Research Reports in Physics

Research Reports in Physics

Nuclear Structure of the Zirconium Region
Editors: J. Eberth, R. A. Meyer, and K. Sistemich

Ecodynamics Contributions to Theoretical Ecology
Editors: W. Wolff, C.-J. Soeder, and F. R. Drepper

Nonlinear Waves 1 Dynamics and Evolution
Editors: A. V. Gaponov-Grekhov, M. I. Rabinovich, and J. Engelbrecht

Nonlinear Waves 2 Dynamics and Evolution
Editors: A. V. Gaponov-Grekhov, M. I. Rabinovich, and J. Engelbrecht

Nonlinear Waves 3 Physics and Astrophysics
Editors: A. V. Gaponov-Grekhov, M. I. Rabinovich, and J. Engelbrecht

Nuclear Astrophysics
Editors: M. Lozano, M. I. Gallardo, and J. M. Arias

Optimized LCAO Method and the Electronic Structure of Extended Systems
By H. Eschrig

Nonlinear Waves in Active Media Editor: J. Engelbrecht

Problems of Modern Quantum Field Theory
Editors: A. A. Belavin, A. U. Klimyk, and A. B. Zamolodchikov

Fluctuational Superconductivity of Magnetic Systems
By M. A. Savchenko and A. V. Stefanovich

Nonlinear Evolution Equations and Dynamical Systems
Editors: S. Carillo and O. Ragnisco

Nonlinear Physics Editors: Gu Chaohao, Li Yishen, and Tu Guizhang

Nonlinear Waves in Waveguides with Stratification By S. B. Leble

Quark-Gluon Plasma Editors: B. Sinha, S. Pal, and S. Raha

Symmetries and Singularity Structures
Integrability and Chaos in Nonlinear Dynamical Systems
Editors: M. Lakshmanan and M. Daniel

Modeling Air-Lake Interaction Physical Background
Editor: S. S. Zilitinkevich

Nonlinear Evolution Equations and Dynamical Systems NEEDS '90
Editors: V. G. Makhankov and O. K. Pashaev

Solitons and Chaos
Editors: I. Antoniou and J. F. Lambert

I. Antoniou F.J. Lambert (Eds.)

Solitons and Chaos

With 44 Figures

Springer-Verlag
Berlin Heidelberg New York London Paris
Tokyo Hong Kong Barcelona Budapest

Dr. Ioannis Antoniou
Professor Dr. Franklin J. Lambert
Vrije Universiteit Brussel, Dienst Theoretische Natuurkunde (TENA), 2, Pleinlaan,
B-1050 Brussels, Belgium

ISBN 3-540-54389-9 Springer-Verlag Berlin Heidelberg New York
ISBN 0-387-54389-9 Springer-Verlag New York Berlin Heidelberg

Typesetting: Camera ready by authors
57/3140-543210 – Printed on acid-free paper

Preface

Solitons and chaos are well-known keywords in a domain of research which is now commonly referred to as "nonlinear science". This cross-disciplinary domain embraces every single field of research involving nonlinear systems.

A concept central to nonlinearity is that of stability, according to which small changes in initial conditions do not produce significant changes in the overall behaviour of the system. Stability upon interaction is the remarkable particle-like characteristic of coherent excitations that we call "solitons". Chaos, on the other hand, refers to incoherence, lack of control, limitations to predictability. The turbulent motion of the atmosphere and the problem of weatherforecast are typical manifestations of the breakdown of stability. The very word "chaos" was introduced twenty years ago into the physics literature in connection with this paradigm.

It is worth emphasizing that both notions were forced upon physicists at a time when computers and supercomputers have become a major exploratory tool for the study of dynamical systems. Quite recently there has been a growing interest in systems and equations which might exhibit some interplay between these complementary aspects of nonlinear dynamics. At the same time there has been an increasing awareness that several mathematical tools, such as perturbation and numerical techniques, Lie group methods and Painlevé analysis, could be used for the study of both aspects. This trend is also reflected by the appearance of recent textbooks which discuss these aspects of nonlinearity, such as R. Sagdeev et al. (1988), M. Tabor (1989), E. Atlee Jackson (1990), E. Infeld and G. Rowlands (1990) and V.E. Zakharov (1991).

In the summer of 1990 we thought of organizing a meeting in the form of an advanced research workshop which would bring together experts in both fields. This idea has been developed at the Theoretical Physics Division (TENA) of the Free University of Brussels (VUB) and has enjoyed the support of the International Solvay Institutes of Physics and Chemistry. About 80 physicists and mathematicians from 14 countries participated at the Brussels Meeting "Aspects of Nonlinear Dynamics: Solitons and Chaos" which was held on the VUB campus. This book contains the proceedings of the Conference including 43 selected contributions by outstanding workers in the field. We believe that these contributions offer an exciting and up-to-date picture of the state of the art.

The wide range of topics covered by this book reflects the very nature of the subject. We did not attempt to provide smooth transitions between articles dealing with very different matters. Each section, however, refers to an explicit link between several papers, as indicated by the title.

A concept repeatedly used throughout the volume is that of integrability. Chaos appears when integrability breaks down, whereas integrability seems to be necessary for the emergence of true solitons. As a result, one may say that integrability concepts, criteria, and related practical tests are at the heart of most studies in nonlinear science. However, it is known that the term "integrable" does not correspond to a unique definition. The ongoing discoveries of systems of ODE's and PDE's which turn out to be integrable from some point of view lead to different working definitions and integrability criteria, the precise interrelations of which are not yet fully understood. We therefore reviewed some of the current notions of integrability used by the present contributors, and listed them in an introductory note at the end of this preface. For the sake of clarity, we also indicated which contributions refer to each notion.

In section 1 we collected a series of papers which discuss particular links between solitons (soliton equations) and chaos (1.2, 1.3), the relation between chaos and irreversibility (1.1, 1.4, 1.5, 1.6) and a generalized normal form transformation method (1.7). A contribution in section 4 relating the Hénon-Heiles system to known IST solvable equations was classified together with other papers on Hamiltonian systems.

Section 2 contains contributions on physical soliton systems, exact or perturbed (2.1, 2.3, 2.5), and on related mathematical techniques (2.2, 2.4).

Several papers dealing with dissipative systems (fluid dynamics, discrete Boltzmann models and chemical reactions) are collected in section 3.

The contributions of section 4 deal with various aspects of Hamiltonian systems, such as long time tails (4.1), chaotic pulsations in variable stars (4.3) and bi-Hamiltonian systems (4.4).

Section 5 is devoted to discrete dynamical systems. An interesting relation between simple chaotic maps and Brownian motion is discussed in 5.1. Papers 5.2 to 5.6 are concerned with chaotic maps and predictability. Integrable maps and their connection with discretized soliton equations are discussed in 5.7 and 5.8.

In section 6 we collected papers on the development and applications of direct methods for the investigation of nonlinear systems, such as singularity analysis (6.1), similarity reductions (6.2) and papers related to Hirota's method of bilinear forms (6.3 to 6.6).

Section 7 deals with inverse methods applicable to nonlinear PDE's which can be "linearized" in some way (by transformation as in paper 7.3 or by association with a linear eigenvalue problem as in papers 7.1, 7.2 and 7.4).

Promising results on integrable (or quasi-integrable) systems with more than one space dimension are presented in section 8. They are related to two-space dimensional solitons (8.2, 8.3), skyrmions (8.4) and localized solutions in $N+1$ dimensions (8.1).

We are particularly grateful to Prof. J. Reignier (Director of the Theoretical Physics Division, VUB) for his encouragement and continuous support. We also thank Prof. I. Prigogine for his enthusiastic support and Prof. G. Nicolis for his interest. We wish to express our thanks to Prof. R. Van Aerschot,

President of the University Council, for his stimulating address at the opening of the Conference. We thank Dr. E. Hefter of Springer-Verlag for his active participation and useful advices.

We are grateful to the following Institutions which provided the financial support to the Meeting: the International Solvay Institutes of Physics and Chemistry, the Free University of Brussels (R&D-IR) and more particularly the Theoretical Physics Division and the Commission of the European Communities DGXII.

Special thanks are due to J. Broekaert, R. Conte, D. MacKernan, M. Musette, M. Peyrard, B. Van Bogaert and particularly to R. Willox, for their help in organizing the Conference. The technical skills of R. Vereecken contributed to the success of the Meeting.

Integrability concepts used in the volume.

1. Integrability through linearization. (sometimes called C-integrability) means that the nonlinear equation may be linearized through a local transformation. The standard prototype is Burgers equation which becomes linear by the Cole-Hopf transformation. This concept is used in contributions 7.3 and 6.2. Integrability through linearization is close to the original and somehow vague concept of an integrable dynamical system as a system for which one can find a suitable transformation to another system with known solutions.

2. Liouville integrability. Liouville's theorem (1855) relates integrability to the existence of constants or invariants or integrals of the motion for Hamiltonian systems. For a thorough discussion see Arnold (1978). Integrals of motion can also be obtained for dissipative systems such as the Lorenz equation in contribution 3.1.

3. Poincaré integrability. demands that the integrating transformation be analytic so that perturbation methods provide actual integration. Poincaré integrability extended to infinite systems is the starting point in Prigogine's (1980) work on irreversibility, see contribution 1.1.

4. Normal form integrability. is a generalization of Poincaré's idea referring to the possibility of reducing an equation to the simplest possible form. Normal forms are discussed by V. Arnold (1988) and used in contributions 1.7 and 4.4.

5. Painlevé integrability. refers to local analytic properties of differential equations (singularity structure). This notion has the advantage of allowing investigations to be made algorithmically (Painlevé test) and of providing criteria of "partial integrability", see R. Conte, N. Boccara (1990). This concept is referred to in contributions 2.2, 4,3, 6.1 and 6.2.

6. IST-solvability of PDE's. (sometimes called S-integrability) corresponds to the fact that the system is associated with a linear eigenvalue (scattering) problem and that it can be "linearized" through an inverse scattering transform, see F. Calogero and A. Degasperis (1982). This notion appears in contributions 2.1, 2.4, 4.2, 7.1, 7.4 and 8.2.

7. Lax integrability of PDE's. is closely related to the concepts 5, 6, 8 and refers to the possibility of expressing the equation as the consistency condition for a suitable pair of linear equations (Lax pair), see P. Lax (1968). This integrability is referred to in contributions 2.2, 2.4, 2.5, 4.2, 7.2 and 8.2.

8. N-soliton integrability of PDE's, $N = 1, 2, \ldots$
refers to the existence of multiple parameter families of special solutions (multisoliton solutions). This property is closely related to algebraic prop-

erties of the equation, such as the existence of an infinite sequence of conserved quantities and can also be tested (to some extent) in the framework of Hirota's bilinear forms. This criteria is referred to in contributions 6.3, 6.4, 6.5, 6.6, 8.1, 8.2, 8.3 and 8.4.

Etymological note.

Chaos is the English transliteration of the Greek word ΧΑΟΣ which comes from the verbs ΧΑΙΝΩ (I gape open, I have gaps) or ΧΑΣΚΩ (I yawn).

Chaos appears in Hesiod's Theogony (Genesis of Gods) as the first primordial entity:

> "Verily, first of all did Chaos come into being and then the broad-bosomed Gaia ..." (Theogony II 6).

Hesiod's Theogony was composed in the early 7th century BC, see Kirk et al. (1983).

Chaos appears also in the Fragmenta of Orpheus. Chaos and Ether were born from Chronos (Time), the first principle of the Orphic Cosmogony. The Orphic oral tradition was written down in Athens during 527-517 BC, by a committee headed by Onomacritus, who also collected Homer's epics, as mentioned by Herodotus VII 6.

It is generally accepted, therefore, that Chaos was mentioned by Hesiod for the first time. However, according to a recent study of the astronomer Chassapis (1967), the astronomical events reported in the Orphic Fragmenta point out that the Orphic tradition goes back to the 15th century BC. Thus, we consider it more appropriate to trace the origin of Chaos back to the teachings of Orpheus.

According to Kirk et al. (1983) the fluidic-diffusive property was introduced into chaos in connection with the verb ΧΕΩ (I pour) by Thales from

Miletus and Pherecydes, the teacher of Pythagoras, in the 6th century BC. The mixing character of chaotic processes was emphasized by Anaxagoras in the 5th century BC as pointed out by Rössler (1985). Anaxagoras' physics is reprinted and translated by Kirk et al. (1983).

- E. Atlee Jackson (1990) "Perspectives of nonlinear dynamics", Cambridge University Press, New York.
- E. Infeld, G. Rowlands (1990) "Nonlinear waves, solitons and chaos", Cambridge University Press, New York.
- R. Sagdeev, O. Usikov, G. Zaslavsky (1988) "Nonlinear physics. From the pendulum to turbulence and chaos", Harwood Acad. Publishers, New York.
- M. Tabor (1989) "Chaos and non-integrability in nonlinear dynamics", Wiley, New York.
- V.E. Zakharov (1991) "What Is Integrability?", Springer-Verlag, Berlin, Heidelberg.
- V. Arnold (1978) "Mathematical Methods of Classical Mechanics", Springer-Verlag, Berlin, Heidelberg.
- V. Arnold (1988) "Geometrical Methods in the Theory of Ordinary Differential Equations", Springer-Verlag, Berlin, Heidelberg.
- I. Prigogine (1980) "From Being to Becoming", Freeman, New York.
- R. Conte and N. Boccara (1990) "Partially Integrable Evolution Equations in Physics", NATO ASI Series, Kluwer, The Netherlands.
- F. Calogero and A. Degasperis (1982) "Spectral Transform and Solitons I", North Holland, Amsterdam.
- P.D. Lax (1968) Comm. Pure Applied Math. 21, 467.
- R. Hirota (1980) "Direct Methods in Soliton Theory" in "Solitons" ed. R.K. Bullough, P.J. Caudrey, Springer-Verlag Berlin, 157.
- C. Chassapis (1967)"The Greek Astronomy of the 2nd Millenia BC according to the Orphic Hymns", Doctorate thesis, University of Athens, Department of Mathematics.
- G. Kirk, J. Raven and M. Schofield (1983) "The Presocratic Philosophers" 2nd ed., Cambridge University Press, London.
- O. Rössler (1985) "Long Time Attractors" in Springer L.N.M. 1165, pp.149-160.

Brussels,
July 1991

I. Antoniou
F. Lambert

Contents

Part I

General Questions on Chaos and Integrability

Integration of Non-Integrable Systems

I. Prigogine* **, T. Petrovsky*, H. Hasegawa*, and S. Tasaki**

*Center for Studies in Statistical Mechanics and Complex Systems
The University of Texas at Austin, Austin, TX 78712 USA.
**International Solvay Institutes for Physics and Chemistry
CP 231, Bd. du Triomphe, 1050 Brussels, Belgium.

I. Introduction

Integrability and chaos are antinomic concepts [1]. This is specially clear for classical dynamics, where "complete integrability" means the existence of tori. It is also apparent in Poincaré's classification [2] into "integrable" and "non-integrable" systems. As was shown by the KAM theory [1,3], non-integrability leads to the *appearance* of random trajectories. In this paper, we summarize our work on "large Poincaré systems" (hereafter refered to as LPS), i.e., the systems which have a continuous spectrum and continuous sets of resonances. This implies that in LPS, almost all trajectories become random. LPS are of special interest as they have a wide range of generality. The concept of LPS is also valid in quantum mechanics and it includes the systems studied in kinetic theory, as well as problems such as radiation damping, interacting fields and so on.

Poincaré's classification of dynamical systems has important consequences concerning the eigenvalue problems of the Liouville operator in classical mechanics and both of the Hamiltonian and Liouville operators in quantum mechanics. These eigenvalue problems have been solved only for a few simple dynamical systems. When there appears the small denominator problem and the system becomes non–integrable, the eigenvalue problems cannot be solved through perturbation analytic with respect to the coupling constant (for details, see section III and also [4]). We want to show that we can overcome this difficulty and obtain perturbation expansions analytic in the coupling constant by the use of non–unitary transformations. This leads as we shall see to an extension of very concept of integrability, which encompasses both integrable and large class of non–integrable systems.

One of the reasons for interest in LPS is the "time paradox": On one side, the importance of irreversibility on the "macroscopic" level of description is now well recognized [5]. On the other side, there is no reference to any privileged direction of time on the "microscopic" level. How then can irreversibility appear on the macroscopic level? In the 20th century, because of the discovery of the constructive role of irreversibility [5], this "time paradox" has become more and more apparent. It is therefore an important point that our method leads to a formulation of dynamics which incorporates irreversibility on the microscopic level (i.e., to a solution of the "time paradox").

Our new integration method is developed in terms of distribution functions (the Gibbs ensemble). We now define a generalized problem of integration as the construction of a complete set of eigenfunctions and eigenvalues for the Liouville–von Neumann operator. For classical integrable systems, a solution of this generalized problem reduces to trajectories [6] and for quantum ones to wave functions. The solution of the generalized problem remains meaningful even for LPS, but then it is no more reducible to trajectories or wave functions. For LPS, the situation is similar to that of radioactive decay, where lifetime has a meaning only for an ensemble of unstable atoms.

The basic idea of our method is the introduction of a suitable *time ordering* of the dynamical states. This leads to a constructive method for the eigenvalue problems of the Liouville operator. In section II, we summarize the Poincaré's theorem and define large Poincaré systems. For LPS, as a consequence of Poincaré's theorem, there exists no perturbative solutions of the eigenvalue problems both for quantum mechanics and for classical mechanics which are analytic in the coupling constant. This is exactly what our method achieves. In section III, after discussing the relation between eigenvalue problems and Poincaré's theorem, we will study the Friedrichs model of quantum theory for the interaction between matter and light as a simple example. In section IV, our general methods will be formulated. In section V, we will make some concluding remarks.

In classical mechanics, LPS correspond to chaotic systems. Therefore our method takes into account the modification of dynamics as a result of chaos. In addition, our results can be extended to quantum theory. The most important single result in quantum theory is that the concept of wave functions is lost in LPS. Our presentation will be a qualitative one emphasizing the physical ideas. For proofs and details, consult the original papers [7,8,9].

II. Poincaré's theorem and the science of chaos

—Large Poincaré systems —

In 1889, Poincaré asked a fundamental question [2,10]; "Can we eliminate interactions ?" This is indeed a very important question. If Poincaré's answer had been yes, the physical universe could be isomorphic to a system of noninteractiong units and there would be no coherence in the universe. So it is very fortunate that he proved that you cannot eliminate interactions in general. Moreover, he gave the reason of this result, namely the existence of the resonances among the various units.

Let us formulate Poincaré's question more precisely. We start with a Hamiltonian of the form

$$H = H_0(J) + \lambda\, V(J, \alpha) \,, \tag{2.1}$$

where λ is the coupling constant and V is the potential energy which depends both on the momenta J (so-called action variables) and the coordinates α (so-called angle variables). For a system with two degrees of freedom the potential can be expanded in a Fourier series

$$V(J_1, J_2, \alpha_1, \alpha_2) = \sum_{n_1, n_2} V_{n_1, n_2}(J_1, J_2)\, e^{i(n_1\alpha_1 + n_2\alpha_2)} \,, \tag{2.2}$$

where n_1 and n_2 are integers. We then ask the question if we can reduce it to the form

$$H = H(J) \,, \tag{2.3}$$

which depends only on the momenta. To perform the transformation from (2.1) to (2.3) Poincaré considered the class of transformations which preserve the structure of the Hamiltonian theory (so-called canonical or unitary transformations). The application of the perturbation techniques lead to the expressions of the form

$$\frac{V_{n_1,n_2}}{n_1\omega_1 + n_2\omega_2} , \tag{2.4}$$

with the frequency ω_i defined as $\omega_i = \partial H_0/\partial J_i$. Here we see the dangerous role of resonances (or "small denominators")

$$n_1\,\omega_1 + n_2\,\omega_2 = 0 . \tag{2.5}$$

Obviously we expect difficulties when (2.5) vanishes while the numerator in (2.4) does not. This has been called by Poincaré [2] the "fundamental difficulty of dynamics". We come in this way to Poincaré's classification of dynamical systems [2,4,6]. If there are "enough" resonances, the system is non-integrable. A decisive progress in our understanding of the role of the resonances has been achieved in the 50's by Kolomogorov, Arnold and Moser (the so-called KAM theory [1,3]). They have shown that if the coupling constant λ in (2.1) is small enough (and also other conditions which we shall not discuss here), "most" trajectories remain periodic as in integrable systems. This is not astonishing. Formula (2.5) can be written as

$$\frac{\omega_1}{\omega_2} = -\frac{n_2}{n_1} \quad \text{a rational number} .$$

Now rationals are "rare" as compared to irrationals. However, whatever the value of the coupling constant λ, there appear now in addition, random trajectories characterized by a positive Lyapounov exponent and therefore by "chaos". This is indeed a fundamental result, since it is quite unexpected to find randomness at the heart of dynamics, which was always considered to be stronghold of a deterministic description. However, it should be emphasized that the KAM theory has not solved the problem of the integration of Poincaré's non-integrable systems. The statement by Arnold that dynamical systems with even only two degrees of freedom lie beyond our present mathematics has been widely quoted. But curiously there is a class of dynamical systems we call large Poincaré systems (LPS), for which we may indeed "integrate" a class of Poincaré's "non-integrable" systems.

A large Poincaré system is a system with a "continuous" spectrum. For example, the Fourier series in formula (2.2) has now to be replaced by a Fourier integral. The resonance conditions then take a new form. The resonance conditions for a small system with an arbitrary number of degrees of freedom are (see (2.5))

$$n_1\omega_1 + n_2\omega_2 + \ldots = 0 \,, \tag{2.6}$$

where n_i are integers. As mentioned, the resonance conditions express the existence of rational relations among frequencies. For LPS, conditions (2.6) have to be replaced by

$$k_1\omega_1 + k_2\omega_2 + \ldots = 0 \,, \tag{2.7}$$

where k_i are *real* numbers. Then resonances occur "everywhere", or in other words the system has a continuous set of resonances. The situation becomes similar to that of the K–flows such as the Baker transformation [10,16], where also almost all motions are random motions. Let us emphasize that the idea of LPS remains meaningful in quantum mechanics. The frequencies ω_i become the energy levels. For small systems, the resonance condition (2.6) would correspond to accidental "degeneracies". But for LPS, we have a continuous spectrum and the situation becomes quite similar to that in classical mechanics. Large Poincaré systems have a surprising generality. We meet them everywhere both in classical and in quantum physics. They involve the situations studied in the kinetic theory, in matter-radiation interaction, in interacting fields and in collision processes.

Large Poincaré systems are not integrable in the usual sense because of the Poincaré's resonances, but as we mentioned we can integrate them through new methods eliminating Poincaré's divergences. This leads to a new "global" formulation of dynamics on the level of distribution functions (classical or quantum). As we deal here with chaotic systems, we may expect new features in this formulation of dynamics. Indeed, we shall find, as compared with the dynamics of integrable systems, an increased role of randomness, and above all a breaking of time symmetry and therefore the emergence of irreversibility at the heart of this new formulation of dynamics. We, in a sense, invert the usual formulation of the time paradox. The usual attempt was to try to deduce the arrow of time from a dynamics based on time reversible equations. In contrast, we now generalize dynamics to include irreversibility.

III. Poincaré's theorem and the quantum mechanical eigenvalue problem —Friedrichs model —

In the quantum mechanical eigenvalue problem we have to find the eigenfunctions and eigenvalues of a given operator. Let us consider the case of a Hamiltonian H, whose eigenfunctions $|u_n\rangle$ and eigenvalues ϵ_n satisfy the relation

$$H\ |u_n\rangle = \epsilon_n\ |u_n\rangle\ . \tag{3.1}$$

Once we have a complete set of eigenfunctions and eigenvalues, we have the "spectral" representation associated with H

$$H = \sum_n \epsilon_n\ |u_n\rangle\langle u_n|\ . \tag{3.2}$$

Finding the spectral representation (or solving the eigenvalue problem) is the central problem of quantum mechanics; however, this problem has only been solved in a few simple situations and most of the time we have to resort to perturbation techniques. We may start, as in Poincaré's theorem, with a Hamiltonian of the form

$$H = H_0 + \lambda V\ , \tag{3.3}$$

where we suppose that the eigenvalue problem can be solved for the "unperturbed" Hamiltonian H_0. We look then for eigenstates and eigenvalues of H, which we could expand in powers of the coupling constant λ. It is here that contact with Poincaré's classification can be made. For non-integrable Poincaré systems, the expansion of eigenfunctions and eigenvalues in powers of the coupling constant leads to Poincaré's catastrophe due to the divergence associated with the small denominators. The relation between Poincaré's theorem and the eigenvalue problem has been studied in a recent paper by two of us [4].

This difficulty leads naturally to the search of non-unitary transformations to eliminate Poincaré's divergences. The study of such non-unitary transformations has been at the forefront of our work during the previous years (see, e.g., [10,11]) but it is only recently that we have succeeded in the construction of a unique non-unitary transformation which leads to the elimination of Poincaré's divergences and permits the use of perturbation techniques for the construction of a spectral representation for LPS. The basic idea which leads to the integration of LPS is the use of a "natural" time ordering of dynamical states.

As a concrete example, let us consider the Friedrichs model [7] which describes the quantum transition from a discrete unstable state $|1\rangle$ to the continuum $|\mathbf{k}\rangle$. Its Hamiltonian in the usual notation is

$$\begin{aligned} H &= H_0 + \lambda V \\ &= \omega_1|1\rangle\langle 1| + \sum_{\mathbf{k}} \omega_k|\mathbf{k}\rangle\langle\mathbf{k}| + \lambda\sum_{\mathbf{k}} V_{\mathbf{k}}\left(|\mathbf{k}\rangle\langle 1| + |1\rangle\langle\mathbf{k}|\right). \end{aligned} \tag{3.4}$$

As is well known since Friedrichs, the spectral representation of H takes the form

$$H = \sum_{\mathbf{k}} \omega_k \, |\phi_{\mathbf{k}}^F\rangle\langle\phi_{\mathbf{k}}^F| \, . \tag{3.5}$$

The important point is that the state $|1\rangle$ has disappeared in (3.5) whatever the value of the coupling constant λ. This leads to a number of difficulties as from the start the "particle" is expressed in terms of the continuum modes [12]. How then can we speak of quantum transitions? Moreover, since the expression (3.5) is not analytic in λ, its calculation is not constructive and no results beyond the Friedrichs model are available.

As mentioned, introducing a natural time ordering of the various states, we can integrate the Friedrichs model keeping the analyticity with respect to the coupling constant λ. At each order of the perturbation calculation, we regularize the propagators *according to the process with which they are associated.* In the decay process there exists first an unstable atomic state which leads later to the emission of radiation. Thus,transitions from 1 to $\mathbf{k}$ (as well as from $\mathbf{k}$ to $\mathbf{k}'$) are considered as "future oriented", while transitions from $\mathbf{k}$ to 1 are considered as "past oriented"[7]. It is remarkable that this natural time ordering of the dynamical states eliminates all Poincaré's divergences and leads to a new complex spectral representation of H [7]

$$H = (\tilde{\omega}_1 - i\gamma)|\varphi_1\rangle\langle\tilde{\varphi}_1| + \sum_{\mathbf{k}} \omega_k \, |\varphi_{\mathbf{k}}\rangle\langle\tilde{\varphi}_{\mathbf{k}}| \, , \tag{3.6}$$

where $\tilde{\omega}_1$ is the (renormalized) frequency of the unstable state and γ^{-1} is its lifetime. Hence, the unstable particle appears explicitly in the spectral representation. The transformation from the unperturbed states $|1\rangle$ and $|\mathbf{k}\rangle$ to the perturbed ones $|\varphi_1\rangle$ and $|\varphi_{\mathbf{k}}\rangle$ is a non-unitary one. The states $|\varphi_\alpha\rangle$ and $\langle\tilde{\varphi}_\alpha|$ are respectively right and left eigenvectors of H and form a complete basis:

$$H|\varphi_1\rangle = (\tilde{\omega}_1 - i\gamma)|\varphi_1\rangle, \qquad \langle\tilde{\varphi}_1|H = (\tilde{\omega}_1 - i\gamma)\langle\tilde{\varphi}_1| \, , \tag{3.7a}$$

$$H|\varphi_{\mathbf{k}}\rangle = \omega_k|\varphi_{\mathbf{k}}\rangle, \qquad \langle\tilde{\varphi}_{\mathbf{k}}|H = \omega_k\langle\tilde{\varphi}_{\mathbf{k}}| \, , \tag{3.7b}$$

and

$$\sum_\alpha |\varphi_\alpha\rangle\langle\tilde{\varphi}_\alpha| \equiv |\varphi_1\rangle\langle\tilde{\varphi}_1| + \sum_{\mathbf{k}} |\varphi_{\mathbf{k}}\rangle\langle\tilde{\varphi}_{\mathbf{k}}| = 1, \qquad \langle\tilde{\varphi}_\alpha|\varphi_\beta\rangle = \delta_{\alpha\beta}. \tag{3.8}$$

It is important that the Hamiltonian H in the new representation (3.6) remains hermitian in spite of the appearance of the complex eigenvalues (or equivalently the states with a broken time symmetry). Indeed, we can show [7]

$$\begin{aligned} H &= (\tilde{\omega}_1 - i\gamma)|\varphi_1\rangle\langle\tilde{\varphi}_1| + \sum_{\mathbf{k}} \omega_{\mathbf{k}} \, |\varphi_{\mathbf{k}}\rangle\langle\tilde{\varphi}_{\mathbf{k}}| \\ &= (\tilde{\omega}_1 + i\gamma)|\tilde{\varphi}_1\rangle\langle\varphi_1| + \sum_{\mathbf{k}} \omega_{\mathbf{k}}|\tilde{\varphi}_{\mathbf{k}}\rangle\langle\varphi_{\mathbf{k}}| = H^+ \, . \end{aligned} \tag{3.9}$$

This is possible because the new basis states are expressed by complex distributions (generalizations of Schwartz's distributions). For example, we have

$$|\varphi_1\rangle = N_1^{1/2}\Big[|1\rangle - \sum_{\mathbf{k}} \frac{\lambda V_{\mathbf{k}}}{(\omega_k - \widetilde{\omega}_1 - z)^+_{-i\gamma}}|\mathbf{k}\rangle\Big], \tag{3.10a}$$

$$\langle\widetilde{\varphi}_1| = N_1^{1/2}\Big[\langle 1| - \sum_{\mathbf{k}} \frac{\lambda V_{\mathbf{k}}}{(\omega_k - \widetilde{\omega}_1 - z)^+_{-i\gamma}}\langle\mathbf{k}|\Big], \tag{3.10b}$$

with N_1 the normalization constant. In (3.10), the function $1/(\omega_k - \widetilde{\omega}_1 - z)^+_{-i\gamma}$ is the complex distribution and is defined through the integration over ω_k with a test function $g(\omega_k)$:

$$\int_0^\infty d\omega_k \, \frac{g(\omega_k)}{(\omega_k - \widetilde{\omega}_1 - z)^+_{-i\gamma}} \equiv \lim_{z\to -i\gamma}\Big[\int_0^\infty d\omega_k \, \frac{g(\omega_k)}{\omega_k - \widetilde{\omega}_1 - z}\Big]_{z\in C^+}, \tag{3.11}$$

where the integration is first carried out by fixing z in the upper-half plane ($\mathbf{C}^+$) and, after that, the limit of $z \to -i\gamma$ is taken. The appearance of this complex distribution is also a natural consequence of the time ordering and introduces the non-unitarity in the transformation which diagonalizes H.

Since there exists a standard solution for the Fredrichs model (see (3.5)), we can compare our results with it and see if our approach makes sense. We indeed recover all known results. For example, with the aid of our states, the evolution of the state $|1\rangle$ is described by

$$\begin{aligned} e^{-iHt}|1\rangle = e^{-i(\widetilde{\omega}_1 - i\gamma)t}\Big[|1\rangle - \sum_{\mathbf{k}} \frac{\lambda\, V_{\mathbf{k}}}{(\omega_k - \widetilde{\omega}_1 - z)^+_{-i\gamma}}|\mathbf{k}\rangle\Big] \\ + \sum_{\mathbf{k}} \frac{\lambda\, V_{\mathbf{k}}}{(\omega_k - \widetilde{\omega}_1 - z)^+_{-i\gamma}} e^{-i\omega_k t}|\mathbf{k}\rangle \,, \end{aligned} \tag{3.12}$$

where the first and second terms are the contributions respectively from $|\varphi_1\rangle$ and $|\varphi_{\mathbf{k}}\rangle$. This evolution as a whole is the same as that obtained from the Friedrichs states $|\phi^F_{\mathbf{k}}\rangle$. However, the states in our description have a broken time symmetry, as is manifest from the exact expression of the time evolution of a state $|\Psi\rangle$

$$e^{-iHt}|\Psi\rangle = e^{-i(\widetilde{\omega}_1 - i\gamma)t}|\varphi_1\rangle\langle\widetilde{\varphi}_1|\Psi\rangle + \sum_{\mathbf{k}} e^{-i\omega_k t}\, |\varphi_{\mathbf{k}}\rangle\langle\widetilde{\varphi}_{\mathbf{k}}|\Psi\rangle \,, \tag{3.13}$$

which is a superposition of the decaying state (the first term) and oscillating states (the second term). Moreover, as a result of the broken time symmetry, we can introduce a functional $\mathcal{H}$ which plays the role of a Lyapounov function

$$\mathcal{H} \equiv |\widetilde{\varphi}_1\rangle\langle\widetilde{\varphi}_1| \ . \tag{3.14}$$

This operator satisfies the commutation relation

$$i[H, \mathcal{H}] = i\{H|\widetilde{\varphi}_1\rangle\langle\widetilde{\varphi}_1| - |\widetilde{\varphi}_1\rangle\langle\widetilde{\varphi}_1|H\} = -2\gamma\mathcal{H} \ , \tag{3.15}$$

and thus decreases monotonically when the unstable particle decays

$$\mathcal{H}(t) \equiv e^{+iHt}\mathcal{H}e^{-iHt} = e^{-2\gamma t}\mathcal{H} \ . \tag{3.16}$$

Its expectation value reaches the minimum when the quantum state can be expressed entirely in terms of the continuum states $|\varphi_{\mathbf{k}}\rangle$. As we can associate a Lyapounov function $\mathcal{H}$ with the particle decay the decay becomes an irreversible process.

We may summarize what we have done as follows: To avoid Poincaré's catastrophe we have introduced more general transformations leading to complex eigenvalues while Poincaré considered only canonical (or unitary) ones which among other properties keep the eigenvalues of H real. The specific choice of this non-unitary transformation follows from our time ordering of the dynamical states. It should be emphasized that the broken time symmetry as expressed by the existence of the complex eigenvalues is *not* induced by the non-unitary transformation but is an intrinsic character of the system. The only role of the non-unitary transformation is to make manifest its broken time symmetry.

The example we have treated is a very simple one as we could introduce a natural time ordering in the frame of the Hilbert space. In general though, this is impossible (think about scattering where all states play a symmetrical role). We then have to introduce a natural time ordering on the level of the statistical description as we shall see in the next section. This leads to the integration of LPS in quite general situations and to a new form of dynamics which breaks radically with the past.

IV. Poincaré's theorem and a statistical formulation of dynamics

From the example studied in the previous section, it should be clear how we may avoid Poincaré's catastrophe. It is by introducing into the theory a time ordering of dynamical states which leads to well-defined "regularization" procedures for the small denominators. But how can we introduce this time ordering? Here as we shall see now, we have to turn to the statistical description.

In the early days of statistical mechanics, Gibbs introduced a quite fundamental concept, the "Gibbs ensembles". Instead of considering single dynamical systems, he considered a large number of dynamical systems with the same structure. They evolve in the phase space associated with the coordinates $q_1, q_2, \ldots, q_N$ and momenta $p_1, p_2, \ldots, p_N$ of the particles contained in each dynamical system. The description is then in terms of the probability distribution ρ in the phase space

$$\rho(q_1 \cdots q_N, p_1 \cdots p_N; t) . \tag{4.1}$$

This description remains also meaningful for quantum systems. The probability distribution is then called the "density matrix", which is specified by its matrix elements like

$$\rho(p_1 \cdots p_N : p'_1 \cdots p'_N; t) . \tag{4.2}$$

Once we know ρ, we can calculate both the velocity distribution of the particles as well as the correlations between the particles.

How then does time ordering enter into this description?

Let us consider a classical gas. Particles collide and these collisions give rise to correlations. First we have binary correlations, then ternary correlations, and time going on, correlations involving more and more particles.

The formation of correlations is somewhat reminiscent to that of two friends who have a conversation (this would correspond to a collision). Even when the partners go away, the memory of their conversation remains. The information associated with this conversation is, time going on, spreading out to more and more participants.

Suppose we look at a glass of water. In this glass of water there is an arrow of time that will persist essentially forever and corresponds to the creation of new correlations involving an ever-increasing number of particles. According to the correlations which exist between the molecules, we can distinguish "young" water from "old"! Recent computer experiments shows that binary correlations appear very rapidly, that ternary correlations involve longer time scales and so on. This time oriented flow of correlations breaks the symmetry involved in the classical description. Suppose we go from state A (of a many-body system) with no correlations at $t = 0$ to a state B at time t involving multiple correlations. Obviously, the transition from A to B involves quite different physical processes than the inverse transition from B to A [18].

The time ordering of correlations has to be introduced into dynamics to avoid Poincaré's catastrophe. Binary correlations come before ternary ones and so on. We have therefore to describe dynamics in terms of the time evolution of correlations.

This corresponds to a different point of view from that of classical (or quantum) dynamics. The question is no more to study the time dependence of the position and momentum of each particle but to follow the evolution of the relations between the particles [13]. In this conceptual framework we can avoid Poincaré's catastrophe by treating transitions to higher correlations as "future–oriented" and transitions to lower correlations as "past–oriented" exactly as we have done before.

As mentioned, Gibbs' ensemble theory leads to a description by the density matrices ρ, which can be regarded as vectors in a linear space (so-called "Liouville space"). We shall call the density matrix ρ a "supervector" and denote it as $|\rho\rangle\rangle$ according to the bra-ket notation. Also a linear operator acting on the density matrix will be called a "superoperator". In this Liouville space, the time evolution of the density matrix ρ is given by

$$i\frac{\partial}{\partial t}\rho(t) = L_H\ \rho(t)\ , \tag{4.3}$$

which is formally quite similar to the Schrödinger equation. The superoperator L_H is the so-called Liouville operator. It can be expressed in terms of the Hamiltonian. In classical mechanics, it is given by the Poisson bracket

$$L_H\ \rho = i\{H,\rho\}_{PB} = -i\sum_j\left(\frac{\partial H}{\partial p_j}\frac{\partial \rho}{\partial q_j} - \frac{\partial H}{\partial q_j}\frac{\partial \rho}{\partial p_j}\right)\ , \tag{4.4a}$$

and, in quantum mechanics, by the commutator

$$L_H\ \rho = H\ \rho - \rho\ H\ . \tag{4.4b}$$

For classical integrable systems, as the Hamiltonian is the function only of the action variables J_j, i.e. $H = H(J_1,\cdots,J_N)$, the Liouville operator becomes

$$L_H = -i\sum_j \omega_j\frac{\partial}{\partial \alpha_j}\ , \tag{4.5}$$

where α_j is the angle variable and $\omega_j = \partial H/\partial J_j$ is the frequency. Therefore its eigenvalue problem has the following solutions

$$u_{\mathbf{n}}(\alpha_1,\cdots\alpha_N) = c_{\mathbf{n}}\ e^{i(\mathbf{n},\alpha)}\ , \quad \text{and} \quad L_H\ u_{\mathbf{n}} = (\omega,\mathbf{n})u_{\mathbf{n}}\ , \tag{4.6}$$

where $c_{\mathbf{n}}$ is the normalization constant, α, ω and $\mathbf{n}$ are respectively angle variable, frequency and integer vectors. From the solution of (4.6), the evolution of the distribution function

$$\rho_0(J_1,\cdots,J_N;\alpha_1\cdots,\alpha_N) = \sum_{\mathbf{n}} a_{\mathbf{n}}(J_1,\cdots,J_N)u_{\mathbf{n}}(\alpha_1,\cdots,\alpha_N)\ , \tag{4.7}$$

becomes

$$\begin{aligned}\rho(J_1,\cdots,J_N;\alpha_1,\cdots,\alpha_N;t) &= \sum_{\mathbf{n}} a_{\mathbf{n}}(J_1,\cdots,J_N)e^{-i(\omega,\mathbf{n})t}u_{\mathbf{n}}(\alpha_1,\cdots,\alpha_N)\\ = \sum_{\mathbf{n}} a_{\mathbf{n}}(J_1,\cdots,J_N)c_{\mathbf{n}}e^{-i(\omega,\mathbf{n})t}e^{i(\mathbf{n},\alpha)} &= \rho_0(J_1,\cdots,J_N;\alpha_1-\omega_1 t,\cdots,\alpha_N-\omega_N t)\ ,\end{aligned} \tag{4.8}$$

which implies that the evolution of the distribution function reduces to that of the trajectories.

For quantum integrable systems, suppose we have a complete set of eigenvectors of H

$$H|\phi_\alpha\rangle = E_\alpha|\phi_\alpha\rangle \ , \quad \sum_\alpha |\phi_\alpha\rangle\langle\phi_\alpha| = 1 \ , \quad \langle\phi_\alpha|\phi_\beta\rangle = \delta_{\alpha\beta} \ , \tag{4.9}$$

then we immidiately obtain an eigen supervector of L

$$L|\phi_\alpha\rangle\langle\phi_\beta| = (E_\alpha - E_\beta)|\phi_\alpha\rangle\langle\phi_\beta| \ . \tag{4.10}$$

Because of the completeness in (4.9), any density matrix ρ can be expressed as

$$\rho = \sum_{\alpha\beta}\langle\phi_\alpha|\rho|\phi_\beta\rangle \ |\phi_\alpha\rangle\langle\phi_\beta| \ . \tag{4.11}$$

This implies that there are no other eigen–supervectors than those given in (4.10) and thus that the eigenvalue problem of L reduces to that of H.

Now we define "integration" of the system as the very determination of eigenfunctions and eigenvalues of the Liouville operator L. As we mentioned, Poincaré's theorem deals with Hamiltonians of the form

$$H = H_0 + \lambda V \ , \tag{4.12a}$$

which leads to a decomposition of the Liouville operator

$$L_H = L_0 + \lambda L_V \ . \tag{4.12b}$$

The problem of integration is closely related to the construction of a set of projection operators. First, let us consider the unperturbed Liouville operator L_0. As before, we suppose that we can find a complete set of the eigenvectors $|\alpha\rangle$ and the eigenvalues ω_α of the "unperturbed" Hamiltonian H_0. And the eigen supervector of L_0 is given by in a bra-ket notation

$$L_0|\nu\rangle\rangle = l_\nu|\nu\rangle\rangle \ , \tag{4.13}$$

where

$$|\nu\rangle\rangle \equiv |\alpha\rangle\langle\beta| \ , \quad l_\nu \equiv \omega_\alpha - \omega_\beta \ . \tag{4.14}$$

Then the above mentioned projection operators $\overset{(\nu)}{P}$ are given by

$$\overset{(\nu)}{P} \equiv |\nu\rangle\rangle\langle\langle\nu| \ , \tag{4.15}$$

which satisfy

$$\begin{gathered} L_0 \overset{(\nu)}{P} = \overset{(\nu)}{P} L_0 \ , \quad \sum_\nu \overset{(\nu)}{P} = 1 \ , \\ \overset{(\nu)}{P} \overset{(\nu')}{P} = \overset{(\nu)}{P} \delta_{\nu\nu'} \ , \quad \overset{(\nu)}{P} = \overset{(\nu)}{P}{}^{+} \ . \end{gathered} \tag{4.16}$$

In the Liouville space, an inner product of two supervectors $|\rho_1\rangle\rangle$ and $|\rho_2\rangle\rangle$ corresponding respectively to density matrices ρ_1 and ρ_2 is defined by

$$\langle\langle\rho_1|\rho_2\rangle\rangle \equiv tr(\rho_1^+ \ \rho_2) \ . \tag{4.17}$$

The first relation of (4.16) corresponds to the fact that $|\nu\rangle\rangle$ is the eigenvector of L_0, the second one to their completeness and the third one to their orthonormality. The last relation is a direct consequence of the definition of $\overset{(\nu)}{P}$. Therefore, the construction of a set of projectors $\overset{(\nu)}{P}$ is closely related to the eigenvalue problem of the Liouville operator L_0.

Now we turn to the eigenvalue problem of L_H. First we shall look for projection operators similar to $\overset{(\nu)}{P}$ since their construction corresponds to the partial solution of the eigenvalue problem. For integrable systems, there is no problem and we can obtain the projection operators which satisfy relations similar to those (4.16) for $\overset{(\nu)}{P}$. However, the problem changes radically for non-integrable systems. Then the Liouville equation describes the emergence of chaos due to the destruction of the invariants of motion associated with the unperturbed system.

Again Poincaré's theorem prevents us from finding projectors satisfying (4.16) which could be expanded in powers of the coupling constant λ [14]. A path to obtain a generalized spectral representation of L is through the construction of a complete set of eigenprojectors $\overset{(\nu)}{\Pi}$ which satisfy (4.16) except the hermiticity condition

$$\overset{(\nu)}{\Pi} \neq \overset{(\nu)}{\Pi}^{+} . \tag{4.18}$$

In short, they satisfy the following relations

$$\begin{aligned} L_H \overset{(\nu)}{\Pi} = \overset{(\nu)}{\Pi} L_H , \quad \sum_{\nu} \overset{(\nu)}{\Pi} = 1 , \\ \overset{(\nu)}{\Pi} \overset{(\nu')}{\Pi} = \overset{(\nu)}{\Pi} \delta_{\nu\nu'} , \quad \overset{(\nu)}{\Pi} \to \overset{(\nu)}{P} \text{ (as } \lambda \to 0) . \end{aligned} \tag{4.19}$$

Let us explain more precisely our natural time ordering through the construction of the projector $\overset{(\nu)}{\Pi}$. From the first equation of (4.19), each term of the perturbation expansion

$$\overset{(\nu)}{\Pi} = \sum_{0}^{\infty} \lambda^n \overset{(\nu)}{\Pi}_n , \tag{4.20}$$

is found to satisfy

$$\overset{(\nu')}{P} \overset{(\nu)}{\Pi}_n \overset{(\nu'')}{P} = \frac{1}{l_{\nu'} - l_{\nu''} + i\epsilon_{\nu'\nu''}} \overset{(\nu')}{P} [\overset{(\nu)}{\Pi}_{n-1} L_V - L_V \overset{(\nu)}{\Pi}_{n-1}] \overset{(\nu'')}{P} , \tag{4.21}$$

where the infinitesimal real number $\epsilon_{\nu'\nu''}$ has been introduced to regularize the small denominator. As a first step, we assign a nonnegative integer to each correlation ν (see e.g., [14]) which we shall call the "degree of correlation" based on the concepts of the flow of correlations described before. For example, in case of an interacting gas, the uniform distribution has degree zero, the binary correlation has degree one, the ternary correlation has degree two and so on. According to the degree of correlation we determine the sign of $\epsilon_{\nu'\nu''}$. Transitions from correlation with lower degree to that with higher degree (as well as between the correlations with the same degree) are considered as "future–oriented" and the retarded branch will be assigned to them. On the other hand, transitions from correlation with higher degree to that with lower degree are considered as "past–oriented" and the advanced branch will be assigned [8,14]. Succinctly then, our result is

$$\epsilon_{\nu'\nu''} = \begin{cases} +\epsilon & \text{if } d_{\nu'} < d_{\nu''}, \\ -\epsilon & \text{if } d_{\nu'} > d_{\nu''}, \\ -\epsilon & \text{if } d_{\nu'} = d_{\nu''}. \end{cases} \tag{4.22}$$

Once we know the projectors $\overset{(\nu)}{\Pi}$, we can solve the eigenvalue problem for the total Liouville operator L_H [8]. For this purpose, we introduce three auxiliary operators

$$\overset{(\nu)}{A} \equiv \overset{(\nu)}{P}\,\overset{(\nu)}{\Pi}\,\overset{(\nu)}{P} \;, \quad \overset{(\nu)}{C}\,\overset{(\nu)}{A} \equiv \overset{(\nu)}{Q}\,\overset{(\nu)}{\Pi}\,\overset{(\nu)}{P} \quad \text{and} \quad \overset{(\nu)}{A}\,\overset{(\nu)}{D} \equiv \overset{(\nu)}{P}\,\overset{(\nu)}{\Pi}\,\overset{(\nu)}{Q} \;, \tag{4.23}$$

where $\overset{(\nu)}{Q} \equiv 1 - \overset{(\nu)}{P}$ is a projector orthogonal to $\overset{(\nu)}{P}$. The operator $\overset{(\nu)}{C}$ expresses the creation of correlation different from ν and $\overset{(\nu)}{D}$ the destruction of it. The evolution of the density matrix in each $\overset{(\nu)}{\Pi}$-subspace is then given by [14]

$$\overset{(\nu)}{\rho(t)} \equiv \overset{(\nu)}{\Pi}\,\rho(t) = e^{-iL_H t}\,\overset{(\nu)}{\Pi}\,\rho(0) \;, \tag{4.24}$$

where

$$e^{-iL_H t}\,\overset{(\nu)}{\Pi} = \overset{(\nu)}{\Pi}\,e^{-iL_H t} = (\overset{(\nu)}{P} + \overset{(\nu)}{C})\,e^{-i\overset{(\nu)}{\theta}\,t}\,\overset{(\nu)}{A}\,(\overset{(\nu)}{P} + \overset{(\nu)}{D}) \,. \tag{4.25}$$

The non-hermitian kinetic operator $\overset{(\nu)}{\theta}$ in each subspace is defined by

$$\overset{(\nu)}{\theta} \equiv l_\nu \overset{(\nu)}{P} + \lambda \overset{(\nu)}{P}\,L_V\,\overset{(\nu)}{C}\,\overset{(\nu)}{P} \;, \tag{4.26}$$

which satisfies

$$L_H(\overset{(\nu)}{P} + \overset{(\nu)}{C}) = (\overset{(\nu)}{P} + \overset{(\nu)}{C})\,\overset{(\nu)}{\theta} \;. \tag{4.27}$$

This relation implies an interesting relation between the spectral problems of the Liouville operator L_H and kinetic operator $\overset{(\nu)}{\theta}$: Their eigenvalues are identical. Indeed when we know a solution of the eigenvalue problem of $\overset{(\nu)}{\theta}$

$$\overset{(\nu)}{\theta}\,|\overset{(\nu)}{u}_\alpha\rangle\rangle = \overset{(\nu)}{Z}_\alpha\,|\overset{(\nu)}{u}_\alpha\rangle\rangle \;, \tag{4.28}$$

we obtain from (4.27) a right–eigenvector of L_H with an identical eigenvalue $\overset{(\nu)}{Z}_\alpha$

$$L_H|\,\overset{(\nu)}{F}_\alpha\rangle\rangle = \overset{(\nu)}{Z}_\alpha\,|\,\overset{(\nu)}{F}_\alpha\rangle\rangle \;, \tag{4.29}$$

where

$$|\,\overset{(\nu)}{F}_\alpha\rangle\rangle = (\overset{(\nu)}{A}_\alpha)^{1/2}(\overset{(\nu)}{P} + \overset{(\nu)}{C})|\overset{(\nu)}{u}_\alpha\rangle\rangle \;, \tag{4.30}$$

with $\overset{(\nu)}{\mathcal{A}_\alpha}$ the normalization constant. Similarly starting from $\overset{(\nu)}{D}$ we obtain left–eigenvectors $\langle\langle \overset{(\nu)}{\widetilde{F}_\alpha} \mid$

$$\langle\langle \overset{(\nu)}{\widetilde{F}_\alpha} \mid L_H = \overset{(\nu)}{Z_\alpha} \langle\langle \overset{(\nu)}{\widetilde{F}_\alpha} \mid . \tag{4.31}$$

These left– and right--eigenvectors are found to form a complete orthonormal set in the Liouville space

$$\sum_{\nu,\alpha} \mid \overset{(\nu)}{F_\alpha} \rangle\rangle\langle\langle \overset{(\nu)}{\widetilde{F}_\alpha} \mid = 1 \ , \quad \langle\langle \overset{(\nu)}{\widetilde{F}_\alpha} \mid \overset{(\nu')}{F_\beta} \rangle\rangle = \delta_{\nu\nu'}\delta_{\alpha\beta} \ , \tag{4.32}$$

and the eigenvectors corresponding to eigenvalues with non-vanishing imaginary parts have zero norm

$$\langle\langle \overset{(\nu)}{F_\alpha} \mid \overset{(\nu)}{F_\alpha} \rangle\rangle = 0 \ . \tag{4.33}$$

This complete set of eigenvectors lead to the new complex spectral representation of L_H

$$L_H = \sum_{\nu,\alpha} \overset{(\nu)}{Z_\alpha} \mid \overset{(\nu)}{F_\alpha} \rangle\rangle\langle\langle \overset{(\nu)}{\widetilde{F}_\alpha} \mid , \tag{4.34}$$

and thus to the following time evolution of a density matrix ρ

$$e^{-iL_H t} \, |\rho\rangle\rangle = \sum_{\nu,\alpha} e^{-i \overset{(\nu)}{Z_\alpha} t} \mid \overset{(\nu)}{F_\alpha} \rangle\rangle\langle\langle \overset{(\nu)}{\widetilde{F}_\alpha} \, |\rho\rangle\rangle \ . \tag{4.35}$$

As the result of our time ordering, we can show $\mathrm{Im}\overset{(\nu)}{Z_\alpha} \le 0$ [8]. We see that any regular initial state† will be projected on the manifold of eigenstates characterized by real eigenvalues. In particular, when zero is the only real eigenvalue, the system approaches equilibrium. This means that the system is mixing and the equilibrium state is an attractor. Since a pure state can be prepared as the initial state and the attracting equilibrium state is a mixed state, this also implies the existence of the loss of wave functions.

---

† Here a "regular state" means a superstate $|f\rangle\rangle$ where $\langle\langle \nu|f\rangle\rangle$ is a regular function of l_ν at the complex eigenvalue $\overset{(\nu)}{Z_\alpha}$ in the analytically continued complex plane.

Reflecting this broken time symmetry, we can introduce a Lyapounov functional $\mathcal{H}$, as in the case of the Friedrichs model,

$$\mathcal{H} = \sum_{\nu,\alpha} | \overset{(\nu)}{\widetilde{F}_\alpha} \rangle\!\rangle \langle\!\langle \overset{(\nu)}{\widetilde{F}_\alpha} | \,. \tag{4.36}$$

Its expectation value for any state $|\rho\rangle\!\rangle$ decreases monotonically

$$\langle\!\langle \rho(t)|\mathcal{H}|\rho(t)\rangle\!\rangle = \langle\!\langle \rho(0)|e^{+iL_H t}\mathcal{H}e^{-iL_H t}|\rho(0)\rangle\!\rangle = \sum_{\nu,\alpha} e^{-2\gamma_{\nu\alpha}t} |\langle\!\langle \overset{(\nu)}{\widetilde{F}_\alpha} |\rho(0)\rangle\!\rangle|^2 \,, \tag{4.37}$$

where $\gamma_{\nu\alpha} = -\mathrm{Im}\,\overset{(\nu)}{Z_\alpha}$. However, because eigenvectors $\langle\!\langle \overset{(\nu)}{\widetilde{F}_\alpha} |$ and $| \overset{(\nu)}{F_\alpha} \rangle\!\rangle$ contain complex distributions, the Liouville operator remains hermitian and the evolution remains unitary. Moreover, we can introduce a non–unitary transformation Λ connecting the eigenvectors of L_0 with the complex eigenvectors of L_H. But it should be emphasized that the dissipative nature *is not* introduced by the non–unitary transformation Λ but is an intrinsic nature of the Liouville space!

Here we have outlined the construction of the eigenvectors through non–hermitian projection operators. As it was shown in [8], we can also construct them by the Brillouin-Wigner perturbation method or directly from the resolvent operator of the Liouville operator L_H.

As was shown above, the spectrum of the Liouville operator L_H is essentially determined by the collision process as expressed by the kinetic operator $\overset{(\nu)}{\theta}$. The latter contains more than two interactions L_V and expresses a process which spreads over space and time. This implies a radical deviation from the usual methods of dynamics valid for integrable systems where the evolution can be resolved into a succession of instantaneous space-time events (remember Feynman diagrams). For this reason the dynamics of LPS can only be formulated on the statistical level, as we cannot reduce it neither to trajectories in the classical case nor to wave functions in the quantum case. This deviation from the great traditions of dynamics is not so astonishing; we deal here with an aspect of dynamics that is totally absent in integrable systems. It is, however, already present in the KAM theory but there the behavior is so complex that it defies any quantitative description (we have to use qualitative criteria for the collapse of resonant tori as the result of the coalescence of resonances). It is precisely the main progress realized by the study of LPS to present a simple description of the physical processes due to resonances which lead to Poincaré's non-integrability. Already our method has been applied to two examples of LPS, i.e., the quantum mechanical scattering problem [8] and classical chaotic systems [9].

V. Concluding remarks

In the conventional formulation of dynamics, time appears as a parameter for the motion of the dynamical object (in Newton's equation of motion) or the wave function associated with it (in Schrödinger equation). In addition to this "first" time, our new formulation of dynamics has a "second" internal time describing the flow of correlations as a "natural time ordering" of dynamical states. As is shown above, it is only on the level of distribution functions (the Liouville space) where we can define this "second" time. Moreover, our formulation is not reducible to trajectories in classical mechanics or to wave functions in quantum mechanics. We expect that nature uses these new eigen-distributions and therefore violates the usual formulation of quantum or classical theory. This belief is based on various general arguments:

1) Our dynamical theory includes time symmetry breaking, appearance of "damping" through complex eigenvalues, and therefore; the second law of thermodynamics.
2) It agrees with previous work on nonequilibrium statistical mechanics (starting with the Van Hove $\lambda^2 t$-limit and leading to the non-Markovian master equation), which aimed at deriving dissipative effects on the microscopic level.
3) It agrees with extensive numerical experiments [9,17].

It is interesting that dynamical instability incorporated in LPS leads to the solution of several fundamental paradoxes.

1) Time paradox: How can irreversibility appear?

   In our method, thanks to the time ordering, the Liouville operator may have complex eigenvalues with nonpositive imaginary parts. This implies the broken time symmetry reflecting irreversibility.

2) Equilibrium paradox: How can a physical system approach its equilibrium state by following dynamics?

   As the Liouville operator now has complex eigenvalues, the eigenstates corresponding to them will die out in the limit of $t \to \infty$ and only the components corresponding to real eigenvalues survive. In short, the system approaches an attractor. When zero is the only real eigenvalue, this attractor corresponds to an equilibrium state.

In short, our method leads to an effective way to construct the spectral representation of both classical and quantum systems. We intend to apply our method to a wide range of problems, e.g., soliton problems in non-integrable nonlinear equations and the problem of interacting fields (including electrodynamics). We are also preparing the extension of our approach to systems with constraints which forces the systems to tend asymptotically to *nonequilibrium* ensembles.

Acknowledgement

We thank Professor E.C.G. Sudarshan and Dr. F. Mayné for fruitful discussions and suggestions while preparing this paper. We also acknowledge the U.S. Department of Energy, Grant N° FG05-88ER13897, the Robert A. Welch Foundation, and the European Communities Commission (contract n° PSS*0143/B) for support of this work. This work has been partially supported by the Belgian Government under the contract "Pole d'attraction interuniversitaire".

References

1) See the excellent introduction by M. Tabor, "Chaos and Integrability in Nonlinear Dynamics", (Wiley, New York) 1989.

2) H. Poincaré, "Méthodes nouvelles de la mécanique céleste", vol. 1 (1892) (Dover, New York, 1957).

3) A.N. Kolmogorov, Dokl. Akad. Nauk **98** (1954) 527, English translation: Los Alamos Scientific Laboratory Translation No. LA-TR-71-67; J. Moser, Nachr. Akad. Wiss. Goettingen Math. Phys. Kl. **21** (1962) 1; V.I. Arnold, Usp. Mat. Nauk **18** (1963) 9 [Russ. Math. Surv. **18** (1963) 85].

4) T.Y. Petrosky and I. Prigogine, Physica **147A** (1988) 439.

5) G. Nicolis and I. Prigogine, "Exploring Copmplexity", (Freeman, New York) 1989.

6) I. Prigogine, "Non-equilibrium Statistical Mechanics", (Wiley, New York) 1962.

7) T. Petrosky, I. Prigogine and S. Tasaki, Physica **A** in press.

8) T. Petrosky and I. Prigogine, to appear in Physica **A**.

9) H. H. Hasegawa and W. C. Saphir, to appear and the contribution to this conference.

10) I. Prigogine, "From Being to Becomming", (Freeman, New York) 1980.

11) Cl. George, F. Mayné and I. Prigogine, Adv. in Chemical Physics, **61** (1985), 223.

12) T. Petrosky and I. Prigogine, Can. J. Phys. **68** (1990), 670.

13) The idea of "dynamics of correlations" was introduced in [6].

14) T. Petrosky and H. Hasegawa, Physica **160A** (1989), 351.

15) B. Misra and E.C.G. Sudarshan, J. Math. Phys. **18** (1977) 756.

16) e.g., B. Misra and I. Prigogine, in "Long Time Predictions in Dynamic Systems", ed. by E.W. Horton, L.E. Reichl and V.G. Szebehely, (Wiley, New York) 1982.

17) T. Petrosky, W. Saphir and I. Prigogine, to appear.

18) I. Prigogine, lecture presented at Nobel Conference XXVI (1990, to be published).

Order and Chaos in the Statistical Mechanics of the Integrable Models in 1 + 1 Dimensions

R.K. Bullough[1,2], Yu-zhong Chen[2] and J. Timonen[3]

[1]LAMF, The Technical University of Denmark,
DK-2800 Lyngby, Denmark.
[2]Department of Mathematics, UMIST
PO Box 88, Manchester M6O 1QD, UK.
[3]Physics Department, University of Jyväskylä,
SF-40100, Jyväskylä, Finland.

1. Introduction

This paper was presented at the meeting under this title. But, originally, the more cumbersome 'Quantum chaos - classical chaos in k-space: thermodynamic limits for the sine-Gordon models' was proposed. Certainly this covers more technically the content of this paper.

Since the early 1980's we have developed, e.g. [1-17], a quantum and classical statistical mechanics of integrable models in 1+1 dimensions like the classical sine-Gordon (s-G) model [1]

$$\phi_{xx} - \phi_{tt} = m^2 \sin \phi ; \tag{1}$$

$m > 0$ is a "mass" and $\phi(x,t) \in \mathbb{R}$ is a classical field in 1+1. The s-G model (1) has soliton solutions - the kinks and antikinks $\phi = 4\tan^{-1}\exp\pm(x - Vt)/(1-V^2)^{\frac{1}{2}}$ $(c = 1)$, and breather solutions $\phi = 4\tan^{-1}[\tan\theta \sin\theta_I \operatorname{sech} \theta_R]$ (where $\theta_I = (m\cos\theta)(t - Vx)(1 - V^2)^{-\frac{1}{2}}$, $\theta_R = (m\sin\theta)(x - Vt)(1 - V^2)^{-\frac{1}{2}}$). There are also "radiative" solutions of harmonic type which we call "phonons" in this paper.

The sinh-G model

$$\phi_{xx} - \phi_{tt} = m^2 \sinh \phi \tag{2}$$

is an integrable model in 1+1 related to s-G by continuation in the coupling constant $\gamma_o > 0$ (see eqn. (4)). It has no soliton solutions only phonon solutions. There are very many other classical integrable models in 1+1 and we have worked out the one-body statistical mechanics (calculation of the free energy F) for some half dozen of them [1-5,8,9,17]. We have called this '*soliton* statistical mechanics '(SM) [5,6]. To each classical SM is a quantum SM of the quantum integrable models and some of these are worked out [2-5]. Generic examples of quantum integrable models are (1) and (2) with the ϕ now quantum fields and the equations interpreted in normal order, e.g.: $\phi_{xx} - \phi_{tt} = m^2 \sin \phi$: and : : means normal order.

The classical *repulsive* nonlinear Schrödinger (NLS) model is

$$- i\phi_t = \phi_{xx} - 2c\, \phi^* \phi^2 , \quad c > 0 \tag{3}$$

with $\phi \in \mathbb{C}$, and this becomes the attractive NLS by continuation in c, $c \to -c$. The repulsive NLS is like sinh-G and has only radiative (phonon) solutions: the SM of both models is in [2]. The attractive NLS ($c<0$) has in addition breather-like soliton solutions. The quantum normally ordered forms use $-2c\phi^\dagger\phi^2$ as nonlinearity. Their quantum SM is in [2] ($c > 0$) and, formally, in [4,5], $c < 0$.

The SM, quantum and classical, of the integrable models in 1+1 (the 'soliton SM') is a generic example of the interaction of integrable nonlinearity (of 'coherent structures' - solitons) with infinite dimensional Hamiltonian chaos. Conceptually the latter is provided by a heat bath at finite temperatures T with infinitely many degrees of freedom. This paper addresses the problem of the *contrast* between the quantum SM and the classical SM in the cases when the classical models have breather-like soliton solutions. This makes the paper particularly relevant to the subject matter of this meeting.

This contrast appears in the following terms:- The now famous spectral transform method (e.g. [18]) solves s-G, eqn. (1) for $x \in \mathbb{R}$, with boundary conditions (b.c.s) vanishing fast enough as $|x| \to \infty$. The solution is made up of the kinks, antikinks, breathers and phonons. But in the classical SM only the kinks, antikinks and phonons contribute to F., In the quantum SM there are really only quantum breathers which, *alone*, contribute to F.

Coupling to the heat bath breaks the integrability of the models. The classical solitons interact only through 2-particle phase shifts [18]. In equilibrium with the heat bath they become "dressed" by both phonons and other solitons. They continue to interact through 2-particle phase shifts, but, really, new excitations, substantially different from the "bare" solitons are created. The simple solitons which have only one degree of freedom are disordered by Brownian motion derived from other solitons and the phonons. They have Gibbsian entropy S and their trajectories are chaotic in this motion. The classical breathers have *two* degrees of freedom - of translation and internal oscillation. We show how the classical breathers *break up* in equilibrium and contribute to the Brownian motion of the chaotic phonons. We contrast this with the quantum breathers. Although the translational degree of freedom is chaotic, the quantised internal degree of freedom forms a structure ordered in momentum space which persists unchanged throughout the motion. This structure is imposed by quantum mechanics. In this sense quantum mechanics imposes order on the classical chaos of the classical breather break-up. This can be thought of as an exotic example of quantum chaos.

Dynamical studies of chaos at this meeting typically look at few degrees of freedom and their language is of Hamiltonian chaos in terms of Poincaré sections, KAM theory, Lyapunov exponents, Arnold webs, of low dimensional chaos and strange attractors, of bifurcation theory, Haussdorf measures and fractal models. We do not attempt in this paper to follow any detailed dynamics. There is a large infinity of degrees of freedom, we work in thermodynamic limit, and the natural language is that of Gibbsian statistical mechanics. The chaos provided by the heat bath is (we believe) Hamiltonian chaos - which we take to be ergodic in that limit.

2. The simplest model - the sinh-G model

Before developing the classical and quantum SM of the s-G models it is helpful to study the simplest model - the sinh-G model. It might appear that the non-relativistic quantum repulsive NLS model (the bose-gas model [19]) is still simpler. However because of the finite mass m, eqn. (2), the classical sinh-G model has a low temperature asymptotic expansion for its free energy which the classical repulsive NLS model does not have [2,4,5].

The integrable models are Hamiltonian. The classical sinh-G is integrable with Hamiltonian

$$H[\phi] = \gamma_o^{-1} \int_{-\infty}^{\infty} dx[\tfrac{1}{2}\gamma_o^{2}\Pi^2 + \tfrac{1}{2}\phi_x^{2} + m^2(\cosh\phi - 1)] \tag{4}$$

and $\gamma_o > 0$ is a dimensionless coupling constant. There is a bracket {.} and a symplectic manifold M coordinatised by Π, ϕ: $\{\Pi, \phi\} = \delta(x - x')$.

Like all of the classical integrable models the sinh-G model is "completely integrable" [6,8-12,14,15,18,20,21], and there exist, for Π, ϕ ε $\mathbb{R}$ and vanishing b.c.s at ∞, action-angle variables $P(k), Q(k)$: $0 \leq \dot{P}(k) < \infty$; $0 \leq Q(k) < 2\pi$ such that $\{P(k), Q(k') = \delta(k-k')$; $k, k' \varepsilon \mathbb{R}$. The $H[\phi]$, eqn. (4), transforms under the canonical transformation which is the spectral transform to [2,20]

$$H[p] = \int_{-\infty}^{\infty} \omega(k)\ P(k)dk \tag{5}$$

in which $\omega(k) = (m^2 + k^2)^{\frac{1}{2}}$. The repulsive NLS has the same form of H[p], but $\omega(k) = k^2$ and H[p] is the non-relativistic limit of (5) with $m = \frac{1}{2}$.

Note that (5) is a large bunch of harmonic oscillators: $H = \omega P$ for one mode k means $\dot{Q} = \omega$, $\dot{P} = 0$, $Q = \omega t + \delta$, $P = \text{constant}\ (= E\ \omega^{-1})$. Thus (5) apparently describes a linear theory, the linear Klein-Gordon (K-G) model. This is illusory since the k-space is unusual. Discretization to the "allowed" modes $\tilde{k}_n$ shows they satisfy the set of *constraints*

$$\tilde{k}_n = k_n - L^{-1} \sum_{m \neq n} \Delta(\tilde{k}_n, \tilde{k}_m)P_m ; \tag{6}$$

L is the 'volume' (a one dimensional box of length L) and $k_n = 2\pi m_n L^{-1}$, m_n an integer; Δ is a 2-body phase shift of S-matrix type – a phonon-phonon shift, and P_m is a discretized action variable. The P_m and canonical Q_m satisfy $\{P_m, Q_n\} = \delta_{mn}$ and Hamiltonian (5) derives from

$$H[p_n] = \sum_{n=1}^{N} \omega(\tilde{k}_n)P_n \to \int_{-\infty}^{\infty} \omega(\tilde{k})P(\tilde{k})d\tilde{k} \tag{7}$$

as $L \to \infty$ in 'thermodynamic limit'. In this thermodynamic limit N particles in L yield the *finite density* (> 0) $\bar{n} = \lim_{L\to\infty} NL^{-1}$. Thus $L^{-1}H[p\ ]$ $\to \int \omega(\tilde{k})P(\tilde{k})d\tilde{k}$ where $\rho(\tilde{k})$ is $\lim_{L\to\infty} L^{-1}P(\tilde{k})$. Since $P_n = O(1)$ in (6), (7), and $d\tilde{k} = O(L^{-1})$, $P(\tilde{k})$ in (7) is O(L). Thus $\rho(\tilde{k}) > 0$ and $P(\tilde{k})$ is *not* the familiar action variable P(k) [20,21] for ϕ defined on $\mathbb{R}$ with vanishing b.c.s at $|x| = \infty$. In fact [1,2,5,7,11] we evaluate the thermodynamic limits under periodic b.c.s and these generate constraints (6). But, because of (6), (7) shows the $\omega(\tilde{k}_n)$ depend on the P_n and the oscillators are *nonlinear*. Quantum mechanics imposes further *quantum constraints* (see below) on the P_m.

The motion of many quantum oscillators labelled by $\tilde{k}$ in a heat bath is a completely solved problem. In principle the density matrix ρ may be solved for the dynamics [5,16]. The equilibrium solution is $\rho = e^{-\beta\hat{H}}/\mathrm{Tr}e^{-\beta\hat{H}}$, and the partition function is $Z = \mathrm{Tr}e^{-\beta\hat{H}}$ and $\hat{H}$ is the quantum Hamiltonian; β^{-1} is the temperature. This can be written as the functional (path) integral [2,5,16] on the symplectic manifold M

$$Z \quad = \mathrm{Tr}\int \mathfrak{D}\mu \, \exp S[p] \tag{8a}$$

with

$$S[p] = \hbar^{-1}\int_0^{\beta\hbar} dt\left[i\int P(k)Q(k),_t dk - H[p]\right] \tag{8b}$$

in which H[p] is (5): $\mathfrak{D}\mu$ is a measure given below; it is constrained by (6).

Evidently Z can also be defined on the manifold M as [13,24]

$$Z \quad = \mathrm{Tr}\int \mathfrak{D}\Pi\mathfrak{D}\phi \, \exp S[\phi] \tag{9a}$$

$$S[\phi] = \hbar^{-1}\int_0^{\beta\hbar} dt\left[i\int \Pi\phi_t dx - H[\phi]\right] \tag{9b}$$

and $H[\phi]$ is (4). Integration on Π takes this expression to Feynman's more familiar form for Z [13]. This is the trace of the Green's function of a (large dimensional) second order *diffusion* equation. Thus, Brownian motion (chaos) is locked into expressions (9). Then canonical transformation on M via the spectral transform takes eqns. (9) to (8). However the classical action $S[\phi]$ must be defined on a 'space-time torus' $\{x,t\}$. For the Tr means periodicity, period $\beta\hbar$, in 'time' t; and the thermodynamic limit means (conveniently) *periodic* b.c.s in x of period L (with $L \to \infty$ at finite density). It is this which distinguishes the action variables from the familiar ones.

Notice now that, since H[p] is a constant of the motion, $\hbar^{-1}\int_0^{\beta\hbar} dt\, H[p] = \beta H[p]$ and only the *pure phase*

$$i\hbar^{-1}\int_0^{\beta\hbar} dt\int P(k)Q(k),_t dk \equiv i\,\Phi[p] \tag{10}$$

contains the quantum mechanics. We shall replace $\Phi[p]$ by an equivalent set of constraints - *quantum constraints* [2,13-15]. In the classical cases $\Phi[p] \equiv 0$, and $S[p] = -\beta H[p]$.

To impose the quantum constraints we need to know first of all that the quantum integrable models have a 'deep' quantisation vested in the quantum groups [5,12,15 and their references]. The elements of a quantum group are the quantum monodromy matrices $\hat{T}(\lambda)$ [12]. The 'group' is a Hopf algebra [12,22] with co-multiplications [22] $\hat{T}(\lambda)\otimes\hat{T}(\mu)$ and $T(\hat{\mu})\otimes\hat{T}(\lambda)(\lambda,\mu \in \mathbb{C})$: $\otimes$ means Kronecker product. They are related by

$$R(\lambda,\mu)\hat{T}(\lambda)\otimes\hat{T}(\mu) = \hat{T}(\mu)\otimes\hat{T}(\lambda)R(\lambda,\mu), \tag{11}$$

and if $\hat{T}(\lambda)$ is 2x2, $R(\lambda,,\mu)$ is a 4x4 quantum R-matrix [6,23]. The matrix trace yields a large number of commuting operator constants $\hat{\Delta}(\lambda) \equiv \mathrm{Tr}\hat{T}(\lambda)$ and this makes the model 'quantum integrable' [6,15].

Evidently (11) are the *commutation relations* for the elements of $\hat{T}(\lambda)$. These impose quantum forms of the constraints (6) on the quantum functional integrals (8) [4,7,12]. These constraints determine the quantum measures $\mathfrak{D}\mu$. In the same way (6) determines $\mathfrak{D}\mu$ in the classical cases: for classical sinh-G and repulsive NLS (with $\hbar = 1$)

$$\mathfrak{D}\mu = \lim_{N\to\infty} \prod_{n=-\frac{1}{2}N}^{\frac{1}{2}N} (2\pi)^{-1}dP_n dQ_n \tag{12}$$

for N independent lattice points in period L, and only the labels n which satisfy (6) enter (12). If the classical case has soliton solutions $\mathfrak{D}\mu$ is more complicated [1].

In the quantum theories [2,13-15] $\mathfrak{D}\mu$ also involves the quantum constraints which we derive from the phase $\Phi[p]$, eqn. (10): $\Phi[p]$ discretizes to

$$\Phi[p_n] = \hbar^{-1}\int_0^{\beta\hbar} dt \sum_n P_n Q_{n,t}; \tag{13}$$

there is periodicity period $\beta\hbar$ in t and we set

$$\hbar^{-1}\int_0^{\beta\hbar} P_n Q_{n,t} dt = \hbar^{-1}\oint P_n dQ_n = 2\pi m_n \tag{14}$$

in which m_n is a positive integer. There is a winding number ν_n: if ν_n = 1, Q_n moves through 2π round the classical torus labelled by n. Since P_n = constant, $P_n = m_n\hbar$, m_n = +ve integer.

There are two obvious choices for the m_n:

$$m_n = 0,\ 1\ \text{(fermions)};\quad m_n = 0,\ 1,\ 2,\ \ldots\ \text{(bosons)}\ . \tag{15}$$

We adopt *both* choices as fermi-bose *equivalent* descriptions [2,5-7,

11-15]. There are evidently other choices [13-15]. Moreover if $\nu_n = 2$, 3, ... one finds forms of *fractional* quantum statistics [13-15] - an aspect still being worked out.

The results of this analysis are that, for quantum sinh-G and repulsive NLS, $F = -\beta^{-1}\ln Z$ is given in boson description (with chemical potential $\mu = 0$) by [2]

$$\lim_{L\to\infty} FL^{-1} = (2\pi\beta)^{-1} \int_{-\infty}^{\infty} \ln(1 - e^{-\beta\varepsilon(k)})dk \tag{16a}$$

$$\varepsilon(k) = \omega(k) + (2\pi\beta)^{-1} \int_{-\infty}^{\infty} dk' (d\Delta_b(k,k')/dk)\ln(1 - e^{-\beta\varepsilon(k')}) , \tag{16b}$$

where Δ_b is the S-matrix phase shift appearing in (6) which (by (15) when $\hbar = 1$) now takes the form of (6) with $P_m = 0, 1, 2, \ldots$. The wholly equivalent *fermion* form has excitation energies $\tilde{\varepsilon}(k)$ relating to (16b) by $\ln(1 + e^{-\beta\tilde{\varepsilon}(k)}) = -\ln(1-e^{-\beta\varepsilon(k)})$ and Δ_b becomes $\Delta_f = \Delta_b - 2\pi\theta(k' - k)$ with $\theta(x)$ the unit step. The jump at $k = k'$ guarantees the equivalence bose to fermi. A point for repulsive NLS is that $\Delta_f(k, k') = -2\tan^{-1}(c(k-k')^{-1})$ and the *smooth branch* $-2\pi < \Delta_f < 0$ is taken; sinh-G is similar [2]. The fermion result equivalent to (16a,b) was first given for the repulsive NLS model (namely for the 'bose gas') in [19]. The equivalent bose form is in [2] and references. The classical limits of the boson forms (16a,b) for both sinh-G and repulsive NLS are also in [2]: $\ln(1 - e^{-\beta\varepsilon(k)}) \to \ln(\beta\varepsilon(k))$ and $\Delta_b \to \Delta_c$ the classical phonon-phonon shift. This classical result is also found for classical Maxwell-Boltzmann particles.

All these systems, quantum and classical, are pure phonon systems. They are disordered (chaotic) and have a finite Gibbsian entropy S. The different entropies are now distinguished by the different statistics rather than the quantum constraints (14).

We define

$$S = \int_{-\infty}^{\infty} dk\big[(f + \rho)\ln(f + \rho) - f\ln f - \rho\ln\rho\big] \tag{17}$$

where $\rho(k)$ is the phonon density, fermion, boson or Maxwell-Boltzmann: $f(k)$ is the corresponding density of states: $f(\tilde{k}) = (2\pi)^{-1}(dk(\tilde{k})/d\tilde{k})$. The constraints like (6) mean for $\Delta = \Delta_f$, Δ_b or Δ_c, that

$$\tilde{k} = k - \int_{-\infty}^{\infty} \Delta(\tilde{k},\tilde{k}')\rho(\tilde{k}')d\tilde{k} \tag{18}$$

and $f(\tilde{k})$ and $\rho(\tilde{k})$ are then connected [4,5,7] by

$$f(\tilde{k}) = (2\pi)^{-1} + (2\pi)^{-1}\int (d\Delta(\tilde{k},\tilde{k}')/d\tilde{k})\rho(\tilde{k})d\tilde{k} \ . \tag{19}$$

From (5), and (5) constrained by (15) in the quantum cases, the energy densities are $\lim_{L\to\infty} EL^{-1} = \int_{-\infty}^{\infty}\omega(\tilde{k})\rho(\tilde{k})d\tilde{k}$. Minimising $\lim_{L\to\infty}(E - \beta^{-1}S)$ with respect to $\rho(\tilde{k})$ yields (16a,b) for bosons or the fermion result or the classical result. The excitation energies are defined as [7]

$$e^{-\beta\tilde{\varepsilon}(k)} = \rho f^{-1} \ (\text{fermions}); \quad \frac{e^{-\beta\varepsilon(k)}}{1 - e^{-\beta\varepsilon(k)}} = \rho f^{-1} \ (\text{bosons}) \ . \tag{20}$$

The classical limit of the boson description has $\rho \gg f$ so that $f\rho^{-1} = \beta\varepsilon(k)$. There is nothing remarkable about these chaotic systems: they are disordered in the usual way for Brownian particles.

3. The classical sine-Gordon model

Instead of the classical limit of (16a) one finds for classical s-G [1] that

$$\lim_{L\to\infty} FL^{-1} = (2\pi\beta)^{-1}\int_{-\infty}^{\infty} \ln(\beta\varepsilon(k))dk - 2(2\pi\beta)^{-1}\int_{-\infty}^{\infty} e^{-\beta\tilde{E}(k)}dk \ . \tag{21}$$

The excitation energies $\varepsilon(k)$ (phonons) and $\tilde{E}(k)$ (solitons) are coupled in a pair of nonlinear integral equations of the form of the classical limit of (16b) [1]. The $\exp(-\beta\tilde{E}(k))$ may be considered [3,7] to derive from fermion-like kinks and antikinks: $\ln(1+e^{-\beta\tilde{E}(k)}) \to e^{-\beta\tilde{E}(k)}$.

The remarkable feature, given that the solution of the initial value problem for $\phi(x,t)$ on $x \in \mathbb{R}$ consists of kinks, antikinks, phonons *and* breathers, is that no breather contributions appear. This is confirmed by on the one hand iterating eqns. (21) for s-G to obtain a low temperature asymptotic expansion in $t \equiv (8m\gamma_o^{-1}\beta)^{-1}$, a result which is also found by the transfer integral method applied to Z eqn. (9a) [1], with the corresponding iteration of the classical limit of (16a,b) for sinh-G [2]. The two series differ by $\gamma_o \to -\gamma_o$ and additional kink and antikink contributions in the s-G case. There are no breather contributions!

Although we have still to follow the details of the actual dynamics the classical breathers of s-G disappear (ie. break up) into the classical chaos which is the phonons - in thermodynamic limit. As noted we construct this limit under periodic b.c.s period L : $L \to \infty$ at finite density. We demonstrate [5,7]:-

(i) The classical action variables P(k) are *bounded above*: thus $\rho(\tilde{k}) = \lim_{L\to\infty} L^{-1}P(\tilde{k}) \to 0$ so there are no phonons!

(ii) The kink (antikink) momenta p_n *pack* with spacing $O(L^{-1})$, $p_n = 2\pi nL^{-1}$, and become dense for $L \to \infty$.

(iii) The breathers have translational momenta p_ℓ and internal action variables $4\gamma_o^{-1}\Theta_\ell$; $0 \le \Theta_\ell < \frac{1}{2}\pi$ with conjugate variables Φ_ℓ: $\{4\gamma_o^{-1}\Theta_\ell, \Phi_{\ell'}\} = \delta_{\ell\ell'}\}$. The $p_\ell = 2\pi\ell L^{-1}$, and become dense for $L \to \infty$.

(iv) The Θ_ℓ "collapse" and the breathers become large amplitude phonons with action variables $\bar{P}_\ell = 16\gamma_o^{-1}\sum_i \Theta_{\ell i}$:- $\Theta_{\ell i}$ is the *set* of Θ's associated with each label ℓ and we choose to pack these like the p_ℓ with spacing $O(L^{-1})$ and in a narrow "cone" of angle $O(L^{-1})$. It is then easy to see [5,7,13] that $\frac{1}{4}L^{-1}\sum_i\Phi_{\ell i} = \bar{Q}_\ell$ is canonical to $\bar{P}_\ell$: $\{\bar{P}_\ell, \bar{Q}_{\ell'}\} = \delta_{\ell\ell'}$, while $L^{-1}\bar{P}_\ell = L^{-1}O(1) \to L^{-1}\bar{P}(k)d\tilde{k} = \rho(\tilde{k})d\tilde{k}$ and $\rho(\tilde{k}) > 0$ since there is no upper bound on the $\bar{P}(\tilde{k})$. The phonon density $\rho(\tilde{k})$ is thus created by the breathers in thermodynamic limit: the $\Theta_{\ell i}$ lower their energies by collapsing into the narrow cone of angle $O(L^{-1})$ at the bottom of their energy band.

4. The quantum s-G model

The quantum breathers are stabilised by quantum mechanics and do not break up:- The quantum 'phonons' (oscillator contribution) disappear: work on the quantum s-G [4] and work on the fermi-bose equivalent quantum massive Thirring model [25-30] shows they go first of all (at zero β^{-1}) as renormalisations in the problem: $\gamma_o \to \gamma_o'' = \gamma_o[1 - \gamma_o/8\pi]^{-1}$, $M = 8m\gamma_o^{-1} \to 8m(\gamma_o'')^{-1}$ in which $m = \frac{1}{8}M\gamma_o''$ is renormalised and M is renormalised.

At finite temperatures the quantum breather translational momenta are chaotic (follow chaotic trajectories) in the same way that the quantum phonons of sinh-G are chaotic. So far [4] we work only in fermion description for the translational momenta. However the quantum internal coordinates $\Theta_{\ell i}$ exhibit structure which persists at finite temperatures. They are not confined to a narrow "cone" of angle $O(L^{-1})$ in thermodynamic limit: they define instead $N_b - 2$ discrete and different fermion-like quantum particles. The contribution of $\Theta_{\ell i}$, $\Phi_{\ell i}$ to the quantum phase $\Phi[p]$ proves to be $\hbar^{-1}\int_0^{\beta\hbar}dt \sum_\ell \sum_i 4(\gamma_o")^{-1} \Theta_{\ell i}\Phi_{\ell i,t}$. The equivalent quantum constraints are therefore

$$\hbar^{-1} \int_0^{\beta\hbar} dt\, \Theta_{\ell i}\Phi_{\ell i,t} = \hbar^{-1}\oint 4(\gamma_o")^{-1}\Theta_{\ell i}d\Phi_{\ell i} = 2\pi m_{\ell i} \tag{22}$$

in which $m_{\ell i}$ is a positive integer. Then (for $\hbar = 1$)

$$\Theta_{\ell i} = m_{\ell i}\gamma_o"/16;\; m_{\ell i} = 1, 2, \ldots, N_b - 1\,. \tag{23}$$

The number $N_b - 1 = [8\pi/\gamma_o"] = [8\pi/\gamma_o] - 1$ in which $[x]$ is the integral part of x. There are thus $N_b - 2$ *distinct* quantum breathers labelled by quantum numbers $m_{\ell i} = 1, 2, \ldots, N_b - 2$ for each ℓ and translational momentum p_ℓ. Their masses are $M_{\ell i} = 2M \sin(m_{\ell i}\gamma_o"/16)$ and there is in addition one quantum kink-antikink *pair* of mass 2M [3-6,26-30]. The free energy density then proves to be [4,5] (with zero chemical potentials)

$$\lim_{L\to\infty} FL^{-1} = -(2\pi\beta)^{-1}\sum_{\ell=1}^{N_b-2}\int_{-\infty}^{\infty} d\alpha_\ell M_\ell \cosh\alpha_\ell \ln(1 + e^{-\beta\tilde{\varepsilon}_\ell})$$

$$- 2(2\pi\beta)^{-1}\int_{-\infty}^{\infty} d\alpha_s M_s \cosh\alpha_s \ln(1 + e^{-\beta\tilde{\varepsilon}_s})\,; \tag{24}$$

$M_\ell \equiv 2M \sin(\ell\gamma_o"/16)$, $M_s = M$; $\varepsilon_\ell = \varepsilon_\ell(\alpha_\ell)$, etc. and rapidities α are being used; $N_b - 1$ coupled integral equations determine the $N_b - 2$ excitation energies ε_ℓ and the ε_s. Thus in the quantum theory $N_b - 1$ quantum breather-like particles contribute to the free energy: there are no phonon contributions, in total contrast with the classical case (§3).

In the *semi*-classical limit $\gamma_o \to 0$ of this theory [3], the $N_b - 2$ quantum breathers become one collective boson; the remaining kink-antikink pair survives as fermion-like contributions. In further *classical* [3,31] limit, the single collective boson becomes the large amplitude chaotic phonons of §3 while the kink-antikink fermion pair becomes the 'classical fermions' of §3. In short at those temperatures where the classical limit is applicable the $N_b - 2$ distinct quantum breathers with label ℓ 'break up' into phonons in equilibrium labelled by chaotic translational momentum p_ℓ.

So what is the quantum imposed structure which survives at finite temperature? From the point of view of the quantum Bethe Ansatz applied to the fermi-bose equivalent Thirring model the quantum breather with quantum number n and mass $M_n = 2M \sin(n\gamma_o''/16)$ is an "n-string" [25-30]. The bare fermion-like particles of mass m_o have complex rapidities β_ℓ. The bare particles, which are of phonon type and which have complex rapidities $\beta_\ell = \alpha_\ell + i\pi$, ($\alpha_\ell \in \mathbb{R}$) have energy $m_o \cosh(\alpha_\ell + i\pi) = - m_o \cosh \alpha_\ell$. These fill the negative energy states of the Dirac sea. The n-string has $\beta_\ell = \alpha_\ell + i\omega\ell_n$, $\ell_n = -(n - 1), -(n - 2), \ldots, (n - 1)$; $\omega = \frac{1}{8}\gamma_o$. Associated with these is $A(n) \le n$ 'holes'. These $A(n) \le n$ holes are holes in the Dirac sea located at $\alpha_\ell + i\pi$. The omission of the A(n) particles in the Dirac sea causes a 'back-flow' in the packing of the α_ℓ [25,26]. The n-string plus $A(n) \le n$ holes plus back flow becomes a complex particle. The remaining holes become kinks and antikinks in pairs. The n-string complex is the quantum breather indexed by n with mass (now renormalised) M_n. The rapidities are also renormalised.

Hole pairs have translational momentum only and follow chaotic trajectories. The n-strings have chaotic translational momenta plus internal structure (namely the n-string + A(n) holes + back flow) which is preserved at finite temperatures. There are no other particles. Quantum mechanics thus stabilises the classical chaotic breathers and imposes the Bethe ansatz structure through the quantum constraints (22). This situation is generic and includes at least [4,5] the cases of the attractive NLS model, Landau-Lifshitz model, Heisenberg magnet, the Heisenberg magnetic in a longitudinal field [17] and the classical massive Thirring model with commutative bracket [4]. A somewhat equivalent account to that of this paper will appear in [32].

References

1. J. Timonen, M. Stirland, D.J. Pilling, Yi Cheng and R.K. Bullough, Phys. Rev. Lett. 56, 2233 (1986).
2. R.K. Bullough, D.J. Pilling and J. Timonen, J. Phys. A: Math. Gen. 19, L955 (1986).
3. J. Timonen, R.K. Bullough and D.J. Pilling, Phys. Rev. B34, 6525 (1986).
4. Yu-Zhong Chen, 'Classical and quantum statistical mechanics of the 1+1 dimensional integrable models', Ph.D. thesis, University of Manchester, July 1989.
5. R.K. Bullough and J. Timonen in: "Nonlinear World" V.G. Bar'yakhtar et al. eds., World Scientific, Singapore (1990) pp.1377-1422, and references.

6. R.K. Bullough, D.J. Pilling and J. Timonen in: "Solitons", M. Lakshmannan ed., Springer-Verlag, Heidelberg (1988), pp.250-281, and references.
7. J. Timonen, Yu-zhong Chen and R.K. Bullough, Nucl. Phys. B (Proc.Suppl.) 5A, 58 (1988).
8. R.K. Bullough, Yu-zhong Chen, S. Olafssen and J. Timonen in: "Integrable Systems and Applications", M. Balabane et al. eds., Springer-Verlag, Heidelberg (1988), pp.12-26.
9. R.K. Bullough, J. Timonen, Yu-zhong Chen, Yi Cheng and M. Stirland in: "Nonlinear evolution equations: integrability and spectral methods", A Degasperis et al eds., Manchester University Press, Manchester (1990), pp.605-617.
10. R.K. Bullough and S. Olafsson, in: IXth Intl. Congress of Math. Phys. 17-27 July, 1988, Swansea, Wales, B. Simon, A. Truman and I.M. Davies eds., Adam Hilger, Bristol (1989), pp.329-334.
11. R.K. Bullough and S. Olafsson, in: "Proc. 17th Intl. Conf. on Diff. Geom. Methods in Theor. Phys." A.I. Solomon ed., World Scientific, Singapore (1989), pp.295-309.
12. R.K. Bullough, Yu-zhong Chen and J. Timonen, in: "Proc. XVIII Intl. Conf. on Diff. Geom. Methods in Theor. Phys." Ling-Lie Chau and Werner Nahm eds., Plenum, New York (1991) and references.
13. R.K. Bullough and J. Timonen, in: "Microscopic Aspects of Nonlinearity in Condensed Matter", A.R. Bishop and V. Tognetti eds., Plenum, New York. To appear 1991.
14. R.K. Bullough and J. Timonen, in: "Nonlinear Science: the Next Decade", A.R. Bishop and D.K. Campbell eds., Physica D: Nonlinear Phenomena. To appear 1991.
15. R.K. Bullough and J. Timonen, in: Proc. XIX Intl. Conf. on Diff. Geom. Methods in Theor. Phys., C. Bartocci, U. Bruzzo and R. Cianci eds., Springer-Verlag, Heidelberg. To appear 1991.
16. J. Timonen, D.J. Pilling and R.K. Bullough, in: "Coherence, Cooperation and Fluctuations", F. Haake et al. eds., C.U.P., Cambridge (1986) pp.18-34.
17. R.K. Bullough, Yu-zhong Chen, J. Timonen, V. Tognetti and R. Vaia, Phys. Lett. A145, 54 (1990).
18. "Solitons", TCP 17, R.K. Bullough and P.J. Caudrey eds., Springer-Verlag, Heidelberg, 1980, especially Chapter I.
19. C.N. Yang and C.P. Yang, J. Math. Phys. 10, 1115 (1969).
20. R.K. Dodd and R.K. Bullough, Physica Scripta 20, 514 (1979).
21. L.D. Faddeev and L.A. Takhtadjan "Hamiltonian Methods in the Theory of Solitons", Springer-Verlag, Berlin (1987).
22. S. Majid, Intl. J. Mod. Phys. A. To appear as of 1989.
23. P.P. Kulish and E.K. Sklyanin, in:"Integrable Quantum Field Theories", J. Hietarinta and C. Montonen eds., Springer-Verlag, Heidelberg (1982).
24. R.K. Bullough in: "Nonlinear Phenomena in Physics", F. Claro ed., Springer-Verlag, Heidelberg (1985), pp.70-103 and references.
25. H.B. Thacker, Rev. Mod. Phys. 53, 253 (1981).
26. H. Bergknoff and H.B. Thacker, Phys. Rev. D 19, 3666 (1979).
27. M. Fowler and X. Zotos, Phys. Rev. B24, 2634 (1981); 25 5806 (1982).
28. M. Imada, K. Hida and M. Ishikawa, J. Phys. C 16, 35 (1983).
29. S.G. Chung and Yia-Chung Chang, J. Phys. A: Math. Gen. 20, 28-75 (1987).
30. J. Timonen and R.K. Bullough, to be published.
31. Nu Nu Chen, M.D. Johnson and M. Fowler, Phys. Rev. Lett. 56, 507 (1986); 1427 (E) (1986).
32. R.K. Bullough and J. Timonen, in: "Nonlinearity with disorder", F.Kh. Abdullaev, A.R. Bishop and S.T. Pneumaticos eds., Springer-Verlag, Berlin (1991). To appear.

Soliton Dynamics and Chaos Transition in a Microstructured Lattice Model

M.K. Sayadi and J. Pouget

Laboratoire de Modèlisation en Mécanique (associé au CNRS),
Université Pierre et Marie Curie, Tour 66,
4, Place Jussieu, 75252 Paris Cedex 05, France.

The nonlinear dynamics of solitons in a lattice model involving internal degrees of freedom in rotation is presented. The emphasis is placed especially on the behavior of solitons under the influence of a constant or time-dependent applied field and damping. The basic lattice model is made of a one-dimensional deformable atomic chain equipped with rotatory molecules. The propagation of coupled solitons in molecule rotation and elastic deformations is placed in evidence and two main classes of solutions are examined. The transient motion of coupled solitons under an applied constant field is presented by means of a method of reduction leading to an equation of motion for the soliton mass center. The problem of the soliton dynamics under the influence of discreteness effects and a time-dependent applied field allows us to show a transition to chaos of the soliton motion.

1. INTRODUCTION

One of the most fascinating features of nonlinear physical systems is they can exhibit spatial and temporal structures, which can be described both by their **coherent** [1] and **chaotic** natures [2]. We propose to illustrate these ideas by means of a rather simple nonlinear lattice model which possesses all the ingredient leading to interesting phenomena such as **soliton propagation** or **chaos transition**. Our physical system is modelled by a one-dimensional monoatomic chain equipped, at each of its nodes, with a rotatory molecules. Two main types of motion can be distinguished in this picture: (i) the longitudinal and transverse displacements of the mass center of the molecules and (ii) two rigid-body rotational motions of the molecules about the chain axis and perpendicular to the latter [3,4]. Since the rotation may have large amplitudes, the lattice has a nonlinear behavior with respect to the rotational motions, however, the coupling between the molecule rotations and lattice deformations plays a central role in the nature of the nonlinear excitations. The propagation of localized structures is investigated in the framework of the continuum version of the microscopic model. Then, the existence of **coupled solitons in rotation and lattice deformation** is proved for two elementary configurations. Such solitons models are particularly suitable for the study of defect movements in crystals associated with the structures in domains and walls. For instance, moving ferroelectric walls in molecular ferroelectric crystals [5] or propagation of a twist defect or kink twist in a long deformable chain of macromolecules [6] (polymer, biological material DNA).

A special attention is devoted next to the influence of an applied field and damping. The **transient motion of a soliton** under an applied constant field is examined and by means of an average technique we find an effective equation governing the soliton position [7]. A second problem to study is the dynamics of the soliton motion on the discrete model. Using a reductive method based on the notion of collective coordinates we place the role of the discreteness effects in evidence. In fact, in addition to the **discreteness effects** we consider a damping process and sinusoidally time-dependent external field and we find, in the range of the first approximation of the discreteness correction, the equation which governs the mass center associated with the soliton. This equation is merely the damped, sinusoidally driven pendulum. Numerical simulations carried out directly on the lattice model show particularly rich dynamics of the soliton and exhibit **transition to chaos** including a cascade of subharmonics as well as strange attractors [8]. The present work illustrate the dynamics of a rather complex system (i.e. the microsctructured lattice) which can be described both by coherent structures (coupled solitons) and by irregular motions as chaos.

2. LATTICE MODEL

Let us consider a one-dimensional monoatomic chain equipped, at each node, with a molecule (see Fig. 1). We suppose next that each molecule is endowed with a microscopic electric dipole of constant length which indicates the molecule orientation. The forces acting on such a system result from (i) the interactions between the first-nearest neighbors, (ii) the mutual interactions between the microscopic dipoles and (iii)the electrostatic interaction due to the external field **E**. On the other hand, the possible motions of the lattice model are (i) the **longitudinal** and **transverse lattice displacements** of the mass center of the molecules denoted by U_n et V_n, respectively, and (ii) the **rigid-body rotational motions** of the molecules or dipoles. The latter can have two rotation angles (see Fig. 1), θ_n rotation of the plan (Π) about the z-axis and φ_n the angle corresponding to the rotation about an axis perpendicular to (Π). Furthermore, we assume that the lattice displacements are infinitesimal whereas the rotation may have large amplitudes [3,4].

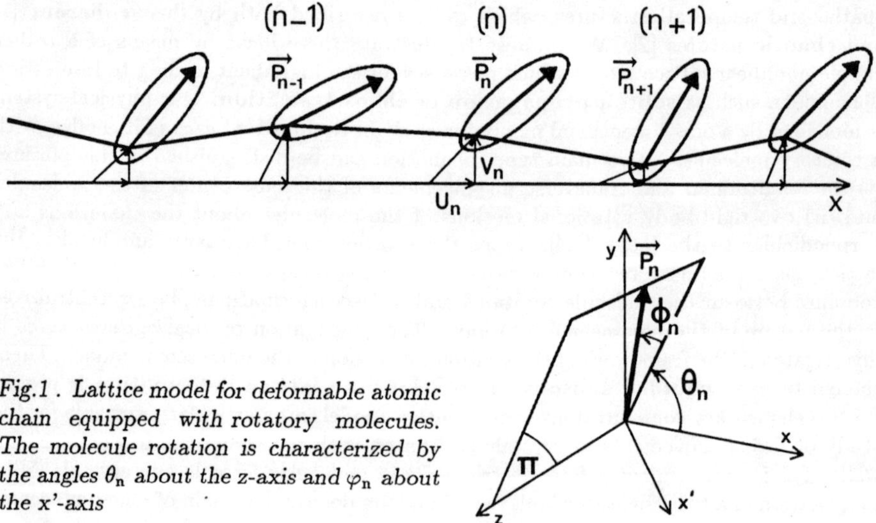

Fig.1 . Lattice model for deformable atomic chain equipped with rotatory molecules. The molecule rotation is characterized by the angles θ_n about the z-axis and φ_n about the x'-axis

Two simple configurations can be examined : (a) configuration (below referred to as A) defined by $\varphi_n = 0$ and $\theta_n \neq 0$, and the dipoles rotate about the z-axis, and (b) configuration B where the dipoles rotate about the chain axis ($\theta_n = \pi/2$).

3. SOLITON PROPAGATION FOR THE CONFIGURATION A

This situation is defined by setting $\varphi_n = 0$ and the Hamiltonian of the discrete system reads as [4]

$$\begin{aligned} H_A &= \sum_n \frac{1}{2}\left(\dot{U}_n^2 + \dot{V}_n^2 + \dot{\Theta}_n^2\right) + \sum_n \frac{1}{2}\left[C_L^2\left(U_{n+1}-U_n\right)^2 + C_T^2\left(V_{n+1}-V_n\right)^2\right] \\ &+ \sum_n \frac{1}{2}\left[\alpha\left(U_{n+1}-U_n\right) cos\Theta_n - \beta\left(V_{n+1}-V_n\right) sin\Theta_n\right] \\ &- \sum_n \left[4cos\left(\frac{\Theta_{n+1}-\Theta_n}{2}\right) - \chi cos\left(\frac{\Theta_{n+1}+\Theta_n}{2}\right)\right] \quad , \qquad (1) \end{aligned}$$

where we have introduced nondimensional notations for sake of convenience and we have set $\Theta_n = 2\theta_n$. The kinetic energy including both translational and rotational motions is defined by the first term of the right hand side of Eq.(1). The second term represents the elastic energy where C_L and C_T are the longitudinal and transverse elastic wave velocities. The third term is the energy of interaction between the lattice of dipoles and lattice deformation and finally the fourth term holds for the dipole interaction.

We suppose now that the discrete displacements and rotations are slowly varying over a crystalline spacing, so that we can get the continuum model of which the equations of motion are [4]

$$U_{tt} - C_L^2 U_{XX} = \alpha(cos\Theta)_X \quad , \qquad (2a)$$

$$V_{tt} - C_T^2 V_{XX} = -\beta(sin\Theta)_X \quad , \qquad (2b)$$

$$\Theta_{tt} - \Theta_{XX} = \chi sin\Theta + \alpha U_X sin\Theta + \beta V_X cos\Theta \quad . \qquad (2c)$$

The structure of the system of nonlinear equations is of special interest. It consists of two **wave equations** for the elastic displacements U and V and a **sine-Gordon equation** for the rotation Θ (twice the physical rotation), each equation being nonlinearly coupled. We look for solutions as functions of the single variable $\xi = X - X_0 - Ct$ where C is the phase velocity. By eliminating the displacements U and V between Eqs (2a), (2b) and (2c) we arrive at

$$\left(1-C^2\right)\Theta_{\xi\xi} = -\chi\left(sin\Theta + \frac{1}{2}\delta(C) sin2\Theta\right) \quad , \qquad (3)$$

where we have defined

$$\chi\delta(C) = \frac{\beta^2}{C_T^2 - C^2} - \frac{\alpha^2}{C_L^2 - C^2} \quad . \qquad (4)$$

We note that the problem of finding solutions to Eqs(2a)-(2c) amounts to solving a nonlinear ordinary differential equation of the **double sine-Gordon** type. Analytical solutions can be obtained for the rotation and elastic deformations but we prefer to illustrate the soliton solution by means of numerical simulations given in Fig. 2. Figure 2.a shows the rotation of the molecules from the region $\theta = -\pi/2$ to the region $\theta = \pi/2$

and we have a kink- like soliton. The elongational deformation generated through the coupling is plotted in Fig. 2.b. At last, the shear deformation is given in Fig. 2.c. Then, these solitons thus computed correspond to the coupled soliton in rotation and lattice deformations.

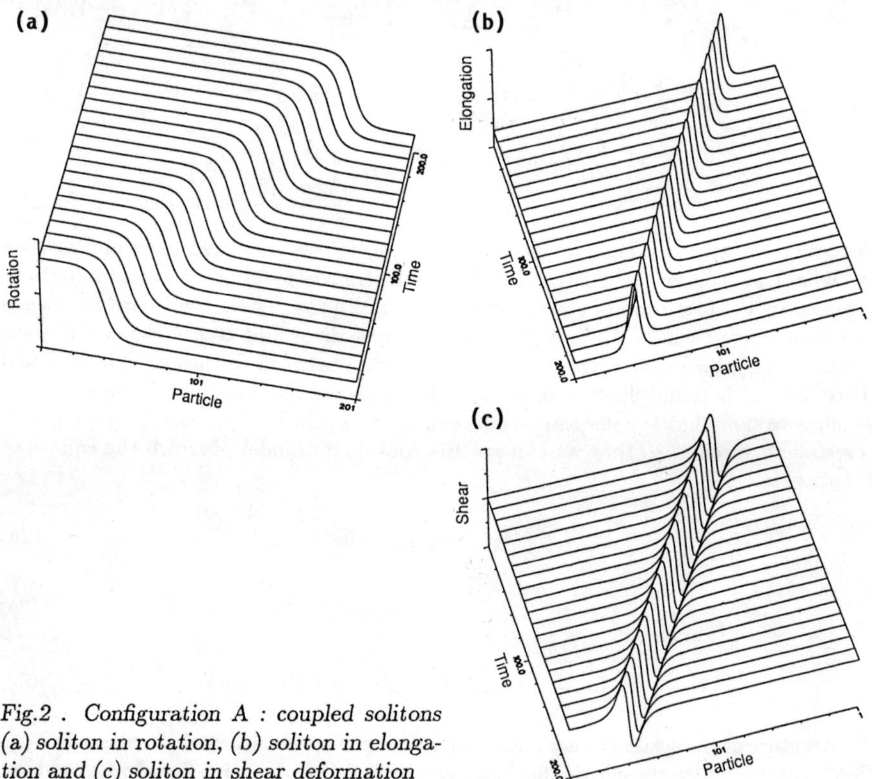

Fig.2 . Configuration A : coupled solitons (a) soliton in rotation, (b) soliton in elongation and (c) soliton in shear deformation

4. SOLITON PROPAGATION FOR THE CONFIGURATION B

For this situation the Hamiltonian of the system is obtained by putting $\theta_n = \pi/2$ and it is written as [4,7]

$$H_B = \frac{1}{2}\sum_n \left(\dot{U}_n^2 + \dot{\varphi}_n^2\right) + \frac{1}{2}\sum_n C_L^2 \left(U_{n+1} - U_n\right)^2$$

$$-\sum_n \left[\cos\left(\varphi_{n+1} - \varphi_n\right) - \mu\cos\left(\varphi_{n+1} + \varphi_n\right)\right] + \kappa\sum_n \left(U_{n+1} - U_n\right)' \cos\left(\varphi_{n+1} - \varphi_n\right) \quad (5)$$

Now we search for solutions as slowly extended waves with respect to the lattice spacing, so that the continuum approximation can be considered. We then find the set of equations

$$U_{tt} - C_L^2 U_{XX} = -\frac{1}{2}\alpha\left(\Phi_X\right)_X^2 \ , \quad (6a)$$

$$\Phi_{tt} - \Phi_{XX} = \lambda sin\Phi - \alpha\,(U_X \Phi_X)_X \quad , \tag{6b}$$

where we have set $\alpha = \kappa/2$, $\lambda = 4\mu$ and $\Phi = 2\varphi$. If the coupling between the displacement and rotation is removed, we recover the usual sine-Gordon equation for the rotation and the wave equation for the longitudinal displacement. However, because of the coupling, the structure of Eqs(6a) and (6b) is much more complicated and its solutions are not trivial. But the one-soliton solution can be obtained in the form of traveling waves.

In order to check the continuum approximation we consider numerical simulations carried out directly on the discrete equations. The results are presented in Fig. 3. The rotation of the molecules is plotted in Fig. 3.a which consists of a kink moving from the domain $\varphi = -\pi/2$ to the domain $\varphi = \pi/2$. The associated elongational deformation is given in Fig. 3.b and shows a sharp hump within the width of the rotation kink.

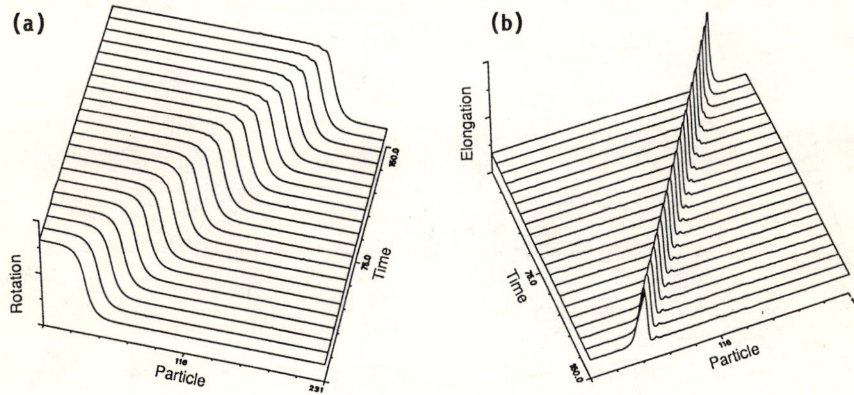

Fig.3 . Configuration B : numerical simulations (a) soliton in rotation and (b) soliton in elongation

5. INFLUENCE OF AN APPLIED CONSTANT FIELD ON A MOVING SOLITON

The problem that we want to solve is the **starting motion of a soliton from rest** when a constant applied external field F is applied to the system. With this in view, we consider the configuration A and the last equation (see Eq.(2.c)) is only modified and reads as

$$\Phi_{tt} - \Phi_{XX} + \gamma\Phi_t = \chi sin\Phi + \alpha U_X sin\Phi + \beta V_X cos\Phi - F cos(\Phi/2) \quad , \tag{7}$$

where the applied field F and dissipative process associated with the rotation have been added. Equations (2.a) and (2.b) are unchanged.

The additional terms (force and damping) to the unperturbed equations (2.a)-(2.c) are considered as small perturbations acting on the coupled soliton in rotation-deformation. The *a priori* effects of the perturbations are : (i) the evolution in time of the soliton location and (ii) modification of the soliton shape (dressing mode) accompanied by radiations. Accordingly, a perturbative scheme is developed and it is based on the hypothesis which consists roughly of assuming that the new variable is $\xi = x - X(t)$ with $dX/dt = C(t)$ where X and C are the location and velocity of the soliton, respectively. That means now that the **soliton position** or the associated mass center is treated as **dynamical variable**. We force the soliton solution to Eqs(2.a)-(2.b) to satisfy a variational principle with respect to the new variable $X(t)$ [3,7]. The Lagrangian principle is, in fact, rewritten for the soliton position including the generalized forces connected with the external force and damping. After some manipulations we find

$$\frac{d}{dt}\left(\frac{C}{\sqrt{1-C^2}}\right) = \hat{F} - \gamma\frac{C}{\sqrt{1-C^2}} \quad , \tag{8}$$

where $\hat{F} = F/2\sqrt{\hat{\chi}}$ with $\hat{\chi} = \chi/(1 - \delta(0)/3)$ is an effective force accounting for the coupling coefficients (see Eq.(4)). It is worthwhile noting that Eq.(8) coincides with that of a relativistic particle of unit rest mass which is uniformly accelerated by $\hat{F}$ and damped.

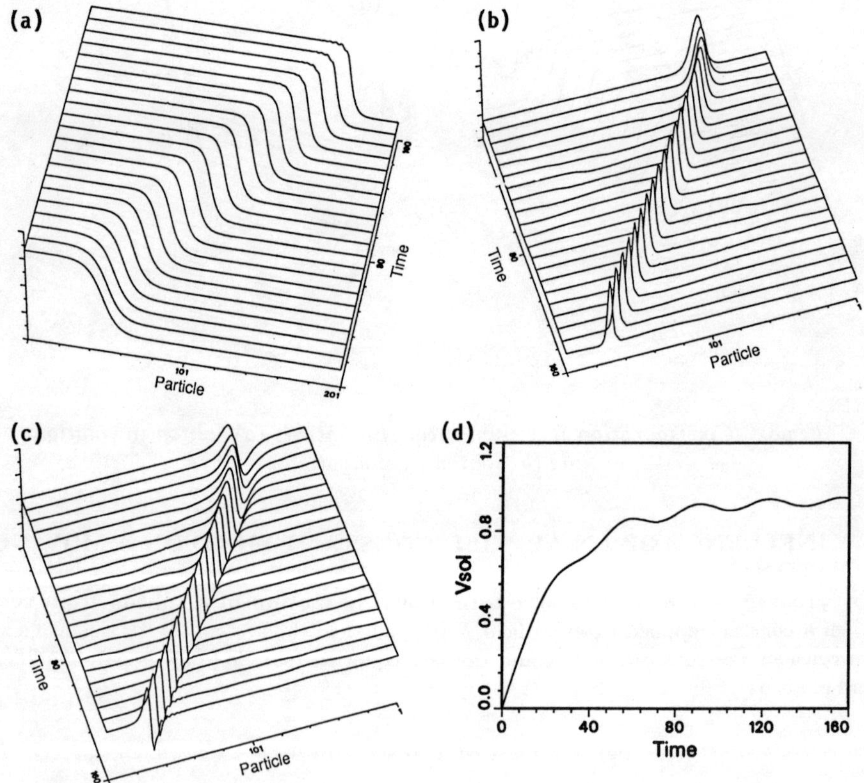

Fig.4 . Influence of an applied constant field and damping on a soliton : soliton in (a) rotation, (b) elongation, (c) shear and (d) soliton velocity vs time

The numerical simulations of this problem are pictured in Fig. 4. Figure 4.a gives the computerized soliton in rotation, we observe that the soliton begins to move and, progressively, gains a constant velocity which depends on the applied field and damping. The corresponding elongational and shear deformations are presented in Figs 4.b and 4.c, respectively. Furthermore, the soliton velocity extracted from the numerical simulations is plotted in Fig.4.d. These numerical results fairly agree with the analytical solution to Eq.(8).

6. DISCRETENESS EFFECTS AND TIME-DEPENDENT APPPLIED FIELD : TRANSITION TO CHAOS

We consider a simplified model extracted from that built in Sections 2 and 3. The lattice displacements are removed and we assume that the intermolecular interactions between the first-nearest neighboring molecules is decomposed into a linear torque and a periodic potential. Therefore, the equations of motion for the rotations take on the following form [8]

$$\ddot{\Phi}_n - (\Phi_{n+1} - 2\Phi_n + \Phi_{n-1}) = \ell_0^{-2} sin\Phi_n - \gamma\dot{\Phi}_n - F(t) cos\,(\Phi_n/2) \quad , \tag{9}$$

where we have set $\Phi_n = 2\theta_n$, and a damping term and external time-dependent field have been added. The single-soliton solution obtained from the continuum approximation and non-relativistic limit is given by

$$\Phi_n(t) = \pi - 4tan^{-1}\left(e^{\xi_n/\ell_0}\right) \quad . \tag{10}$$

The semi-discrete nature of the solution has been considered by setting $\xi_n = n - X(t)$. Because, first, of the **discrete nature of the lattice**, and next, of the influence of the damping and external field, the kink position depends on time in a nontrivial way and it can be considered as a **particle- like object**. On using the same technique as in Section 5, we derive an effective equation for the kink position. On neglecting the dressing correction, this equation reads as [8]

$$\ddot{\varphi} + \Gamma\varphi + sin\varphi = F_D sin\,(\omega_D t) \quad , \tag{11}$$

where we have defined

$$\Gamma = \gamma/\Omega_{PN} \ , \quad F_D = -\pi\ell_0 F_0/\Omega_{PN}^2 \ , \quad \omega_D = \omega_0/\Omega_{PN} \quad , \tag{12a}$$

$$\varphi(t) = 2\pi X(t) \ , \quad \Omega_{PN}^2 = \pi^6 \ell_0 \left(2 - 1/\pi^2\ell_0^2\right)/3sinh\left(\pi^2\ell_0\right) \quad . \tag{12b}$$

Moreover, the time-dependent field is $F(t) = F_0 sin\,(\omega_0 t)$. We notice that the equation for $X(t)$ or $\varphi(t)$ is merely the equation of the **sinusoidally driven and damped pendulum** [8]. The problem that we want to tackle is to ascertain, by carrying out numerical simulations directly on the discrete system, that the motion of the kink position is rather well described by Eq.(11). The control parameters which determine the soliton dynamics on the discrete system is the amplitude of the driving force F_0, the other parameters ω_0, ℓ_0 and γ are kept constant. The characteristic length ℓ_0 (the soliton width is $\pi\ell_0$) is chosen such that the Peierls-Nabarro frequency is not too large but producing rather significant discreteness effects [9]. For small enough F_0, the soliton position motion is periodic and a closed orbit in the phase plane is a limit cycle. When the driving force amplitude F_0 is increased we observe a **cascade of**

odd subharmonic components. The situation depicted in Fig. 5 corresponds to the phase portrait of a closed orbit with a period five times the driving period. By increasing further the amplitude of the driving force, the dynamics of the soliton position becomes more complicated. For this case, the Poincaré section plotted in Fig. 6 exhibits folded structures which is a **strange attractor** [8].

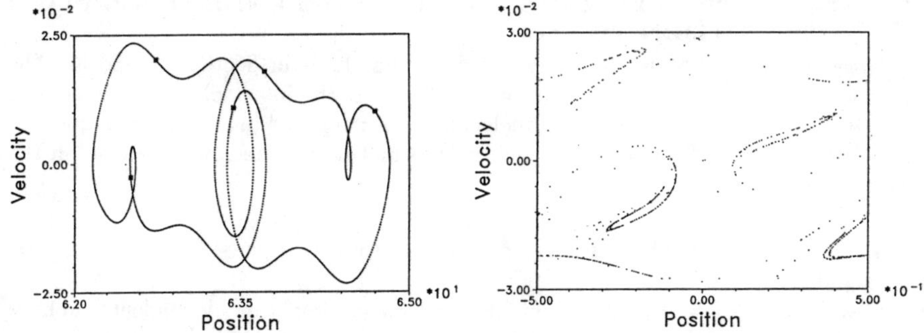

Fig.5 : Motion of the soliton position : the phase portrait shows a motion with a period five times the drive period

Fig.6 . Poincaré section for the chaotic motion of the soliton position

7. CONCLUSIONS

In this paper we have investigated the dynamics of solitons on a discret lattice model, made of a deformable atomic chain equipped with rotatory molecules. On the basis of this model we have studied the propagation of coupled solitons in rotation and deformation which characterize the coherent structures of the physical system. Moreover, the problem of the transient motion of a soliton from rest has been examined. For this problem, we have used a variational principle which has provided the equation of motion for the soliton position. On the other hand, on accounting for discreteness effects of the lattice and sinusoidally time-dependent applied field we have shown that the discrete system gives rise to very rich dynamics which does not exist in the continuum approximation. Then, a cascade of odd subharmonics and transition to chaos have been examined by means numerical simulations. This has proved that the motion of the soliton mass center can be described by the dynamics of the forced and damped pendulum. The latter is, in fact, a classical and simple nonlinear system with a rather complex dynamics [10]. Nevertheless, a possible extension of the present work could be the computation of the fractal dimension of the strange attractor thus obtained in order to compare it to that of the strange attractor corresponding to the pendulum [10].

References

[1] A.C. NEWELL, Trans. of the ASME, *J. Appl. Mech. 50*, 1127 (1983).

[2] P.J. HOLMES and F.C. MOON, Trans. of the ASME, *J. Appl. Mech. 50*, 1021 (1983).

[3] J. POUGET, NATO-ASI Series C, 225 "*Physical and Thermodynamic Behaviour of Minerals*", ed. E.K.H. Salje (D. Reidel, Dordrecht, 1988), p. 359.

[4] M.K. SAYADI and J. POUGET (Submitted for publication in J. Phys. A).

[5] K.N. KINASE, W. MAKINO and K. TAKAHASHI, *Ferroelectrics 64*, 173 (1985).

[6] Ph. L. TAYLOR and A. BANERTEA, *Ferroelectrics 66*, 135 (1986).

[7] J. POUGET and M.K. SAYADI, IUTAM Proc. on *Mechanical Modelling of New Electromagnetic Materials*, ed. R.K.T. HSIEH (Elsevier, Amsterdam, 1990), p. 179.
[8] M.K. SAYADI and J. POUGET (Submitted for publication in Physica D).
[9] R. BOESCH and C.R. WILLIS, *Phys. Rev. B39*, 361 (1989).
[10] E.G. GWINN and R.M. WESTERVELT, *Phys. Rev. A33*, 4143 (1986).

What is the Role of Dynamical Chaos in Irreversible Processes?

P. Gaspard

Faculté des Sciences,
Université Libre de Bruxelles,
Campus Plaine CP 231, 1050 Bruxelles, Belgium.

1. Irreversible Processes

The motion of atoms and molecules in matter is described by Hamiltonians like

$$H = \sum_{i=1}^{N} \frac{p_i^2}{2m} + \sum_{1 \leq i < j \leq N} V(r_{ij}) \; . \tag{1.1}$$

Trajectories in phase space are uniquely determined from the initial positions and velocities of the particles. Because of the time-reversal symmetry of Hamiltonian systems, we can recover these initial conditions, if we reverse the velocities a short time after the beginning of the process. However, when we wait too long before reversing the velocities, a return to the initial conditions is no longer observed. This irreversibility results into dramatic effects on the large scale properties of matter like diffusion, viscosity, heat or electrical conductivities: the phenomenological equations describing these transport properties like the diffusion equation,

$$\partial_t f = \mathcal{D} \nabla^2 f \; , \tag{1.2}$$

are not time-reversal symmetric.

The irreversibility of physical processes has been described for a long time [1]. Nowadays, advances made in the theory of dynamical chaos in nonlinear systems [2-39] and the outcoming debate [2-7] are suggesting a quantitative and intrinsic understanding for this phenomenon in terms of dynamical sensitivity to initial conditions. In this context, we recently derived a formula relating the transport coefficients to the characteristic quantities of dynamical chaos like the Lyapunov exponents and the Kolmogorov-Sinai (KS) entropy per unit time [8].

Our purpose in this communication is to summarize these new results. In Secs. 2 to 4, we recall properties of Lyapunov exponents and of the KS entropy in large systems of statistical mechanics. In Sec. 5, we describe the phenomenon of classical chaotic scattering and the important concept of fractal repeller [9-12]. In Sec. 6, we show how the escape rate of the repeller can be obtained from the Liouville dynamics

[13, 14]. In Sec. 7, we associate fractal repellers to the nonequilibrium states of large dynamical systems and we obtain the relationship between the transport coefficients and the properties of dynamical chaos [8]. Conclusions are drawn in Sec. 8.

2. Sensitivity to initial conditions and Lyapunov exponents

The inability to come back to initial conditions finds its origin in the sensitivity to the initial conditions. Nearby trajectories of typical Hamiltonian systems are exponentially separating along several directions of stretching in phase space. The rate of separation in each stretching direction defines the corresponding Lyapunov exponent (λ_n) [15, 16]. By time-reversal symmetry, a direction of contraction with a negative Lyapunov exponent ($-\lambda_n$) is associated to each stretching direction. As a consequence, typical N-particle systems present $3N$ positive and $3N$ negative Lyapunov exponents in a phase space of dimension $6N$. In spite of stretching and contraction, volumes are preserved according to Liouville's theorem and the sum of all Lyapunov exponents vanishes. In such chaotic systems, typical trajectories are of saddle type.

3. Spectrum of Lyapunov exponents

Lyapunov exponents have been calculated for many-particle systems like the Fermi-Pasta-Ulam (FPU) chain [17] as well as 2D and 3D fluids or solids with a repulsive Lennard-Jones (LJ) interaction between particles [18]. In the thermodynamic limit, the Lyapunov exponents form a spectrum which can be fitted by the empirical rule [17-20]

$$\lambda_n = \lambda_{\max} \left(\frac{n}{n_{\max}} \right)^{\beta} , \tag{3.1}$$

where the index n runs from 0 to $n_{\max} \simeq \nu N$, ν being the dimension of the physical space. The maximum Lyapunov exponent appears independent of the number of particles. $\beta \simeq 1$ for the FPU chain at high temperature [17]; $\beta \simeq 1/3$ for the 3D fluid with a repulsive LJ interaction and $\beta \simeq 1$ for the 3D solid [18].

For hard-sphere gas, a typical Lyapunov exponent can be evaluated from the geometry of a collision [2, 20]. If ℓ is the mean free path, d the particle diameter, and v the mean velocity, we have

$$\lambda_{\max} \sim \frac{v}{\ell} \log \frac{2\ell}{d} \sim 10^{+10} \text{ digits/sec} . \tag{3.2}$$

This microscopic chaos must be distinguished from the recently studied macroscopic chaos arising from the nonlinear coupling between the hydrodynamic modes of far-from-equilibrium physico-chemical systems. The known examples of chaos at the macroscopic scale present Lyapunov exponents of the order of 0.01 to 100 digits/sec [21]. In celestial mechanics, systems have a much lower power of dynamical chaos: for instance, Hyperion has Lyapunov exponents of the order of 10^{-8} digits/sec, without talking about the motion of Pluto or the solar system itself for which $\lambda \sim 10^{-16}$ digits/sec [22]. On the other hand, microscopic chaos takes place on a much shorter time scale of the order of the mean intercollisional time between the atoms or molecules.

4. Entropy per unit time and its density

In bounded systems, the exponential separation of nearby trajectories cannot continue forever in a global way so that folding of phase space volumes rapidly follows the first stage of stretching. This folding mechanism and the reinjection into the initial volume generate dynamical randomness which is characterized by the KS entropy per unit time defined as follows [23].

We construct a partition $(\mathcal{P})$ of phase space into small and arbitrary cells labelled by integers. We denote by the string of integers $\omega_0\omega_1 \cdots \omega_{n-1}$ the set of trajectories which visit successively the corresponding cells at times $(0, \tau, \cdots, n\tau - \tau)$. Let $\mu(\omega_0 \cdots \omega_{n-1})$ be the equilibrium invariant probability measure of this set. The KS entropy is then

$$h_{\mathrm{KS}} \equiv - \mathrm{Sup}_{\mathcal{P}} \lim_{n\to\infty} \frac{1}{n\tau} \sum_{\omega_0 \ldots \omega_{n-1}} \mu(\omega_0\omega_1 \ldots \omega_{n-1}) \log \mu(\omega_0\omega_1 \ldots \omega_{n-1}) \,. \quad (4.1)$$

Because the supremum is taken over all the possible partitions, the resulting KS entropy is independent of these partitions and characterizes an intrinsic property of the system, namely, its dynamical randomness. h_{KS} gives the data accumulation rate required to resolve the time evolution of the system into trajectories [24]. If the measuring device has a data accumulation rate smaller than h_{KS}, the evolution can only be resolved into fractal ensembles of trajectories as shown elsewhere [14].

Dynamical randomness arises from the repetitive stretching and folding of the partition cells. Consequently, the KS entropy of closed dynamical systems is equal to the sum of the positive Lyapunov exponents

$$h_{\mathrm{KS}} = \sum_{\lambda_n > 0} \lambda_n \,, \quad (4.2)$$

according to Pesin's formula [15, 25]. However, in open chaotic systems from which particles can escape, the difference between both members gives the escape rate as will be shown in Sec. 5.

In large systems of statistical mechanics which present a spectrum of Lyapunov exponents like (3.1), the KS entropy is an extensive quantity. Therefore, we introduce the corresponding intensive quantity which is the entropy per unit time and volume [26]

$$h^{(\mathrm{time,\ vol})} \equiv \lim_{V\to\infty} \frac{h_{\mathrm{KS}}}{V} = \lim_{V\to\infty} \frac{1}{V} \int_0^{\nu N} \lambda_n \, dn = \frac{\nu}{1+\beta} \rho \, \lambda_{\max} \,, \quad (4.3)$$

where ρ is the particle density. For a typical dilute gas like Argon at room temperature and one atmosphere, (4.3) takes the value

$$h^{(\mathrm{time,\ vol})} \sim 10^{29} \text{ digits / sec} \cdot \mathrm{cm}^3 \,, \qquad (\mathrm{Ar},\ 300^\circ\mathrm{K},\ 1\ \mathrm{atm.}) \,, \quad (4.4)$$

where we used (3.2) [20]. Recent numerical simulations by Posch and Hoover [18] describe a fluid of Argon near its critical point. Using the numbers provided by these authors, we obtain a more precise evaluation of the space-time entropy in this case

$$h^{(\text{time, vol})} \simeq 1.4\ 10^{34} \text{ digits / sec} \cdot \text{cm}^3 , \qquad (\text{Ar, } 150^\circ\text{K, } 50 \text{ atm.}) . \tag{4.5}$$

The difference of five orders of magnitude between (4.4) and (4.5) comes from the difference in density between both examples. Nevertheless, they are typically large numbers. We conclude that a gas of interacting particles is characterized by a positive entropy per unit time and volume in contrast to ideal gases which have a vanishing entropy per unit time and volume. Indeed, because the mean free path in a dilute gas is given by $\ell \simeq 1/\rho\pi d^2$, ρ and v being independent of the particle diameter d, we infer from (3.2) that

$$h^{(\text{time, vol})} \sim \rho^2\, v\, \pi d^2 \log \frac{2}{\rho\, \pi d^3} \to 0 \quad \text{for} \quad d \to 0 . \tag{4.6}$$

Similarly, the harmonic solid has a vanishing entropy per unit time and volume.

The entropy per unit time and volume characterizes disorder for space-time processes as done by the standard thermodynamical entropy per unit volume for disorder in space. Going back to Boltzmann's formula, $S = k_{\text{B}} \log \Omega$, the entropy per unit volume is known to give the logarithm of the number of instantaneous spatial configurations (Ω) of the system. In analogy, the KS entropy per unit time is an estimation for the logarithm of the number of possible trajectories that a system can take per unit time. Similarly, $h^{(\text{time, vol})}$ multiplied by a time interval T and a volume V is the logarithm of the number of space-time configurations that the gas, liquid, or solid can present in the space-time volume $T \times V$. We have here a generalization of Boltzmann's results from spatial configurations to space-time processes.

5. Chaotic scattering

Systems under nonequilibrium conditions sustain fluxes of particles. For instance, we consider a piece of matter that we model as a finite lattice of hard discs fixed in the plane (Fig. 1). Across this lattice, we impose a concentration gradient of independent point particles undergoing elastic collisions on the discs. Because the discs present a convex surface to the incoming particles, each collision is defocusing and the trajectories are of saddle type with positive Lyapunov exponents. Most particles spend a finite time inside the scatterer and exit at the left or the right hand. This process can be described as classical scattering. A scattering map relates the ingoing trajectories to the outgoing ones

$$S = \lim_{t_1 \to -\infty} \lim_{t_2 \to +\infty} \Phi_0(t_2) \circ \Phi(t_2, t_1) \circ \Phi_0^{-1}(t_1) , \tag{5.1}$$

in analogy with quantum mechanics [11]. Φ is the canonical transformation of the interacting Hamiltonian flow while Φ_0 describes the free flow. Underlying the scattering

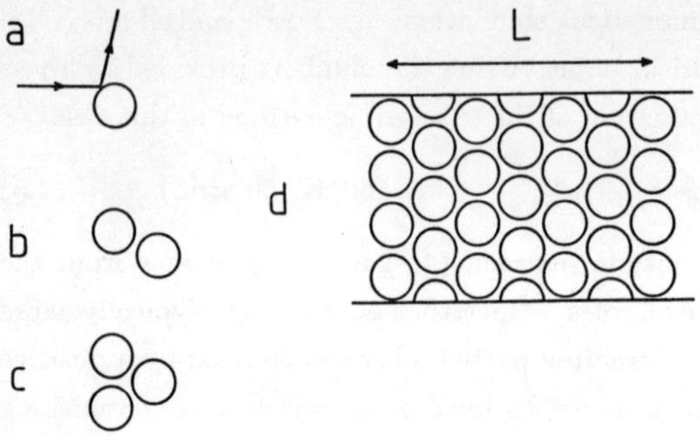

Fig. 1. Different scatterers formed by clusters of discs fixed in the plane (a-d); (d) Finite and open Lorentz gas constructed with a slab of width L in a lattice of discs. Particles escape in free motion at both sides

process, we find the trajectories which are indefinitely trapped between the discs. These trajectories are highly unstable and form a set of zero Lebesgue measure in the phase space. The invariant set of trapped trajectories contains uncountable trajectories forming a fractal object which plays a central role in the nonequilibrium diffusion process.

To see how this fractal repeller arises, let us first consider a two-disc scatterer (Fig. 1b). Here, the invariant set contains a single periodic trajectory along the line joining the centers of the discs. Going to a three-disc scatterer (Fig. 1c), the trajectories are now in one-to-one correspondence with the biinfinite sequences $\cdots\omega_{-2}\omega_{-1}\omega_0\omega_1\omega_2\cdots$ on the integers $\{1, 2, 3\}$ labelling the discs [9, 27]. This symbolic dynamics describes the motion on a fractal repeller, which has a dramatic effect on the scattering process. For instance, the exit time from the scatterer as well as the S-function are singular on the fractal set formed by the stable manifolds of the repeller [9, 11]. Today, the fractal property of many classical scattering processes has been observed not only in disc scatterers but also in Hamiltonian flows or mappings modelling atomic ionizations as well as molecular reactions in chemical kinetics [10, 11, 28]. Because trajectories escape from the fractal repeller, this latter is characterized by an escape rate (γ) which gives the reaction rate in chemical applications [9, 28].

The dynamics on a fractal repeller is chaotic with positive Lyapunov exponents (λ_n). The stretching of the phase space volumes has here two contributions. First, stretching induces folding, cutting, and reinjection inside the scatterer like in closed systems. This first contribution produces a positive KS entropy. In open systems, stretching is also at the origin of the escape from the scatterer. As a consequence, the escape rate is given by the difference between the sum of positive Lyapunov exponents and the KS entropy. Pesin's formula generalizes to open systems in the form [15, 29]

$$\gamma = \sum_{\lambda_n > 0} \lambda_n - h_{\mathrm{KS}} . \tag{5.2}$$

For a closed system, the escape rate vanishes and (4.2) is recovered.

6. Classical resonances and spectral theory of the evolution operator

Beside the escape rate, several decay rates and different kinds of decay modes participate in typical scattering processes [9, 12, 30]. Thanks to newly developed techniques, it is now possible to calculate these scattering resonances directly from the evolution operator of the classical dynamics, also called the Koopman operator,

$$U^t = \exp itL , \tag{6.1}$$

where L is the Liouville operator [31]

$$L = i\,\{H, \cdot\} = i \sum_k \left(\frac{\partial H}{\partial q_k} \frac{\partial}{\partial p_k} - \frac{\partial H}{\partial p_k} \frac{\partial}{\partial q_k} \right) , \tag{6.2}$$

with the Hamiltonian (1.1). For hard-disc systems, U^t is defined using the free-motion Hamiltonian supplemented by boundary conditions describing the elastic collisions. For area-preserving mappings, Φ^t ($t \in \mathbf{Z}$), the evolution operator is defined directly with $(U^t f)(X) = f(\Phi^t X)$.

The scattering resonances are then calculated in terms of the unstable periodic orbits in the system. For chaotic systems with two degrees of freedom, the characteristic spectral equation of the evolution operators is given by [13]

$$\det(I - e^{-iz}\, U) = \frac{1}{\zeta_1(z)} \frac{1}{\zeta_2(z)^2} \frac{1}{\zeta_3(z)^3} \cdots , \tag{6.3}$$

for mappings, and [30]

$$\mathrm{Tr}\left(\frac{1}{z - L} - \frac{1}{z - L_0} \right) = \frac{d}{dz} \ln \frac{1}{\zeta_1(z)} \frac{1}{\zeta_2(z)^2} \frac{1}{\zeta_3(z)^3} \cdots , \tag{6.4}$$

for flows where L_0 is the free-motion Liouvillian. In the right member of these equations, we find Ruelle's zeta functions [32]

$$\zeta_\beta(z) = \prod_{\mathbf{p}} \left(1 - \frac{e^{-izT_{\mathbf{p}}}}{|\Lambda_{\mathbf{p}}|\Lambda_{\mathbf{p}}^{\beta-1}} \right)^{-1} , \tag{6.5}$$

defined as product over the prime periodic orbits of the repeller. We denote their period by $T_{\mathbf{p}}$ and their stability eigenvalue by $\Lambda_{\mathbf{p}}$. The Lyapunov exponent of the periodic orbit $\mathbf{p}$ is $\lambda_{\mathbf{p}} = \ln \Lambda_{\mathbf{p}} / T_{\mathbf{p}}$.

The scattering resonances are defined as the poles of the characteristic equation. Formally, they may be considered as complex eigenvalues of the Liouville or Koopman operators [31]

$$L\,\varphi \;=\; z\,\varphi\,, \qquad \text{or} \qquad U^t\,\varphi \;=\; e^{izt}\,\varphi\,. \tag{6.6}$$

According to the Perron-Frobenius-Ruelle theorem [32], the scattering resonances lie higher in the complex plane than the escape rate, Im $z \geq \gamma$, so that the decay modes have a lifetime shorter than or equal to $1/\gamma$. Therefore, a gap which is empty of resonances is formed above the real axis of the complex plane z in the spectrum of open systems.

Eqs. (6.3) to (6.5) also apply to chaotic dynamical systems which are closed. In mixing systems, the decay mode corresponding to the vanishing escape rate turns into the equilibrium invariant probability measure and the resonance $\gamma = 0$ becomes the unique real eigenvalue [23]. All other resonances are complex and correspond to the relaxation modes of the systems. The classical spectrum of mixing systems has thus a continuous part in contrast to integrable and quasiperiodic systems which have a discrete spectrum. Let us recall that the Hamiltonian of an integrable system only depends on the action variables, $H(J_1, J_2, \ldots, J_N)$, describing independent oscillations with frequencies, $\omega_k = \partial H/\partial J_k$, or quasi-free motions. The eigenvalues of the Liouvillian are given by [31]

$$z \;=\; m_1\,\omega_1 \;+\; m_2\,\omega_2 \;+\; \cdots \;+\; m_N\,\omega_N\,, \qquad (m_k \in \mathbf{Z})\,, \tag{6.7}$$

for each among the continuum of ergodic components [15].

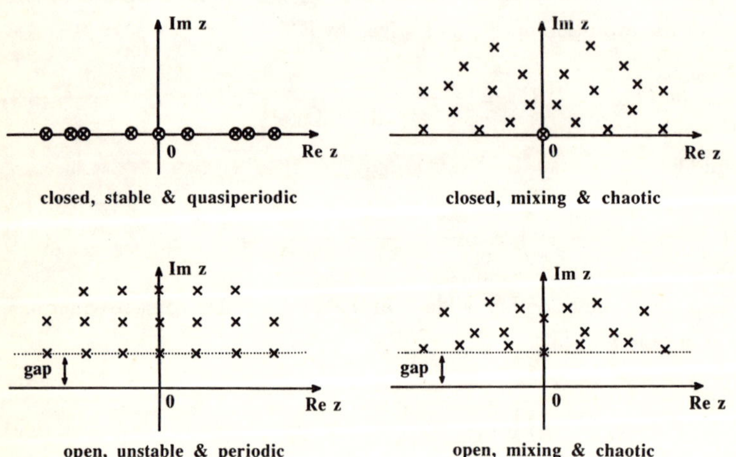

Fig. 2. Spectra of eigenvalues and resonances of the Liouville dynamics in different classical systems

Different types of classical spectra for closed and open, chaotic and non-chaotic systems are depicted in Fig. 2. Let us emphasize that nonintegrable and chaotic systems with a continuous spectrum are typical in classical mechanics.

Classical resonances have been calculated for several dynamical systems such as a simple flow model invented by Ruelle as well as complex mappings [33], the intermittent map [34], the Hénon map [35], and some 1D maps [36]. They are also studied in intramolecular dynamics where they describe the dominant quasiclassical features of photoabsorption molecular spectra [37]. We present in Fig. 3 the classical resonances for the three-disc scatterer of Fig. 1c. The aforementioned gap due to escape appears above the real axis. The escape rate is given by the resonance in A_1 at Re $z = 0$.

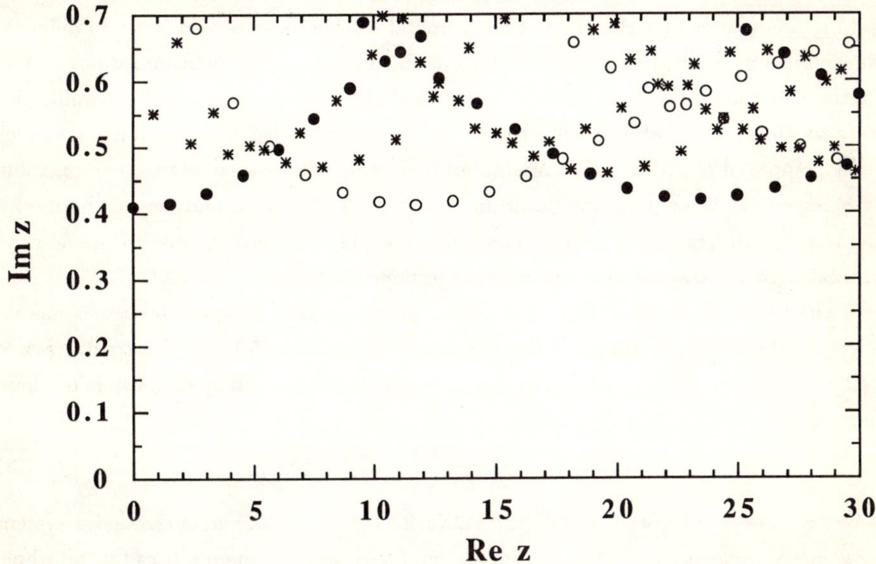

Fig. 3. Scattering resonances for a particle of unit velocity hitting three discs of unit radius fixed at the vertices of an equilateral triangle of side $r = 6$. Because of the three-fold symmetry, the resonances belong to one of the three symmetry representations A_1 (•), A_2 (o), and E (∗) of the group C_{3v} [9]

7. Nonequilibrium states in large systems

When the concentration gradient is removed from our lattice of hard discs described at the beginning of Sec. 5, particles escape from the scatterer at a rate which is controlled by diffusion in the Lorentz gas (Fig. 1). We assume that particles cannot go through the lattice without collision so that the diffusion coefficient $\mathcal{D}$ is finite [38]. The escape rate can be evaluated solving the diffusion equation (2.1) with two absorbing walls separated by a distance L [8, 9]. The slowest decay rate is then

$$\gamma = \mathcal{D}\left(\frac{\pi}{L}\right)^2 , \tag{7.1}$$

which gives us the escape rate from the fractal repeller ($\mathcal{F}_L$) of trapped trajectories. Inserting (7.1) in (5.2), we obtain the diffusion coefficient by

$$\mathcal{D} = \lim_{L\to\infty} \left(\frac{L}{\pi}\right)^2 \left[\sum_{\lambda_n>0} \lambda_n(\mathcal{F}_L) - h_{\mathrm{KS}}(\mathcal{F}_L)\right] , \tag{7.2}$$

in terms of the Lyapunov exponents and the KS entropy of the dynamics on the nonequilibrium fractal $\mathcal{F}_L$ [8].

We can also consider first-passage problems in deterministic systems by placing absorbing boundaries in a closed system. A fractal repeller is thus associated to different nonequilibrium problems. Equivalently, fractal repellers define nonequilibrium states. On the other hand, the Liouville invariant probability measure describes the equilibrium state which is stable because it fills the whole phase space. Each nonequilibrium repeller being composed of saddle-type trajectories is dynamically and spontaneously unstable with respect to the equilibrium state in agreement with the second law of thermodynamics. For this result to hold, dynamical chaos plays the central role to provide the instability of the trajectories participating in nonequilibrium processes.

The probabilistic character of the laws governing the transport processes has its origin in the wild variability of the scattering S-function (5.1) on the fractal set of its singularities. For the 2D open Lorentz gas, the information dimension of these singularities is

$$D_{\mathrm{I}} = 2 - \frac{\mathcal{D}}{\lambda_1(\mathcal{E})}\left(\frac{\pi}{L}\right)^2 + \mathcal{O}(L^{-3}) , \tag{7.3}$$

in terms of the size L of the system, of its diffusion coefficient $\mathcal{D}$, and of the closed-system Lyapunov exponent $\lambda_1(\mathcal{E})$. For a large system ($L \to \infty$), the singularities fill the whole 2D space of incoming trajectories, rendering unpredictable the outgoing trajectories. We have obtained similar behaviors for an exactly solvable model composed of a chain of coupled baker transformations, we called the multibaker map [14]. See also related studies in Ref. 39.

8. Conclusions

Although the results of Sec. 7 have been studied for large systems with only two degrees of freedom like the Lorentz gas or the multibaker map, they can be generalized to many-particle systems such as the hard-sphere or the Lennard-Jones gases which can sustain transport processes like viscosity or heat conductivity beside self-diffusion. We can here place absorbing boundaries in phase space to define a nonequilibrium state where the total energy or momentum in a small subdomain have values higher than the equilibrium values. The escape rate out of this situation can be calculated from the Navier-Stokes and heat equations. On the other hand, the trapped trajectories satisfying the nonequilibrium conditions form a highly unstable fractal repeller. Its escape rate is given by (5.2). This remark suggests a relation like (7.2) for this many-body problem. Similar considerations apply to effusion of a gas through a small hole in a container or to the density fluctuations between two containers joined by a small tube (Fig. 4). When the opening of area A is smaller than the mean free path (Knudsen regime), the escape rate is given by

$$\gamma = \frac{1}{4}\, \rho \, \langle v \rangle \, A \,, \qquad \text{with} \qquad \langle v \rangle = \left(\frac{8 k_{\mathrm{B}} T}{\pi m} \right)^{1/2} , \tag{8.1}$$

where T is the gas temperature, ρ its density, and m the particle mass [39]. Our point is that this escape rate characterizes a fractal repeller in the phase space of the many-body system.

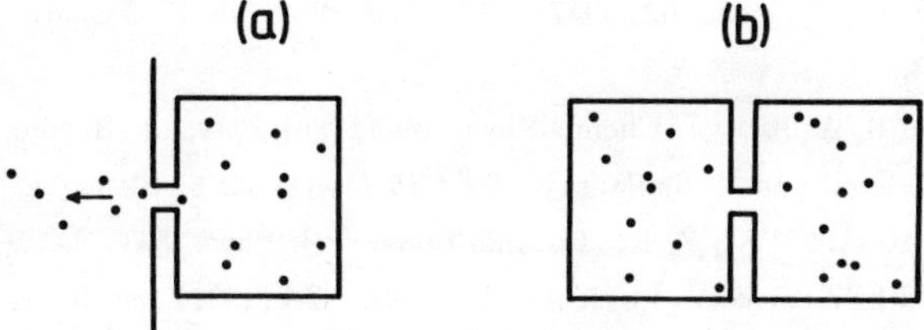

Fig. 4. (a) Effusion of a gas; (b) Density fluctuations between two containers

Let us now summarize the results we reported. We showed that it is possible to solve the Liouville evolution equation and to find its scattering resonances which can be assimilated formally to complex eigenvalues. In large systems, there is a spectral isomorphism between these resonances of the Liouville dynamics and the eigenvalues of the phenomenological irreversible equations like (1.2). Among the resonances, we find the escape rate which is related, on one hand, to the diffusion coefficient, and, on the other hand, to the difference between the Lyapunov exponents and the KS entropy. This result establishes a bridge between the kinetic and the ergodic theories. To conclude, irreversibility in the sense of Secs. 1-2 appears so intimately connected to microscopic chaos that we propose the hypothesis to assimilate both of them and to use the entropy per unit time and volume described in Sec. 4 as indicator of the irreversibility of a many-body Hamiltonian system.

Acknowledgement

I thank G. Nicolis as well as I. Prigogine and S. A. Rice for their support and encouragement. I received nice suggestions from P. Cvitanović, J.-P. Eckmann, R. Kapral, and D. Ruelle. The author is “Chercheur Qualifié” at the National Fund for Scientific Research (Belgium).

References

1. R. P. Feynman, R. B. Leighton, and M. Sands, *Lectures on Physics* (Addison-Wesley, Reading, MA, 1964) vol. I.
2. N. N. Krylov, *Works on the Foundations of Statistical Mechanics* (Princeton University Press, 1979); Ya. G. Sinai, *ibid.* p. 239.
3. Ya. G. Sinai, *On the entropy per particle for the dynamical system of hard spheres*, Preprint 1978; Ya. G. Sinai and N. I. Chernov, Russian Math. Surveys, **42**:3 (1987) 181.
4. I. Prigogine, *From being to becoming* (Freeman, San Francisco, 1980); G. Nicolis and I. Prigogine, *Exploring Complexity* (Freeman, New York, 1989).
5. G. Nicolis and C. Nicolis, Phys. Rev. A **38** (1988) 427.
6. D. Ruelle, in: *L'ordre du chaos* (Pour la Science, Paris, 1984) p. 136.
7. M. C. Mackey, Rev. Mod. Phys. **61** (1989) 981.
8. P. Gaspard and G. Nicolis, Phys. Rev. Lett. **65** (1990) 1693.
9. P. Gaspard and S. A. Rice, J. Chem. Phys. **90** (1989) 2225, 2242, 2255; **91** (1989) 3279.
10. P. Gaspard and S. A. Rice, J. Phys. Chem. **93**, 6947 (1989); S. A. Rice and P. Gaspard, Israel J. Chem. **30** (1990) 23.
11. D. W. Noid, S. K. Gray, and S. A. Rice, J. Chem. Phys. **84** (1986) 2649; M. Hénon, Physica D **33** (1988) 132; G. Troll and U. Smilansky, *ibid.* **35** (1989) 34; S. Bleher, E. Ott, and C. Grebogi, *ibid.* **46** (1990) 87; Z. Kovács and Tamás Tél, Phys. Rev. Lett. **64** (1990) 1617; P. Eckelt and E. Zienicke, J. Phys. A: Math. Gen. **24** (1991) 153; R. A. Jalabert, H. U. Baranger, and A. D. Stone, Phys. Rev. Lett. **65** (1990) 2442.
12. L. P. Kadanoff and C. Tang, Proc. Natl. Acad. Sci. U.S.A. **81** (1984) 1276.
13. S. A. Rice, P. Gaspard, and K. Nakamura, *Signatures of Chaos in Quantum Dynamics and the Controllability of Evolution in a Quantum System*, preprint 1990.
14. P. Gaspard, *Diffusion, Effusion, and Chaotic Scattering: An Exactly Solvable Liouvillian Dynamics*, preprint 1991.
15. J.-P. Eckmann and D. Ruelle, Rev. Mod. Phys. **57** (1985) 617.
16. A. J. Lieberman and A. J. Lichtenberg, *Regular and Stochastic Motion* (Springer, New York, 1983).
17. R. Livi, A. Politi, and S. Ruffo, J. Phys. A: Math. Gen. **19** (1986) 2033.
18. H. A. Posch and W. G. Hoover, Phys. Rev. A **38** (1988) 473; *ibid.* **39** (1989) 2175.
19. C. M. Newman, Commun. Math. Phys. **103** (1986) 121; J.-P. Eckmann and C. E. Wayne, J. Stat. Phys. **50** (1988) 853.
20. P. Gaspard and G. Nicolis, Physicalia Magazine (J. Belg. Phys. Soc.) **7** (1985) 151.
21. J.-P. Eckmann, S. O. Kamphorst, D. Ruelle, and S. Ciliberto, Phys. Rev. A **34** (1986) 4971.
22. J. Wisdom, S. J. Peale, and F. Mignard, Icarus **58** (1984) 137; G. J. Sussman and J. Wisdom, Science **241** (1988) 433; J. Laskar, Nature **338** (1989) 237.

23. V. I. Arnold and A. Avez, *Ergodic Problems of Classical Mechanics* (Benjamin, New York, 1968).

24. P. Billingsley, *Ergodic Theory and Information* (Wiley, New York, 1965).

25. Ya. B. Pesin, Math. USSR Izv. **10**:6 (1976) 1261; Russian Math. Surveys **32**:4 (1977) 55.

26. S. Goldstein, J. L. Lebowitz, and M. Aizenmann, in *Dynamical Systems, Theory and Applications*, J. Moser (Ed.), Lect. Notes in Phys. **38** (Springer, Berlin, 1975) p. 112.

27. B. Eckhardt, J. Phys. A: Math. Gen. **20** (1987) 5971.

28. S. K. Gray, S. A. Rice, and M. J. Davis, J. Phys. Chem. **90** (1986) 3470; S. K. Gray and S. A. Rice, Faraday Discuss. Chem. Soc. **82** (1986) 1; D. W. Noid, S. K. Gray, and S. A. Rice, J. Chem. Phys. **84** (1986) 2649; S. K. Gray and S. A. Rice, J. Chem. Phys. **86** (1987) 2020; G. Casati, B. V. Chirikov, D. L. Shepelyansky, and I. Guarneri, Phys. Rep. **154** (1987) 77.

29. H. Kantz and P. Grassberger, Physica D **17** (1985) 75; T. Bohr and D. Rand, Physica D **25** (1987) 387.

30. P. Cvitanović, *Recycling chaos*, preprint 1990; P. Cvitanović and B. Eckhardt, *Periodic orbit expansions for classical smooth flows*, preprint 1990.

31. I. Prigogine, *Non-Equilibrium Statistical Mechanics* (Wiley, New York, 1962).

32. D. Ruelle, Commun. Math. Phys. **125** (1989) 239.

33. D. Ruelle, Phys. Rev. Lett. **56** (1986) 405; J. Stat. Phys. **44** (1986) 281; C. R. Acad. Sci. Paris **296** Sér. I (1983) 191.

34. V. Baladi, J.-P. Eckmann, and D. Ruelle, Nonlinearity **2** (1989) 119.

35. S. Isola, Commun. Math. Phys. **116** (1988) 343.

36. F. Christiansen, G. Paladin, and H. H. Rugh, Phys. Rev. Lett. **65** (1990) 2087.

37. D. W. Noid, M. L. Koszykowski, and R. A. Marcus, Ann. Rev. Phys. Chem. **32** (1981) 267; J. M. Gomez Llorente and H. S. Taylor, J. Chem. Phys. **91** (1989) 953.

38. J. Machta and R. Zwanzig, Phys. Rev. Lett. **50** (1983) 1959 ; J. Stat. Phys. **32** (1983) 555; L. A. Bunimovich and Ya. G. Sinai, Commun. Math. Phys. **78** (1980) 247, 479.

39. R. S. MacKay, J. D. Meiss, and I. C. Percival, Physica D **27** (1987) 1; I. Dana, N. W. Murray, and I. C. Percival, Phys. Rev. Lett. **62** (1989) 233; I. Dana, Phys. Rev. Lett. **64** (1989) 2339.

40. R. K. Pathria, *Statistical Mechanics* (Pergamon, Oxford, 1972).

A Propositional Lattice for the Logic of Temporal Predictions

H. Atmanspacher

Max Planck Institut für Extraterrestrische Physik, D-8046 Garching, Germany

1. Introduction and Summary

The concept of chaos, apart from its significance with respect to the modelling of specific complex systems and the prediction of their behavior, bears important implications for our general understanding of nature and the natural sciences. One of the central quantities characterizing chaotic systems is the dynamical entropy K. It can be interpreted as a temporal rate of internal information production of a system due to the specific dynamical laws governing its evolution. These laws can be formalized in different ways, using different types of operators acting on a phase space distribution function. Two respective operator formalisms refer to the Liouville operator L and to an information (or entropy) operator M. Both are incommensurable in the sense of a non–vanishing commutator given by K (Sec.2).

Here we intend to interpret this incommensurability, compare it with incommensurabilities we know from quantum theory, and indicate conjectures of the common features of both. The interpretative tool we utilize is lattice theory, the corresponding approach is a logico – algebraic one. It will be illustrated in Sec.3, how propositions based on L and M can be mapped onto elements of a lattice, and how the properties of this lattice can be studied.

The properties of distributivity and modularity will be addressed in particular. With respect to distributivity, it will be shown that the incommensurability of L and M violates the distributive equality as the quantum incommensurability of position and momentum does. Instead, both incommensurabilities satisfy the mutually dual distributive inequalities, thus indicating a formal relation of duality among each other (Sec.3.3). With respect to modularity, it may be conjectured that there are limiting cases in which it is satisfied. An illustrative interpretation of these limits as well as of the general, non–modular case will be given in Sec.4.

2. Temporal Dynamics: Liouvillean and Information Approach

Evolution in the dynamical sense we are addressing here means evolution of variables of a system as it may be described by differential equations.[1] The variables $\mathbf{x} = (x_1, ..., x_n)$ correspond to properties of the constituents of the system, the temporal evolution of which is determined by a matrix $\mathbf{F}$ characterizing interactions among the constituents. To begin with, it will

[1]This kind of description is not taken to be unique. There are of course additional possibilities like Fokker–Planck type equations, Markov chains, symbolic dynamics, etc.

generally be assumed that the variables $\mathbf{x}$ and the interactions $\mathbf{F}$ may be localized in the weak sense of requiring that a boundary for the system may be defined, e.g., by minimal flows (see, however, Sec.4). This informal characterization intentionally allows for the inclusion of thermodynamically open systems.

As has been discussed in detail elsewhere [1, 2] the temporal evolution of the variables $\mathbf{x}$ is more appropriately described by the temporal evolution of a distribution function ρ. It characterizes the state of a system in phase space including a certain degree of unavoidable ignorance (uncertainty). In this framework, two different formalizations of temporal dynamics are possible, which we shall denote as Liouvillean dynamics and information dynamics.

2.1. Liouvillean Dynamics

The temporal evolution of a distribution function ρ characterizing the state of a system in phase space with non–vanishing uncertainty can be described by the Liouville equation

$$i\partial_t \rho = L\ \rho \qquad (1)$$

where L is the Liouville operator acting on ρ and $i = \sqrt{-1}$. The concept of an evolution operator acting on distribution functions in phase space has been introduced and extensively applied by Prigogine [1]. The solution of Eq.(1) is given by:

$$\rho(t) = e^{-iLt}\rho(0) \qquad (2)$$

It represents a unitary evolution based on the unitary operator $U_t = e^{-iLt}$. The Liouville operator L is the generator of the corresponding unitary dynamical group. The existence of an inverse element in this group ensures its invariance under time reversal. A temporal evolution of ρ as given by L is temporally reversible. The dynamics determined by Eq.(2) will be called *Liouvillean dynamics* in the following.

2.2. Information Dynamics

In addition to the dynamical evolution based on L we now consider the information flow in a system. This means that we are interested in studying how much information is accessible about the variables of a given system instead of studying the variables themselves. In this sense our viewpoint is one of meta – observers instead of observers.

Dynamical systems can be regarded to generate information with a temporal rate K, if they give rise to at least one positive Ljapunov exponent λ [3-7]. In case of conservative systems, this condition is satisfied for K – flows, dissipative systems with finite $K > 0$ can be characterized by strange ("chaotic") attractors. K is an operationally accessible quantity [6, 8]. Its determination from experimental data (see, for instance, [9]) provides a sufficient criterion for processes denoted as deterministic chaos.

The information production rate K can be identified [10] with the dynamical (Kolmogorov–Sinai) entropy [11, 12] which may (under certain conditions) be approximated by the sum of positive Ljapunov exponents [13]:

$$K = \sum_i \begin{cases} \lambda_i, & \text{if } \lambda_i > 0 \\ 0, & \text{otherwise} \end{cases} \tag{3}$$

A corresponding formalisation of the resulting information flow would be given by:

$$I(t) = I(0) + Kt \tag{4}$$

This expression describes conservative as well as dissipative systems, since it merely contains the dynamical entropy K as an independent quantity. (It does not include the Hamiltonian of the system as L does.) Two remarks are in order concerning the linearity and the sign of $I(t)$.

The positive sign of $I(t)$ originates in a specific viewpoint concerning the fact that the intrinsic "instability" of the system due to positive Ljapunov exponents enhances initial uncertainties exponentially. From an *internal* (with respect to the system) point of view, this is equivalent with an increasing amount of information within the system. From an *external* (observer) point of view, the information about the actual state of the system decreases to the same degree. It is of general importance to distinguish between these two viewpoints. Following C.F.v.Weizsäcker [14] information in the first sense (on the level of the system) will therefore be called *potential information.* It is addressed in Eq.(4). (Note that potential information is nothing but positive entropy.) It can be gained by an observer who then obtains *actual information* about the state of the system by measurement. As a fundamental idea behind this two–level description one might see an – informal – principle of information conservation, aiming at a description of information transfer between both levels. A more detailed elaboration on these matters in the context of elementary information transfer between two levels has been started elsewhere [15].

The linearity of $I(t)$ in Eq.(4) does not reflect any "realistic" temporal evolution of information in the system. It is thus spurious in the sense that it is merely a consequence of the linear stability analysis providing the Ljapunov exponents. Under certain conditions, Eq.(4) describes a first order approximation to the (nonlinear) mutual information [6], only valid for small $t > 0$. Since it is well known that the information flow is generally nonlinear, the dynamical entropy K is to be considered as an averaged *local* rate of information production in the system. Its inverse, $1/K$, estimates the time interval for which the temporal evolution of the system can reasonably well be predicted. Since $K > 0$ is a sufficient and necessary condition for deterministic chaos, the temporal prediction of chaotic systems is generally limited. As mentioned above, this limitation occurs in a completely deterministic manner due to an intrinsic instability acting on the distribution function ρ.

Defining an information operator M with eigenvalues $I(t)$ according to Eq.(4) provides the eigenvalue equation:

$$M\rho = I(t)\rho = (I(0) + Kt)\,\rho \tag{5}$$

The operator M is time independent. It acts on a time dependent distribution ρ, yielding a temporally increasing eigenvalue $I(t)$ if $K > 0$. (A more detailed discussion of the eigenvalues of information has been given elsewhere [5].) In contrast to L, the information operator M introduces the notion of irreversible evolution in the sense that M provides a semigroup representation (lacking inverse element) whenever $K > 0$. The dynamics according to Eq.(5) will subsequently be denoted as *information dynamics.*

2.3. Incommensurability of L and M

Both approaches can be compared by investigating the commutator of the evolution operators[2] which has turned out [5] to be given by the dynamical entropy K ($\mathbf{I}$ is the identity operator):

$$i[L, M] = K\mathbf{I} \tag{6}$$

This commutation relation expresses the incommensurability of Liouvillean and information dynamics. This incommensurability is important in various respects (the following list is not intended to be complete):

1. It gives a formal account of a symmetry breaking in time, originating in the difference between unitary (L) and semigroup (M) dynamics.
2. There are no common eigendistributions for L and M at $t > 0$, whenever $K > 0$.
3. The commutator is system – specific, it is not a universal constant (as h is in the quantum theoretical uncertainty relations).
4. The commutator can be considered as a temporal rate of information production. For this reason, an interpretation according to "simultaneous" applicability (as in quantum theory) is inappropriate.
5. The commutator is of statistical significance since K is defined as the sum of temporal averages, the Ljapunov exponents.
6. Both dynamical approaches are different (for $K > 0$) with respect to the *prediction* of a future state of the system. This may be interpreted as a consequence of the increasing uncertainty in predicting the state of a system as time proceeds. Whenever $K > 0$ the state $\rho(t + \Delta t)$ of a system cannot be predicted as accurate as initial conditions have been measured (or otherwise fixed) at t. While both L and M refer to a *deterministic* dynamics, a future state is not completely *determinable* by M.[3]

For the purposes of the present article, the last point is of particular importance. It represents a basic point of departure for the following section.

---

[2]This idea originated in Prigogine's commutation relation concerning a Liouvillean L and an entropy operator which dates back to the 70s [16, 17]. An extensive discussion of this approach can be found in the monograph *From Being To Becoming* [2], see also a more recent article [18] adressing information theoretical issues.

[3]"Completely determinable" is here used in the sense of non – increasing uncertainty for a temporal prediction of a future state of the system. This view may be related to Krylov's [19] notion of an inexhaustively complete experiment.

3. Lattice Theory of a Calculus of Temporal Propositions

Before going into details some remarks are necessary with respect to the notions of incommensurability, logical structure, and propositional calculus. Concerning the commutation relation (6), we shall subsequently speak of an incommensurability of propositions, not of operators. In this manner a formal investigation of the logical structure of the propositional calculus related to (6) can be achieved. It has become common practice to study such a logical structure in the framework of lattice theory.

Another possible approach to investigate incommensurabilities is commonly better known, namely that used in the field of quantum logic [20]. The historical paper [21] in this context deals with subspaces of the relevant Hilbert space which can be interpreted as an orthocomplemented, non-distributive lattice. A corresponding attempt for temporal logic in terms of such a formalism could be related to the concept of superspace, as it is for instance discussed by Balian *et al.* [22]. However, the argumentation given in the following appears to be formally simpler than one using techniques of functional analysis. Moreover, it will become clear below that a presupposed equivalence of lattices with geometrical spaces can be misleading.

Talking about propositions implies that (6) can be interpreted in the sense of characterizing an incommensurability of propositions concerning predictions given by dynamical laws, i.e., by theories. (We consider a set of dynamical laws as a subset of a theory.) This incommensurability of propositions as predictions may be contrasted to quantum theoretical incommensurabilities referring to propositions as they express results of measurement. In the framework of a semiotic terminology, the act of *measurement* can be considered as a pragmatic aspect of a *fact*, whereas the act of *prediction* concerns a pragmatic aspect of a *theory*. In this sense one can distinguish incommensurabilities of facts and of theories as two basically different concepts.[4] Some general relations between both of them as well as empirical consequences have been indicated recently [7, 28, 29].

We are now going to study the logical structure of the propositional lattice related to the incommensurability of propositions concerning predictions based on L and M. For a comprehensive textbook on lattice theory, we recommend the extensive monograph of Birkhoff [30]. A compact overview of important definitions and theorems is listed in the appendix of Jammer's book [31]. An excellent discussion of various lattice properties in the context of quantum theory has been given by Primas [32]. The present article deals only with those details relevant for our purposes.

A propositional lattice $V(G, \geq, \leq)$ consists of a set $G = \{a, b, ...\}$ of

---

[4] A relation of these notions to contemporary ideas in the philosophy of science [23-25] is worth to be noted. In this context, facts in our sense can be regarded as being related to the traditional viewpoint of realism, while theories can be associated with the philosophical concept of relativism. We should also point out that the notion of incommensurable theories is much discussed in the historical development of (scientific) theories [26, 27]. Our approach expresses precisely how this notion of incommensurable theories can be formalized mathematically.

propositions for which partial ordering relations are defined as implications "$\rightarrow$", "$\leftarrow$", and "$\leftrightarrow$". Propositions can be combined by the basic logical operations "and" and "or".

Let us now denote propositions (a, b) concerning predictions given by theories related to L and M as (prop$L$, prop$M$). Typical propositions we have in mind are "A system S conserves information", or "Measurement accuracy and prediction accuracy in S are identical for each $t > 0$", or "$K = 0$ in S". All these propositions mean that the number of possible states of a system S in phase space is time – independent for a fixed phase space resolution ϵ. They are true in the context of both Liouvillean as well as information dynamics. Hence they are equally well classified as prop$L$ or prop$M$. However, propositions like "A system S generates information", or "Prediction accuracy is lower than measurement accuracy in S for each $t > 0$", or "$K > 0$ in S" (more specifically, "$K = K_s > 0$" in S) have to be classified as prop$M$ since they belong to the context of information dynamics exclusively. They indicate that the number of possible states in a phase space with fixed ϵ increases exponentially with a rate K.

Since we are interested in interpreting the incommensurability of L and M in Eq.(6) we relate prop$L$ and prop$M$ to the classifying criterion K of a dynamical system S. As discussed above prop$L$ refers to systems with vanishing dynamical entropy, while prop$M$ refers to systems with zero or positive dynamical entropy. Consequently we define sets

$$\mathcal{L} := \{K | K = 0\} \tag{7}$$
$$\mathcal{M} := \{K | K \geq 0\} \tag{8}$$

meaning the set of possible values of K corresponding to possible propositions prop$L$ (prop$M$) which apply to a specific system S characterized by $K = 0$ ($K \geq 0$). As G is the set of all possible propositions with respect to S, $\mathcal{G}$ is the corresponding set of possible values of K.[5] Since systems with negative K are excluded by definition (3) (the dynamical entropy is positive semidefinite), the lowest upper bound of $\mathcal{G}$ is $I = \mathbf{R}^+$, whereas the greatest lower bound of $\mathcal{G}$ is $0 = \emptyset$. Logical implications translate into the set theoretical relations "$<$", "$>$", and "$=$". The logical "and" and "or" are represented by the set theoretical operators meet ($\wedge$) and join ($\vee$) acting on subsets of $\mathcal{G}$.

At this point it should be stressed that the definitions (7) and (8) refer to the sets of *potential* (possible) values of K for a *specific* system S. (These potential values of K in turn refer to *potential* propositions which may be valid for the system.) The general validity of a theory represented by L or M for *all* systems would then have to be achieved stepwise for additional systems. Subsequently we shall restrict ourselves to a specific system and consider the basic lattice properties for propositions of the form "$K = 0$"

---

[5]The assignment of real numbers to propositions resembles a "Gödelization" procedure except that we have to allow for reals instead of natural numbers only. This modification provides problems, for instance if one is interested in the exact boundaries of the set; cf. Sec. 4.

and "$K \geq 0$" (subsets of G) by their set theoretical equivalents $\mathcal{L}$ and $\mathcal{M}$ (subsets of $\mathcal{G}$) as given in (7) and (8).

3.1. Basic Properties

A partially ordered set constitutes a lattice if the following algebraic properties are satisfied:

$$\text{Idempotency}: \quad a \wedge a = a \tag{9}$$

$$\text{Commutativity}: \quad a \wedge b = b \wedge a \tag{10}$$

$$\text{Associativity}: \quad a \wedge (b \wedge c) = (a \wedge b) \wedge c \tag{11}$$

$$\text{Absorption}: \quad (a \wedge b) \vee b = b \tag{12}$$

together with the corresponding dual properties which are obtained by interchanging $\wedge$ and $\vee$.

It is easy to check these properties for sets $\mathcal{L}$ and $\mathcal{M}$ as introduced above. $\mathcal{G}$ is *idempotent* since:

$$\mathcal{L} \wedge \mathcal{L} = \{K|K = 0\} \wedge \{K|K = 0\} = \{K|K = 0\} = \mathcal{L} \tag{9a}$$

$$\mathcal{M} \wedge \mathcal{M} = \{K|K \geq 0\} \wedge \{K|K \geq 0\} = \{K|K \geq 0\} = \mathcal{M} \tag{9b}$$

and the dual properties.

Similarly we show *commutativity* by:

$$\mathcal{L} \wedge \mathcal{M} = \mathcal{L} = \mathcal{M} \wedge \mathcal{L} \tag{10a}$$

$$\mathcal{L} \vee \mathcal{M} = \mathcal{M} = \mathcal{M} \vee \mathcal{L} \tag{10b}$$

(The detailed derivation in terms of $\{K|K...\}$ has been omitted for brevity.)

For *associativity* we have to introduce a third set of values of K, namely a subset $\mathcal{N} \subset \mathcal{M}$. The choice of a specific subset is arbitrary since:

$$\mathcal{L} \vee (\mathcal{M} \vee \mathcal{N}) = (\mathcal{L} \vee \mathcal{M}) \vee \mathcal{N} = \mathcal{M} \tag{11a}$$

$$\mathcal{L} \wedge (\mathcal{M} \wedge \mathcal{N}) = (\mathcal{L} \wedge \mathcal{M}) \wedge \mathcal{N} = \mathcal{L} \wedge \mathcal{N} \tag{11b}$$

where $\mathcal{L} \wedge \mathcal{N}$ is either $\emptyset$ or $\mathcal{L}$.

The self–dual property of *absorption* is shown by:

$$\mathcal{L} \wedge (\mathcal{L} \vee \mathcal{M}) = \mathcal{L} \vee (\mathcal{L} \wedge \mathcal{M}) = \mathcal{L} \tag{12a}$$

$$\mathcal{M} \wedge (\mathcal{M} \vee \mathcal{L}) = \mathcal{M} \vee (\mathcal{M} \wedge \mathcal{L}) = \mathcal{M} \tag{12b}$$

Since (9) - (12) are satisfied, the set $\mathcal{G}$ of K–values constitutes a lattice, and the set G of propositions constitutes a propositional lattice.

3.2. Complementation

A lattice V is *complemented* if for each a an element a' exists such that:

$$a \wedge a' = 0 \tag{13}$$

and (dually)

$$a \vee a' = I \tag{14}$$

Here I (unity) and 0 (zero) represent universal upper and lower bounds in G, respectively.

The complement $\mathcal{L}'$ of the set $\mathcal{L}$ is simply given by:

$$\mathcal{L}' = \{K|K \neq 0\} = \{K|K > 0\} \tag{15}$$

such that

$$\mathcal{L} \wedge \mathcal{L}' = 0 \tag{13a}$$

$$\mathcal{L} \vee \mathcal{L}' = \{K|K \geq 0\} = I \tag{14a}$$

The physical meaning of the complement $\mathcal{L}'$ is that the system under consideration is not adequately described by L. This can be empirically checked by comparing a prediction about the future state of the system given at $t = 0$ with the actually observed state at $t + \Delta t$. The prediction based on L ($K = 0$; the system does not generate information) is significantly incorrect, if the system is not found to be in the predicted state. Then L does not apply, and the K–value of the system is a member of $\mathcal{L}'$.

The situation is more involved for $\mathcal{M}$, since $\mathcal{M}$ contains infinitely many potential values of K. This implies that the comparison of prediction and observation as used above does not provide a comparably easy assignment of K to $\mathcal{M}$ or $\mathcal{M}'$. Let us again assume that a prediction is made at $t = 0$. This time it is based on M, so that $K \geq 0$. (Remember that no particular value of K is specified. The purpose is only to find out whether the system behaves according to M or not.) Then, any positive check procedure (confirmation) for any time interval Δt confirms the proposition prop$M$ corresponding to the set $\mathcal{M}$ of potential values of K. However, such a confirmation for specific time intervals does neither extend nor reduce the set $\mathcal{M}$. Although the check procedure acquires more and more empirical confirmation material for prop$M$ (and thus enhances the "believability" of prop$M$), the set $\mathcal{M}$ remains the same as in (8):

$$\mathcal{M}(\Delta t) = \mathcal{M} = \{K|K \geq 0\} \tag{16}$$

This is not the case for its complement $\mathcal{M}'$. Since the series of confirming checks for prop$M$ excludes potential elements of $\mathcal{M}'$ (which might be candidates for falsification before the check), the size of the set $\mathcal{M}'$ of remaining potential values of K (values which might still be relevant for falsification) decreases as Δt increases. Hence $\mathcal{M}'$ is explicitly time dependent:

$$\mathcal{M}'(\Delta t) = \{K|K < (\Delta t)^{-1}\} \tag{17}$$

$\mathcal{M}$ and $\mathcal{M}'$ satisfy Eq. (14) since $\mathcal{M} \vee \mathcal{M}' = I$. However, satisfying Eq. (13) requires the limit $\Delta t \to \infty$, i.e., the unrealistic limit of an exhaustive check procedure. Only in this limit could $\mathcal{M} \wedge \mathcal{M}' = 0$ be obtained. This indicates that the lattice based on $\mathcal{G}$ is complemented "in principle", but the complement cannot be uniquely determined [33]. This situation is exactly what leads to the property of non–distributivity, which we treat in the following.

3.3. Distributivity

A lattice V is *distributive* with respect to the complement if for each pair $(a, b) \in V$ the relations

$$a = (a \wedge b) \vee (a \wedge b') \tag{18a}$$

$$\vee \quad b = (b \wedge a) \vee (b \wedge a') \tag{18b}$$

and (dually)

$$a = (a \vee b) \wedge (a \vee b') \tag{19a}$$

$$\wedge \quad b = (b \vee a) \wedge (b \vee a') \tag{19b}$$

are satisfied. If V is non–distributive (18) and (19) have still to satisfy the distributive inequalities, obtained by replacing "=" by "$\geq$" in (18) and replacing "=" by "$\leq$" in (19). In order to study the distributivity of V with respect to the complement, we analyze relation (19) using the set $\mathcal{G}$ with its relevant subsets to express "a", "b", and their complements. The algebraic term $(\mathcal{M} \vee \mathcal{L}) \wedge (\mathcal{M} \vee \mathcal{L}')$ is characterized by $\mathcal{M}$ according to (8), $\mathcal{L}$ according to (7), and $\mathcal{L}'$ according to (15):

$$\{K|K \geq 0 \vee K = 0\} \wedge \{K|K \geq 0 \vee K \neq 0\} = \{K|K \geq 0\} \tag{20a}$$

such that

$$(\mathcal{M} \vee \mathcal{L}) \wedge (\mathcal{M} \vee \mathcal{L}') = \mathcal{M} \tag{21a}$$

The second relation to be checked concerns $(\mathcal{L} \vee \mathcal{M}) \wedge (\mathcal{L} \vee \mathcal{M}')$. Here we have to use $\mathcal{L}$ according to (7), $\mathcal{M}$ according to (16) (which effectively equals (8)), and $\mathcal{M}'$ according to (17), and find:

$$\{K|K = 0 \vee K \geq 0\} \wedge \{K|K = 0 \vee K < (\Delta t)^{-1}\} = \{K|0 \leq K < (\Delta t)^{-1}\} \tag{20b}$$

such that

$$(\mathcal{L} \vee \mathcal{M}) \wedge (\mathcal{L} \vee \mathcal{M}') > \mathcal{L} \tag{21b}$$

Relation (21b) satisfies the distributive inequality. Since according to (19) distributivity would require the equality of both (21a) and (21b), a complement as discussed here is non – unique with respect to $\mathcal{G}$.

Here it should be emphasized that we have shown non–distributivity based on (19) only. Of course, this means that the lattice is also non–distributive according to (18) due to the duality of the lattice. However, we have investigated *particular* sets of propositions, namely temporal predictions, in terms of the corresponding sets of numbers assigned to them. The duality of the lattice does not imply that this particular set of propositions, satisfying one of the dually related properties, also satisfies the other one necessarily. It is conceivable that a lattice consists of general (meta–) propositions, and that the dual properties refer to particular propositions of mutually different content. If this is the case, then (18) and (19) can be satisfied "separately", thus indicating an underlying unity of both sets of propositions.

It is the latter situation which we believe to be relevant in the present context. The content to which (18) refers is that of quantum logical propositions, and the lattice containing (18) and (19) is supposed to effectively unify temporal logic as introduced in this article with quantum logic. The type of logic combining both has been discussed to provide basic properties of a four–valued modal logic [28]. Although such a logic still misses the fundamental issue of self–reference, it indicates an improved understanding of the duality of external and internal viewpoints toward the structure and dynamics of systems [29]. This understanding is completely different from the fiction of a never ending (infinitely recursive) construction of meta–viewpoints, each of them including its predecessors, but at the same time requiring its successive extension. In the sense that even the unifying property of the lattice and its duality are ultimately considered to be relevant simultaneously, our approach may be understood as a radically complementary one.

4. Locality, Modularity, and Geometry

A well–known essential property of quantum systems is their fundamental non–locality as implied by EPR–type correlations. This property represents a strong argument against the applicability of the lattice approach in general. The definition of a lattice is based on the definition of a partial order, $a \leq b$, which may become indefinite if there are subsets a and b with an indefinite boundary.

Such an indefinite boundary concerning propositional sets a, b can be caused by nonlocal correlations in a system consisting of subsystems to which a and b refer. In EPR–type situations, nonlocal (with respect to spatial distance) correlations introduce a fundamental inseparability into our description of a system, which renders an adequate decomposition of a system into independent subsystems fundamentally impossible. Consequently, boundaries between subsystems cannot be well–defined, and propositions about those subsystems cannot be treated independently from each other.

From the viewpoint of the duality of quantum logic and temporal logic indicated in Sec.3.3 the propositional lattice corresponding to temporal predictions would have to be limited in an analogous manner. This can be understood using the above argument with respect to temporally nonlocal correlations. It means that *a decomposition of the temporal behavior of a system into subsequent independent temporal intervals is as inadequate as its spatial counterpart.*

These deliberations suggest that a lattice unifying quantum logic and temporal logic is not relevant in general, i.e., for nonlocally correlated composite systems. However, it is worthwhile to identify limiting cases in which boundaries between spatial/temporal subsystems may be approximately defined or become irrelevant, so that the above argument against the definition of a lattice does no longer apply. The boundary of a subset a with respect to another one, c, is certainly irrelevant if $a \equiv c$. This case represents a description of a system with internal correlations, but without any external relationships. This implies that no observation/modelling of the system from an external viewpoint is possible.

The other alternative is to *approximate* the description of a composite system by a description in terms of separable subsystems. This approximation may be reasonable if the influence of nonlocal correlations on the investigated effect is small. In this case, it can be justified to treat the system lattice theoretically. On the level of facts, this second case refers to the situation of a "detached observer", with respect to models it could be interpreted as that of a "detached 'describer'".

At this point, some remarks on the lattice property of modularity are appropriate. A lattice is *modular* if $a \leq c$ implies:

$$a \vee (b \wedge c) = (a \vee b) \wedge c \tag{22}$$

It is easy to see that both limits discussed above satisfy this property. In the first case, $b \leq a \equiv c$, one can think of an internal "participant" of the system characterized by b, and the second case can be covered by characterizing with b an external "detached observer", where $b \wedge (a \equiv c) = \emptyset$.

It is very tempting to identify both limits as limits of the general situation of a "participatory observer" as suggested by Wheeler [34]. They are formally reflected by the simplifying assumption $a \equiv c$ which gives rise to the modular limit of a generally non–modular (e.g., weakly modular, orthomodular, or semimodular) lattice. As mentioned above, the relevance of a modular approximation depends on the separability of subsystems. Although there is no generally accepted good criterion for inseparability at hand, almost all types of systems, quantum or classical, simple or complex, are presently treated in terms of those modular limits (as far as their treatment in the framework of physics is concerned).

Nevertheless, inseparability due to nonlocal correlations, if taken seriously, basically invalidates the lattice approach as such, and consequently the modular property becomes irrelevant. It is only in a non–rigorous (perhaps even self–contradictory) manner that modularity might be regarded as violated without a simultaneous breakdown of the lattice approach.

The modular property is particularly important concerning the geometrical interpretation of the lattice approach. Since any metric lattice is modular [30], non–modularity prevents the unique definition of a metric. Hence, the uniqueness of distances as it is required in metric spaces is lost in such a case. In addition, modularity is a continuity condition [32], such that a non–modular lattice limits the relevance of a representation of a corresponding system in a continuous manner. Two fundamental consequences arise:

1. Any continuous geometrical representation of the structure and dynamics of systems is based on the limit of modularity. Within this

limit, non–distributivity restricts the possible classes of geometries to skew–symmetric ones, e.g., dual quaternionic spacetimes [29]. Distributivity as a further limiting case would allow for symmetric spacetimes or Boolean lattices, respectively.

2. A more general representation appears to be conceivable, if spaces are taken into consideration which do not require a unique definition of a metric [35] and which are not necessarily continuous [34]. This point indicates the relevance of a discrete and topological instead of a continuous geometrical concept of spacetime. A corresponding algebraic approach (Grassmann's extensor algebra which generalizes the lattice approach) with discrete spacetime topology has recently been suggested [36].

Acknowledgment

It is a pleasure to thank I. Antoniou and F. Lambert for the opportunity to present an informal version of this article at the Brussels meeting. Comments by J. Becker, B. Pompe, and H. Scheingraber are gratefully appreciated.

References

1. I. Prigogine, *Non - Equilibrium Statistical Mechanics* (Interscience, New York, 1962).
2. I. Prigogine, *From Being to Becoming*, 2nd ed. (Freeman, San Francisco, 1980).
3. R. Shaw, *Z. Naturforsch.* **36 a**, 80 (1981).
4. J.D. Farmer, *Z. Naturforsch.* **37 a**, 1304 (1982).
5. H. Atmanspacher and H. Scheingraber, *Found. Phys.* **17**, 939 (1987).
6. A.M. Fraser, Ph.D. Thesis, University of Texas at Austin 1988.
7. H. Atmanspacher, in *Parallelism, Learning, Evolution*, eds. J. Becker, F. Mündemann, and I. Eisele (Springer, Berlin, 1991) in press.
8. P. Grassberger and I. Procaccia, *Phys. Rev. Lett.* **50**, 346 (1983).
9. H. Atmanspacher and H. Scheingraber, *Phys. Rev.* **A 34**, 253 (1986).
10. S. Goldstein, *Israel J. Math.* **38**, 241 (1981).
11. A.N. Kolmogorov, *Dokl. Akad. Nauk. SSSR* **119**, 861 (1958).
12. Y. Sinai, *Dokl. Akad. Nauk. SSSR* **124**, 768 (1959).
13. J.B. Pesin, *Russ. Math. Survey* **32**, 455 (1977).
14. C.F. v. Weizsäcker, *Aufbau der Physik* (Hanser, München, 1985) Sec.5.
15. H. Atmanspacher, *Found. Phys.* **19**, 553 (1989).
16. B. Misra, *Proc. Ntl. Acad. Sci. USA* **75**, 1627 (1978).
17. B. Misra, I. Prigogine, and M. Courbage, *Physica* **98 A**, 1 (1979).
18. Y. Elskens and I. Prigogine, *Proc. Ntl. Acad. Sci. USA* **83**, 5756 (1986).
19. N.G. Krylov, *Works on the Foundations of Statistical Physics* (Princeton University Press, Princeton, 1979).

20. J.M. Jauch, *Foundations of Quantum Mechanics* (Addison Wesley, Reading, 1968).
21. G. Birkhoff and J. von Neumann, *Ann. Math.* **37**, 823 (1936).
22. R. Balian, Y. Alhassid, and H. Reinhardt, *Phys. Rep.* **131**, 1 (1986).
23. Y. Elkana, in *Sciences and Cultures. Sociology of the Sciences*, Vol.5, E. Mendelsohn and Y. Elkana, eds. (Reidel, Dordrecht, 1981) pp.1-76.
24. H. Putnam, *Reason, Truth, and History* (Cambridge University Press, Cambridge, 1981).
25. J. Margolis, *Pragmatism Without Foundations* (Blackwell, Oxford, 1986).
26. P. Feyerabend, *Against Method* (New Left Books, 1975).
27. T. Kuhn, *The Structure of Scientific Revolutions* (Univ. Chicago Press, Chicago, 1962).
28. H. Atmanspacher, F.R. Krueger, and H. Scheingraber, in *Parallelism, Learning, Evolution*, eds. J. Becker, F. Mündemann, and I. Eisele (Springer, Berlin, 1991) in press.
29. H. Atmanspacher, in *Information Dynamics*, eds. H. Atmanspacher and H. Scheingraber (Plenum Press, New York, 1991) in press. For a concrete empirical consequence concerning cosmological redshifts we refer to H. Atmanspacher and H. Scheingraber, "An internal observer's view of moving objects in a closed universe", preprint 1991.
30. G. Birkhoff, *Lattice Theory*, 3rd ed. (AMS Coll. Publ., Vol. 25, Providence, 1979).
31. M. Jammer, *The Philosophy of Quantum Mechanics* (Wiley & Sons, New York, 1974).
32. H. Primas, *Chemistry, Quantum Mechanics, and Reductionism* (Springer, Berlin, 1983).
33. D. Finkelstein, in *The Universal Turing Machine - A Half Century Survey*, ed. R. Herken (Oxford University Press, Oxford, 1988).
34. J.A. Wheeler, in *Some Strangeness in the Proportion*, ed. H. Woolf (Addison Wesley, Reading, 1980).
35. See P.A.M. Dirac, *Proc. Roy. Soc. (London)* A**180**, 1 (1942).
36. D. Finkelstein and J. Hallidy, "An algebraic language for quantum spacetime topology", preprint 1990.

Damping, Quantum Field Theory and Thermodynamics

E. Celeghini[1], M. Rasetti[2] and G. Vitiello[3]

[1]Dipartimento di Fisica dell'Università,50125 Firenze, Italy and INFN, Sezione di Firenze.
[2]Dipartimento di Fisica del Politecnico di Torino, 10129 Torino, Italy and CISM, Unità Politecnico di Torino.
[3]Dipartimento di Fisica dell'Universitá, 84100 Salerno, Italy and INFN, Sezione di Napoli.

The purpose of this communication is to discuss the canonical quantization of the damped harmonic oscillator [1]:

$$m\ddot{x} + \gamma\dot{x} + kx = 0 \tag{1}$$

In particular we will show that in the infinite volume limit the set of the states of the damped oscillator splits into disjoint folia, each one parametrized by the time t in such a manner that time evolution is described in terms of trajectories across the folia.

Doubling of phase-space degrees of freedom is required in order to deal with an isolated system, as the canonical quantization scheme prescribes. By introducing the auxiliary variable y to this aim, the Lagrangian L is then written as:

$$L = m\dot{x}\dot{y} + \frac{1}{2}\gamma(x\dot{y} - \dot{x}y) - kxy \tag{2}$$

The y variable is seen to be the time-reversed of the x oscillator. After introduction of creation and annihilation operators for x and y systems and by performing a linear canonical transformation, the Hamiltonian $H = H_O + H_I$ is obtained, with

$$H_O = \hbar\Omega(A\ A - B\ B)\ ; \quad H_I = i\hbar\Gamma(A\ B - AB) \tag{3}$$

where $\Gamma=\gamma/2m$ is the decay constant for the classical variable x(t). The states generated by B^+ represent the sink where the energy dissipated by the damped oscillator flows. It is important to note that $[H_O,H_I] = 0$ which shows that the eigenvalue of H_O, say $(n_A - n_B)$, in the basis $\{|n_A,n_B\rangle\}$ of simultaneous eigenvectors of A^+A and B B is a conserved quantity.

We consider now the time evolution of the vacuum $|0,0\rangle \equiv |0\rangle$: $A|0\rangle = B|0\rangle = 0$, $\langle 0|0\rangle = 1$. Let $|0\rangle$ be the vacuum at t=0; at time t it is

$$\begin{aligned} |0\rangle|_t &= e^{-iHt/\hbar}|0\rangle = e^{-iH_It/\hbar}|0\rangle \equiv |0(t)\rangle \\ |0(t)\rangle &= [\cosh(\Gamma t)]^{-1}\exp[\tanh(\Gamma t)A^+B^+]|0\rangle \end{aligned} \tag{4}$$

namely, a two-modes Glauber coherent state. $|0(t)\rangle$ is a well normalized state at each value of t: $\langle 0(t)|0(t)\rangle=1$; however

$$\lim_{t\to+\infty} \langle 0(t)|0\rangle = \lim_{t\to+\infty} \exp[-\log\cosh(\Gamma t)] \longrightarrow 0 \tag{5}$$

Eq.(5) expresses the instability (decay) of the vacuum under time evolution: as an effect of the damping time evolution brings out of the t=0 Fock space; This is of course a non-acceptable pathology in a proper Quantum Mechanics scheme. Canonical quantization requires infact the operator algebra be fully realized in a specific Hilbert space of states (defined up to unitary transformation). In this sense we may say that dissipation leads to "unquantization" of the original (i.e. at t=0) quantum realization of the operator algebra. We then move to Quantum Field Teory (QFT) where infinite number of degrees of freedom allows infinitely many unitarily inequivalent representations (uir) of the ccr. In such a framework the vacuum instability may be represented as tunneling between uir and a bona fide quantum realization of the operator algebra may be implemented at each value of time t by GNS construction in the C*-algebra formalism [2]. In QFT we have $H = H_0 + H_I$

$$H_0 = \sum_k(\hbar\Omega_k A_k^+ A_k - \hbar\Omega_k B_k^+ B_k)\ , \quad H_I = i\sum_k \hbar\Gamma_k(A_k^+ B_k^+ - A_k B_k) \tag{6}$$

$$[A_k, A_q^+] = [B_k, B_q^+] = \delta_{kq}\ , \quad [A_k, B_q] = [A_k, B_q^+] = 0 \tag{7}$$

where k (q) denotes spatial momentum. We now have

$$|0(t)\rangle = \prod_k [\cosh(\Gamma_k t)]^{-1} \exp[\mathrm{tangh}(\Gamma_k t) J_{+k}]|0\rangle\ , \quad J_{+k} = A_k B_k \tag{8}$$

$$\lim_{t\to+\infty} \langle 0(t)|0\rangle = \lim_{t\to+\infty} \exp[-\sum_k \log\cosh(\Gamma_k t)] = 0 \tag{9}$$

Moreover, by using $\sum_k \to [V/(2\pi)^3]\int dk^3$,

$$\langle 0(t)|0(t')\rangle \xrightarrow[V\to\infty]{} 0 \qquad \text{for any } t\neq t' \tag{10}$$

which shows that as $V\to\infty$ the states split into uir $\{|0(t)\rangle\}$ and these are spanned as t evolves; time evolution can be represented as the inner automorphism of SU(1,1) which carries the system operator algebra across the disjoint folia $\{|0(t)\rangle\}$:

$$\begin{aligned} A_k \longrightarrow A_k(t) &= e^{-iH_I t/\hbar} A_k\, e^{iH_I t/\hbar} = A_k \cosh(\Gamma_k t) - B_k^+ \sinh(\Gamma_k t) \\ B_k \longrightarrow B_k(t) &= e^{-iH_I t/\hbar} B_k\, e^{iH_I t/\hbar} = B_k \cosh(\Gamma_k t) - A_k^+ \sinh(\Gamma_k t) \end{aligned} \tag{11}$$

and their h.c.. It is interesting to notice that eq. (11) are time-dependent Bogoliubov transformations. Moreover, at each time t a new set of annihilation and creation operators is defined : $A_k(t)|0(t)\rangle = B_k(t)|0(t)\rangle = 0$. The number of modes A_k at time t is

$$n_{A_k}(t) = \langle 0(t)|A_k^+ A_k|0(t)\rangle = \sinh^2(\Gamma_k t) \tag{12}$$

and similarly for modes B_k . The automorphism (11) is not bounded in time even at finite volume, which expresses the already mentioned fact that dissipation "carries out" of the original (t =0) Fock space ("unquantization").In the infinite volume limit the transformations (11) are non-unitary transformations. We thus obtain the remarkable result: *At microscopic level the irreversibility of time evolution of the damped oscillator is expressed by the non unitary evolution across the disjoint folia* $\{|0(t)\rangle\}$.

We observe that the state $|0(t)\rangle$ may be written as

$$|0(t)\rangle = \exp(-S/2)\, \exp(\sum_k A_k^+ B_k^+)|0\rangle \tag{13}$$

$$S = -\sum_k\{A_k^+A_k \log \sinh^2(\Gamma_k t) - A_k A_k^+ \log \cosh^2(\Gamma_k t)\} \tag{14}$$

S may also be expressed by replacing A by B in eq. (14). By writing $|0(t)\rangle = \sum_n W_n|n,n\rangle$, $\sum_n W_n = 1$, we have $\langle 0(t)|S|0(t)\rangle = -\sum W_n \log W_n$, which allows us to refer to S as to the "entropy" for the dissipative system. It is interesting to observe that $\frac{\delta}{\delta t}|0(t)\rangle = -\frac{i}{\hbar} H_I|0(t)\rangle$ and $\frac{\delta}{\delta t}|0(t)\rangle = -(\frac{1}{2}\frac{\delta S}{\delta t})|0(t)\rangle$, which shows that $-i(1/2)(\delta S/\delta t)$ acts as the generator of time translations: the fact that S can be interpreted as the "entropy" and at the same time it controls the time translations is related with the irreversibility of time evolution. Dissipation infact implies the choice of a preferred direction in time evolution with a consequent breakdown of time-reversal invariance. Let us also observe that the total entropy $(S_A - S_B)$ is a constant of motion: $[S_A - S_B, H_I] = 0$.

Even if we did not start our analysis by introducing since the beginning statistical notions and quantities, the above description of time evolution in terms of trajectories or paths in the "space of uir of the ccr" seems to lead in a natural way to concepts (e.g. the entropy) which are proper of statistical mechanics and thermodynamics. It may be therefore interesting to analyse more closely such a connection with thermal field theory.

Let us focus our attention on A mode and introduce the functional

$$F_A(t) = \langle 0(t)|\, H_A - \frac{1}{\beta} S_A\, |0(t)\rangle \tag{15}$$

where $H_A = \sum \hbar\Omega_k A_k^+ A_k$. S_A is given by (14) and β is a positive function of time which is to be determined and is assumed to be non-zero for finite time. We minimize $F_A(t)$ with respect to $\vartheta_k(t) = \Gamma_k t$ for each k and each t and obtain $\beta(t)\hbar\Omega_k = -\log\tanh^2(\Gamma_k t)$. We then have

$$n_{A_k}(t) = \sinh^2(\Gamma_k\, t) = \frac{1}{e^{\beta(t)E_k} - 1} \qquad \text{at each t} \tag{16}$$

with $E_k = \hbar\,\Omega_k$. Eq.(16) is the Bose distribution for A_k at t if we take $\beta(t)$ to be the inverse-temperature at t: $\beta(t) = (\kappa\, T(t))^{-1}$, κ is the Boltzmann constant. The functional F (t) is thus recognized to be the A mode free energy and the representation $\{|0(t)\rangle\}$ is nothing else than the thermo field dynamics[3,4] representation $\{|0(\beta)\rangle\}_{at\ t}$. $n_A(t)$ is the statistical average of A modes at the temperature $\beta(t)^{-1}$.

When $(\delta\beta/\delta t) = -(1/\kappa T)(\delta T/\delta t) \approx 0$, then by minimizing the free energy we have

$$dF_A(t) = dE_A(t) - \frac{1}{\beta} dS_A(t) = 0 \tag{17}$$

which expresses the first principle of thermodynamics for a system coupled with the environment at constant temperature β^{-1} and in the absence of exchange of mechanical work.

As observed above, the automorphism (11) is unbounded in t. This may be related also with the fact that the Bose distribution is singular for $\beta = 0$ and $E_k \neq 0$.

Eq.(17) may also be obtained by noting that $dE_A(t) = \sum_k \hbar\,\Omega_k\, \dot{n}_{A_k}(t)dt$ and that $dS_A(t) = d(\langle 0(t)|\, S_A|0(t)\rangle) = \beta dE_A(t)$ (see also ref. 5). The change in energy is thus related to the variations of the number of condensed particles in $|0(t)\rangle$. By defining the heat variation $dQ = \beta^{-1}\cdot dS$, we see that time evolution, controlled by dS , turns out into heat dissipation dQ related with the change dn of particle condensed in the vacuum. As the total number of A and B particles increases and $|0\,(t)\rangle$ becomes "fatter", it decays as $\exp(-\,t\Sigma\Gamma_k)$ for large t. Neverthless, as $n_A - n_B$ = const. in time for any k, $|0(t)\rangle$ remains the lowest energy state for each t (which is the meaning of the first principle of thermodynamics eq.(17)).

We finally observe that the interaction picture in the case of the damped oscillator is exact in the sense that $H_{ip} = e^{iH_0t} H_S e^{-iH_0t} = H_0 + H_I$ where H_{ip} and H_S are the hamiltonian in the interaction picture and in the Schrodinger picture, respectively. In perturbation theory a basic role is played by the "adiabatic hypothesis" by which the interaction may be switched off at $t \longrightarrow \pm\infty$. It is such a possibility which allows the definition and the introduction of the "free", i.e. non-interacting fields. In the case of the damped oscillator the switching off of the interaction, i.e. $\Gamma \to 0$, is not possible since one would completely lose the main feature of the system which is the damping. As matter of fact, we have seen that the set of annihilation and creation operators changes at each time t and the same concept of "non-interacting" field thus loses any meaning. On the other hand, normal ordering of operator fields, which is a crucial tool in perturbation calculations, depends on the representation of ccr and thus needs to be redefined at each t (and for each value of the temperature T)[6]. We thus conclude that damping and dissipation require a non-perturbative approach and perturbation methods may only be used for local (in time and temperature variables) fluctuations; in other words, localizable "free" fields (particles and quasi-particles) can be defined at most only locally, i.e. at each "point" in the space of the representations. An example of such a situation is provided by the description of unstable particles in QFT [7]and by the problem of the quantization of the matter field in the presence of a classical gravitational background, i.e. in curved space-time[8].

We also note that the conventional squeezed states of quantum optics, have been shown[9] to be equivalent, up to elements of the group of automorphism of SU(1,1), to the states of the damped harmonic oscillator.

References

1. E.Celeghini, M.Rasetti and G.Vitiello, Damping and thermal quantum field theory , Proceed. of The Second Inter. Workshop on Thermal Field Theories and Their Applications , Tsukuba, Japan, July 1990
2. O.Bratteli and D.W.Robinson, Operator Algebras and Quantum Statistical Mechanics, Springer, Berlin 1979
3. Y.Takahashi and H.Umezawa, Collective Phenomena 2, 55 (1975)
4. H.Umezawa, H.Matsumoto and M.Tachiki , Thermo Field Dynamics and Condensed States, North-Holland Publ.Co., Amsterdam 1982
5. H.Umezawa, Origin of quantum noise and time dependent Thermo field Dynamics , Proceed. of The Second Inter. Workshop on Thermal Field Theories and Their Applications , Tsukuba, Japan, July 1990
   T.Arimitsu, Nonequilibrium Thermo Field Dynamics and thermal process , ibid.
6. R.Manka and G.Vitiello, Annals of Phys.(N.Y.) 199, 61 (1990)
7. S.De Filippo and G.Vitiello, Lett. Nuovo Cimento 19, 92 (1977)
8. M.Martellini, P.Sodano and G.Vitiello, Nuovo Cimento 48A, 341 (1978)
9. E.Celeghini, M.Rasetti, M.Tarlini and G.Vitiello, Mod. Phys. Lett. B3, 1213 (1989)

Quasi-Monomial Transformations and Decoupling of Systems of ODE's

L. Brenig and A. Goriely

Physique Statistique Plasmas et Optique Nl,
Université Libre de Bruxelles, Campus Plaine CP 231,
Blv du Triomphe, 1050 Bruxelles, Belgium.

Summary: The quasi-monomial representation and transformations of systems of nonlinear ODE's provide new conditions of integrability or/and decoupling. In particular, n-dimensional Poincaré normal forms can be transformed into (n-1) closed systems.

I. Quasi-monomial representation and transformations.

The analysis of linear ODE's rests upon two concepts. A matrix representation:

$$\dot{x}_i = \sum_{j=1}^{n} L_{ij}\, x_j \qquad (1)$$

(where the dot means $\frac{d}{dt}$ and $i=1,\ldots n$) and a group of transformations:

$$x_i = \sum_{j=1}^{n} T_{ij}\, x_j \qquad (2)$$

For nonlinear ODE's we have found a similar couple. Indeed, any system of the form:

$$\dot{x}_i = P_i\,(x_i\,,\ldots,x_n\,);i=1,\ldots,n \qquad (3)$$

where the P_i are general polynomials or sums of real or complex powers of the variables $x_1\,,\ldots,x_n$, can be cast in the following standard shape:

$$\dot{x}_i = x_i \sum_{\gamma=1}^{m} A_{i\gamma} \prod_{k=1}^{n} x_k^{B_{\gamma k}} ; i=1,\ldots,n \qquad (4)$$

where $A_{i\gamma}$ and $B_{\gamma k}$ are time independant real or complex numbers. The integer m is generally different from n. Form (4) includes linear and non-linear terms as well. The x_i factor may be cancelled by a well chozen form of the mxn matrix B. Associated to representation (4) are the quasi-monomial transformations (QMT's):

$$x_i = \sum_{j=1}^{n} x'^{T_{ij}}_j \qquad (5)$$

where T is any nxn invertible real or complex matrix. For linear ODE's representation (1) is covariant under transformations (2) that is:

$$\dot{x}'_i = \sum_{j=1}^{n} L'_{ij} x'_j \qquad (6)$$

with

$$L' = T^{-1} \circ L \circ T \qquad (7)$$

Similarly, for nonlinear ODE's representation (4) is covariant under transformations (5):

$$\dot{x}'_i = x'_i \sum_{\gamma=1}^{m} A'_{i\gamma} \prod_{k=1}^{n} x'^{B'_{\gamma k}}_k \qquad (8)$$

with:

$$A' = T^{-1} \circ A \qquad (9)$$

$$B' = B \circ T \qquad (10)$$

In analogy with the diagonalization of linear systems (1) our couple representation-transformation may lead to simpler forms of the nonlinear systems. These are related to the rank of the two fundamental matrices A and B as shown below. We present here some of these results (see also references (1), (2), (3)).

II. Integrability and rank of A

Theorem 1: Any system of ODE's of the form:

$$\dot{x}_i(t) = \sum_{j=1}^{n} \sum_{\gamma=1}^{m} N_{ij\gamma} x_j \prod_{k=1}^{n} x_k^{B_{\gamma k}} ; i=1,\ldots,n; m \text{ arbitrary} \quad (11)$$

with a tensor N factorized as follows:

$$N_{ij\gamma} = \sum_{k=1}^{n} L^{-1}_{ik} L_{kj} A_{k\gamma} ; \; j=1,\ldots,n \; ; \; \gamma=1,\ldots,m \quad (12)$$

and where rank (A)=r<n, possesses (n-r) first integrals of the form:

$$K_\alpha = \prod_{i=1}^{n} \left[\sum_{j=1}^{n} L_{ij} x_j \right]^{T^{-1}_{\alpha i}} ; \; \alpha = 1,\ldots,n-r \quad (13)$$

The matrix T^{-1} is explicitely built with the (n-r) vectors $\varphi^{(\alpha)}$ of the null-space of A: $\varphi^{(\alpha)} \circ A=0$.

This is a generalization of a result included in ref. (3). It is readily obtained by combining a QMT with the linear transformation generated by the nxn matrix L appearing in (12).

III. Decoupling and rank of B

Theorem 2: Any system of ODE's written in the form (4) with rank(B)=r<n can be decoupled after transforming it by a QMT into a r-dimensional system and a (n-r)-dimensional linear one. The proof proceeds along the same lines as in ref (3): if B is of rank r<n then there exist (n-r) vectors $\varphi^{(\alpha)}$ (α=1,...,n-r) such that :

$B \circ \varphi^{(\alpha)}=0$. An nxn invertible matrix T is then built with these vectors as its (n-r) last columns. Next, one performs on system (4) a QMT (5) with the above T matrix. The resulting equation (8) involves a B' matrix verifying relation (10). Obviously, by our above construction of T, B' has its (n-r) last columns made of zeros. This results in a decoupled r-dimensional system:

$$\dot{x}'_l = x'_l \sum_{\gamma=1}^{m} (T^{-1} \circ A)_{l\gamma} \prod_{k=1}^{r} x'^{(B\circ T)_{\gamma k}}_k ; \; l=1,\ldots,r \tag{14}$$

and a linear (n-r) - dimensional system:

$$\dot{x}'_s = x'_s \sum_{\gamma=1}^{m} (T^{-1} \circ A)_{s\gamma} \prod_{k=1}^{r} x'^{(B\circ T)_{\gamma k}}_k ; \; s=r+1,\ldots,n \tag{15}$$

whose coefficients depend on time via the solutions of (14).

IV. Decoupling of Poincaré normal forms

In order to simplifly the discussion we extract the linear part of (4) and assume it is already in diagonal form:

$$\dot{x}_i = \lambda_i x_i + x_i \sum_{\gamma=1}^{m-1} A_{i\gamma} \sum_{k=1}^{n} x_k^{B_{\gamma k}} \tag{16}$$

We also assume that the $B_{\gamma k}$ are non-negative integers or are equal to -1 and such that $\sum_{k=1}^{n} B_{\gamma k} \geqslant 1$. A monomial $x_i \prod_{k=1}^{n} x_k^{B_{\gamma k}}$ in the i-th equation of (16) is said to be resonnant if $\sum_{k=1}^{n} B_{\gamma k}\lambda_k = 0$.

The Poincaré-Dulac theorem (ref (4)) claims that there exist a formal transformation:

$$x_i = y_i + \sum_{m_1,\ldots m_n=0}^{\infty} c_i(m_1\ldots m_n) y_1^{m_1}\ldots y_n^{m_n}; \quad \sum_{i=1}^{n} m_i \geqslant 2 \qquad (17)$$

which leads (16) to the normal form:

$$\dot{y}_i = \lambda_i y_i + y_i \sum_{\mu=1}^{r} \bar{A}_{i\mu} \prod_{k=1}^{n} y_k^{\bar{B}_{\mu k}} \qquad (18)$$

where only resonant monomials are present.

In our approach, this means that $\sum_{k=1}^{n} \bar{B}_{\mu k}\lambda_k = 0$ for all $\mu=1,\ldots,r$. Thus, the vector $(\lambda_1,\ldots,\lambda_n)$ belongs to the null-space of $\bar{B}$. Hence, theorem 2 is applicable and yields the following theorem :

Theorem 3: Any n-dimensional Poincaré normal form may be decoupled into a (n-1)- dimensional system by a well-chozen QMT built along the prescriptions of theorem 2.

As a simple example, let us take the normal form for a local codimension 2 bifurcation in a 3-D system:

$$\begin{aligned} \dot{Z} &= i\omega\, Z + axZ \\ \dot{x} &= bx^2 + c|Z|^2 \end{aligned} \qquad (20)$$

All the monomials are resonant with eigenvalues $i\omega$, $-i\omega$, 0. The QMT which decouples system (20) is found to be:

$$Z = y_1 y_3^{i\omega} \qquad (21)$$

$$x = y_2 \qquad (22)$$

with $(y_1, y_2, y_3) \in \mathbb{R}^3$.

Putting $y_3 \equiv e^{\frac{\varphi}{\omega}}$, transformation (21) reduces to the polar coordinate transformation. Although here this seems trivial, by no means it is in the general case.

<u>References</u>:

(1) L. Brenig, Phys. Lett. <u>A133</u>, 378 (1988).

(2) L. Brenig and A. Goriely, Phys. Rev. <u>A4</u>, 4119 (1989).

(3) A. Goriely and L. Brenig, Phys. Lett. <u>A145</u>, 245 (1990).

(4) V. I. Arnold, "Geometrical Methods in Theory of ODE's" 2nd ed. Springer-Verlag (Berlin, 1988).

Part II

Physical Systems with Soliton Ingredients

Solitons in Optical Fibers: First- and Second-Order Perturbations

D.J. Kaup

Clarkson University,

Department of Mathematics and Computer Science, and the Institute for Nonlinear Studies.

Potsdam, NY 13699-5815, USA.

Abstract. A review of solitons in optical fibers is given, along with a discussion of various small effects which can be treated as perturbations. We also estimate the second-order long-range phase-indepenedent soltion-soliton interaction and find it to be too small to account for recent experimental results.

1 Introduction

With the current interest of using solitons as pulse bits in long optical fibers for communication purposes [1], it is important for us to re-evaluate the practicality of using analytical techniques for predicting the behavior of such bits. Since such pulses are near a pure soliton solution, it becomes feasible to use analytical soliton techniques and the potential for obtaining useful analytical results becomes quite high.

Although there are many examples in the past of perturbation calculations [2–11], few of these have been applied to solitons in optical fibers. Most recently, this author [12] has demonstrated that perturbative calculations can easily compare with extensive numerical calculations [1]. Hasegawa and Kodama [13–14] have recently applied guiding center ideas from plasma physics to describe the motion of solitons which are periodically amplified. These examples represent applications of old perturbative techniques to solving new problems. What we shall describe here is another example where we extend perturbative techniques into second order in an attempt to understand a recently observed phase-independent interaction between solitons in optical fibers [15]. One suggested mechanism [15] is the exchange of continuous spectra between solitons.

Here we use perturbation theory to calculate the continuous spectra. Which is generated by the non-soliton-like terms arising from the Raman pumping of the soliton. This Raman pumping creates a spatially periodic non-soliton perturbation. As a result of this perturbation, a long wavelength continuous spectra is generated which slowly evolves away from the soliton where it was generated. This continuous spectra eventually reachs and interacts with the next soliton (if present), giving a mechanism for solitons to mutually interact over distances much larger than their individual soliton widths. However when we finally estimate the strength of such an interaction, we find that it is in general too small (by at least a factor of 100) to account for the observed shifts. Thus we can assert that one should look elsewhere for an explanation of the experimental observations [15].

We shall now detail how one makes such calculations.

2 The Second-Order Perturbed NLS

As shown in Ref. 12, one may expand solutions of the perturbed NLS

$$i\partial_t q = -\partial_x^2 q - 2q^* q^2 + \epsilon R[q, q^*] \tag{1}$$

in the singular perturbation expansion about a single soliton as

$$q = q_0 + \epsilon q_1 + \epsilon^2 q_2 + \cdots \tag{2}$$

where

$$q_0 = \frac{Ae^{i\alpha}}{cosh\theta} \tag{3}$$

$$\theta = 2\eta(x - \bar{x}) \tag{4}$$

$$\alpha = -2\xi(x - \bar{x}) + \bar{\alpha} \tag{5}$$

In the above, A is the amplitude of the soliton, η is the inverse of the width, $\bar{x}$ is its position, and $\bar{\alpha}$ is its phase at its center. As in Ref. 12, we allow the soliton amplitude and width to be independent variables. However if the perturbations is small, we do expect A to be very close to the value of 2η, which is the pure soliton value.

We also introduce the multiple time scales whence

$$\tau_n = \epsilon^n t \tag{6a}$$

$$\frac{\partial}{\partial t} = \frac{\partial}{\partial \tau_0} + \epsilon\frac{\partial}{\partial \tau_1} + \epsilon^2\frac{\partial}{\partial \tau_2} + \cdots \tag{6b}$$

We take $A = A(\tau_1, \tau_2, \cdots)$ and $\eta = \eta(\tau_1, \tau_2, \cdots)$. For simplicity and to avoid certain secular terms, we also take $\bar{x}$ and $\bar{\alpha}$ to be independent of $\tau_0(= t)$ and to be dependent only on the slow time scales. Consequently, in order to have the proper zeroth-order behavior, we expand $\bar{x}$ as

$$\bar{x} = \bar{x}_{-1}\frac{1}{\epsilon} + \bar{x}_0 + \bar{x}_1\epsilon + \cdots \tag{7}$$

and similarly for $\bar{\alpha}$.

For the higher order parts of the solution, define the column matrix

$$v = \begin{bmatrix} e^{-i\alpha}(q_1 + \epsilon q_2 + \cdots) \\ e^{i\alpha}(q_1^* + \epsilon q_2^* + \cdots) \end{bmatrix} \tag{8}$$

then the evolution of v is given by

$$i\partial_t v + 4\eta^2 L v = F \tag{9}$$

where the operator L is

$$L = \sigma_3(\partial_\theta^2 - 1) + \frac{2}{cosh^2\theta}(2\sigma_3 + i\sigma_2), \tag{10}$$

with σ_1, σ_2 and σ_3 being the Pauli spin matrices. The eigenstates and closure of L are summarized in appendix A.

The Source F is of the form

$$F = \begin{bmatrix} \mathcal{R} \\ -\mathcal{R}^* \end{bmatrix} \tag{11}$$

where from all of the above, we obtain

$$\begin{aligned} \epsilon\mathcal{R} = \epsilon e^{-i\alpha} R - \frac{i}{cosh\theta}\partial_t A - \left(\frac{A\theta}{\eta cosh\theta} + \frac{\epsilon\theta}{\eta}\mu\right)\partial_t\xi \\ +(4\eta^2 - A^2)\left[\frac{2A}{cosh^3\theta} + \frac{2\epsilon}{cosh^2\theta}(2\mu + \mu^*)\right] \\ +i\theta\left[\frac{A sinh\theta}{\eta cosh^2\theta} - \frac{\epsilon}{\eta}\partial_\theta\mu\right]\partial_t\eta \\ +\left[\frac{2\xi A}{cosh\theta} - 2iA\eta\frac{sinh\theta}{cosh^2\theta} + 2i\eta\epsilon\partial_\theta\mu + 2\epsilon\xi\mu\right](\partial_t\bar{x} + 4\xi) \\ +\left(\frac{A}{cosh\theta} + \epsilon\mu\right)(\partial_t\bar{\alpha} - 4\eta^2 - 4\xi^2) - 2\epsilon^3(\mu^2\mu^*) \\ -\frac{2A\epsilon^2}{cosh\theta}(\mu)^2 - \frac{4\epsilon^2 A}{cosh\theta}(\mu^*\mu) \end{aligned} \tag{12}$$

where μ is the first component of v, which by (8) is

$$\mu = e^{-i\alpha}(q_1 + q_2\epsilon + q_3\epsilon^2 + \cdots) \tag{13}$$

The terms of order ϵ in (12) were given in Ref. 12. Here we have given the general expression which can be expanded to any order desired. One would only have to use (6), (7), and (13) in order to expand (12) to the order desired.

One should note that the coefficients of $\partial_t\xi$ and $\partial_t\eta$ contain terms secular in θ at second-order since, in general, μ will approach a plane wave for large θ. The origin of these terms and their handling will be described elsewhere [16]. We shall not need to be concerned with them here.

The last two terms in (12) give the second-order nonlinear coupling between the soliton and the continuous spectrum (radiation). In these terms we shall find the phase-independent long-range interaction.

As discussed in Ref. 12, the correct expansion for v is to expand in terms of the continuous eigenfunctions of L. Thus we take

$$v = \eta \int_{-\infty}^{\infty} dk \left[\, g(k,t)|\psi, k> + \bar{g}(k,t)|\bar{\psi}, k> \,\right] \tag{14}$$

Consequently v, to all orders, is orthogonal to the four states $|\phi_e>, |\phi_0>$, $|\chi>$, and $|\theta\phi_e>$. Inner products of (12) with these states determine the evolution of the soliton parameters. In first order [12], these are

$$\partial_t\xi = \frac{\epsilon\eta}{2A} < \phi_0|\sigma_3|F_{ext}> \tag{15}$$

$$4\partial_t A - 2\frac{A}{\eta}\partial_t\eta = -i\epsilon < \phi_e|\sigma_3|F_{ext}> \tag{16}$$

$$\partial_t\bar{x} + 4\xi = -\frac{i\epsilon}{4A\eta} < \theta\phi_e|\sigma_3|F_{ext}> \tag{17}$$

$$\partial_t\bar{\alpha} - 4\eta^2 - 4\xi^2 = \frac{\epsilon}{2A\eta} < \chi|\sigma_3|F_{ext}> \tag{18}$$

where

$$|F_{ext}> = \begin{bmatrix} Re^{-i\alpha} \\ -R^* e^{i\alpha} \end{bmatrix} \tag{19}$$

is the external forcing. The density of the radiation, g and $\bar{g}$, is given in first-order by

$$\partial_t g - 4i\eta^2(k^2+1)g = \frac{1}{2\pi\eta a^2} < \bar{\phi}|\sigma_3|F_{ext}> \tag{20}$$

$$\partial_t \bar{g} + 4i\eta^2(k^2+1)\bar{g} = \frac{-1}{2\pi\eta a^2} < \phi|\sigma_3|F_{ext}> \tag{21}$$

where a is given by (A.19) and $<\phi|$ and $<\bar{\phi}|$ by (A.11) and (A.12). This is the radiation which will evolve away from the soliton and will propagate as a forced wave.

Let us now consider the evolution under the following general perturbation.

$$R = -i\gamma q + iC_1\partial_x^3 q + iC_2\partial_x^2 q + C_3 q\partial_x(q^* q) \tag{22}$$

where γ is the coefficient of linear damping [2,8,12,17,18,19], C_1 is the coefficient of the third-order dispersion [17,20], C_2 is the coefficient of the second-order delayed Raman effect [17,18] (the first-order delayed Raman effect is just a shift in the group velocity), and C_3 is the coefficient of the self-frequency shift [8,18,21,22,23]. Using (3) to evaluate (22) in lowest order, we have

$$|F_{ext}>= 4\eta A(12i\eta^2 C_1 - A^2 C_3\sigma_3)\frac{tanh\theta}{cosh^3\theta}\begin{pmatrix}1\\1\end{pmatrix}$$

$$+8\eta A(3i\xi^2 C_1 - C_2\xi\sigma_3 - i\eta^2 C_1)\frac{tanh\theta}{cosh\theta}\begin{pmatrix}1\\1\end{pmatrix}$$

$$+4\eta^2 A(iC_2 + 6C_1\xi\sigma_3)(\frac{1}{cosh\theta} - \frac{2}{cosh^3\theta})\begin{pmatrix}1\\1\end{pmatrix}$$

$$-A(i\gamma + 4i\xi^2 C_2 + 8\xi^3 C_1\sigma_3)\frac{1}{cosh\theta}\begin{pmatrix}1\\1\end{pmatrix} \tag{23}$$

Evaluating the inner products in (15) - (18) gives

$$\partial_t\xi + \frac{16}{3}\epsilon\eta^2 C_2\xi = -\frac{16}{15}\epsilon A^2\eta^2 C_3 + \cdots \tag{24}$$

$$\frac{4}{A}\partial_t A - \frac{2}{\eta}\partial_t\eta = -4\epsilon\gamma - \frac{16}{3}\epsilon C_2(\eta^2 + 3\xi^2) + \cdots \tag{25}$$

$$\partial_t\bar{x} + 4\xi = 4\epsilon C_1(\eta^2 + 4\xi^2) + \cdots \tag{26}$$

$$\partial_t\bar{\alpha} - 4(\eta^2 + \xi^2) = 2(A^2 - 4\eta^2) + 16\epsilon\xi C_1(\eta^2 - \xi^2) + \cdots \tag{27}$$

Of course, the first two results are exactly the same as those derived from the conservation laws[17,18], since the conservation laws are a consequence of the equations of motion[7,8]. However, the evolution of the phases and the radiation densities g and $\bar{g}$, cannot be obtained from the conservation laws.

The soliton self-frequency shift is given by (24), where $-\xi$ is the relative frequency. Since C_2 and C_3 are normally positive, the frequency always downshifts, but is stabilized by a nonzero C_2 [18]. From (25), once a relationship between A and η is chosen[24], one can determine how the soliton's amplitude will vary. It is driven only by the damping and the delayed Raman effect. Eq

(26) gives how the soliton's center will evolve. In zeroth order, it is driven only by the relative frequency, $-\xi$, which is normally zero or at least very small. Thus small corrections can be important in the evolution of $\bar{x}$, such as the higher-order dispersion in (26). This correction is nothing more than what one would obtain from a linear theory. Eq (27) gives the evolution of the soliton's phase. Note that it has a zeroth-order part of $4\eta^2$ (since ξ is normally zero or small and $A \approx 2\eta$ to within an order of ϵ).

Before giving quantitative estimates of the constants, first we need to determine the units in which (1) is written. The procedure for obtaining these has been given in other references [see for example the appendix of Ref. 1]. Here we shall simply summarize the results. First, one has one arbitrary unit which one could choose to be the pulse width, or some convenient unit of time on the order of the pulse width. This unit of time, t_c, converts time quantities into the unitless coordinate x where

$$x = t_{lab}/t_c \tag{28}$$

and t_{lab} is the time coordinate in the laboratory frame. Once t_c is fixed then the characteristic length, z_c, is determined by the condition that the coefficient of $\partial_x^2 q$ be unity. This leads to [1]

$$z_c = \frac{4\pi c}{D\lambda^2} t_c^2 \tag{29}$$

where D is the dispersion coefficent and λ is the wavelength. Note that (29) differs by a factor of 2 from that in Ref. 1. This is because of the different normalizations used in (1). The only real difference is that our z_c will be twice that of Ref. 1. One uses z_c to convert spatial distances into the unitless coordinate t via

$$t = z_{lab}/z_c \tag{30}$$

where z_{lab} is the spatial distance in the laboratory.

For convenience, we shall choose $t_c = 50ps$, since this is the typical soliton width in optical fibers. And we also take $D = 17ps/nm/km$ and $\lambda = 1.56\mu m$ as typical values[15], although a value for D as small as 2ps/nm/km seem to be feasible[25]. Once these parameters are specified, then the characteristic length is calculated to be z_c = 228km. We shall refer to these parameters as the "standard case". Note that for lower D values, z_c increases. Thus if $D = 2ps/nm/km$, then z_c increases by a factor of about 10 to around 2000km. Similarly if t_c is increased, z_c increases quadatically.

With these values now specified, we may now estimate the (unitless) size of the various perturbations. Starting with the damping, from Ref. 1 we have that the Raman gain is typically $\alpha_g \approx 0.07/km$ whereas the net gain, which is periodic, has a maximum amplitute of $(\alpha_g - \alpha_s) \approx 0.02/km$. To determine the damping constant γ in (22), we simply multiply by the value of z_c. Thus the amplitude of the periodic damping (or gain) is $\gamma_{max} \approx 4.56$. Now this is larger than unity and therefore is not a small perturbation. In fact it suggests that the damping will be just as important, if not more important, than the dispersion or the nonlinearity. That is certainly the case if one would use a lower value for D, such as 2ps/nm/km. In this case, the larger value of z_c will increase the unitless amplitude of the periodic damping to ≈ 40. Now clearly the periodic damping dominates both the dispersion and nonlinearity, and attemping to describe this system with a perturbation theory based on a soliton would not be advisable. Rather the linear theory would dominate. Still one could recover the soliton case by reducing the solitons's width until z_c was sufficiently small. However for $D = 2ps/nm/km$, one would require a pulse width of less than 1 ps in order to reduce the value of the unitless periodic damping below unity.

Next let us look at the coefficient of the soliton self-frequency shift, C_3. This coefficient is basically the delay time of the Stoke's response, which is of the order of 5 fs[18]. To obtain the unitless constant C_3, we divide this delay time by the characteristic time t_c. Whence $C_3 \approx 10^{-4}$ for the standard case and is very small. However according to (24), this small value can have a cumulative effect if C_2 is not too large. Thus we also need to know this value as well. It can be obtained by expanding the Raman gain, α_g, in a Taylor series about a zero time delay. Assuming a simple exponential decay, one has $C_2 \approx \frac{1}{2}(\alpha_g z_c)(t_d/t_c)^2$ where t_d is the time delay of around 5 fs. One can obtain the same form from a bandwidth-limited amplification argument also [18]. Thus the unitless value of C_2 is of the order of 8×10^{-8} for the standard case. Consequently for realistic values of $\eta(\sim 1)$ and $\xi(|\xi| << 1), C_2$ is never important and C_3 is only important over extremely long distances. For the standard case a 4000km cable, one would have t in (24) ranging from 0 to 18, giving a net change in ξ of only 7×10^{-3}. By (26), this would allow a maximum shift in $\bar{x}$ of around .02 which is only a small fraction of the width of a soliton. Thus for these effects to be important at all, narrower pulses or longer distances or higher dispersion would have to be used.

The size of the higher-order dispersion coefficient, C_1, can be estimated from experimental dispersion curves [25]. The typical slope in Fig. 36 of Ref. 25 is $40ps/km/nm/\mu m$. Since $D = -2\pi ck''/\lambda^2$, this gives $|\frac{1}{6}k'''| \approx 10^{-2}(ps)^3/km$ which upon using the characteristic lengths of the standard case gives $C_1 = \frac{1}{6}z_c k'''/t_c^3 \approx 2 \times 10^{-5}$. Even with the lower dispersion of $2ps/km/nm$, one can increase this part by only a factor of 10. And since z_c is proportional to t_c^2, to dramatically increase C_1, one would require pulse widths on the order of 10fs. So higher-order dispersion effects are also very small.

In conclusion, the dominate perturbation is the periodic gain (damping). In fact, it is so large that it is almost <u>not</u> a perturbation. Instead, it is a dominate feature of the evolution and is as important as the dispersion or nonlinearity. The one saving feature in the optical fiber case is the short periodicity of the periodic gain. Because of this, the average or first-order effects are smaller than one would normally expect. Thus one still can use a soliton based perturbation theory.

To see this, let us model the Raman-compensated case with

$$\gamma = \gamma_m \cos(\frac{2\pi t}{l}) \tag{31}$$

where $l = L/z_c$ and L is the laboratory distance between repeaters. The important quantity is the integral of (31),

$$\Gamma = \int_0^t \gamma dt = \frac{\gamma_m l}{2\pi} \sin(\frac{2\pi t}{l}) \tag{32}$$

which for small l, does give Γ as being small. In fact, for the standard case and repeaters spaced 40 km apart, the maximum amplitude of Γ is 0.13, which is indeed small. (However for the low dispersion case, the maximum amplitude of Γ is 1.10. So for this case, the damping is just as important as the nonlinearity!)

Let us now look at the radiation generated by the periodic damping, and let us use the constraint $\partial_t \eta = 0$[24]. Then (25) gives

$$A = 2\eta exp(-\epsilon\Gamma) \tag{33}$$

With this, upon expanding, (21)-(23), we have

$$\partial_t \bar{g} + 4i\eta^2(k^2+1)\bar{g} = -4i\eta^2 \frac{(k+i)^2\Gamma}{cosh(\frac{\pi}{2}k)} + \cdots \tag{34}$$

If there is no resonance, then the amplitude of $\bar{g}$ will be on the order of $\Gamma/cosh(\frac{\pi}{2}k)$, and thereby small if Γ is small. In this case, $\bar{g}$ will be composed mostly of long wavelengh radiation where $|k|\tilde{<}1$ (and therefore slowly moving). The higher values of k will be present, but their amplitudes will be seriously reduced due to the exponential nature of the $cosh(\frac{\pi}{2}k)$ factor.

However, when Γ is periodic, then a resonance at certain values of k can occur whereby $\bar{g}$ will grow secularly in t. Using the model (31), this is at

$$k_0^2 = \frac{\pi}{2\eta^2 l} - 1 \tag{35}$$

For the standard case, a solution always exists for (35) for real k whenever $L < 716km$. However for large k_0, the amplitude of the radiation density will be reduced because of the factor $1/cosh(\frac{\pi}{2}k)$ in (34). In the standard case, this resonance occurs at $k_0 = 2.82$ for which $cosh(\frac{\pi}{2}k_0) = 42$. Thus the amount of radiation produced should be small. Still, the secular nature of the resonance could allow a growth to occur.

To analyze this, we integrate (34) using the model (31). The solution for $\bar{g}$ is found to consist of three parts. There is one part of the continuous spectrum which moves with the soliton, and has no dispersive time dependence. This part contains the reshaping of the soliton due to the perturbation. There is another part that is a transient part of the continuous spectrum. It is generated because we started with a soliton which was initially unperturbed. Over a long time, this part evolves similar to the solution of the free Schrodinger equation: phase-mixing to zero with an amplitude vanishing like $t^{-\frac{1}{2}}$.

Lastly there is a part which does not vanish and does travel away from the soliton and is given by

$$\bar{g}_p = \frac{i\eta^2(k+i)^2\gamma_m l}{\pi cosh(\frac{\pi}{2}k)} \cdot \frac{[e^{-4i\eta^2(k^2+1)t} - e^{-2\pi it/l}]}{[4\eta^2(k^2+1) - 2\pi/l]} \tag{36}$$

Due to the resonance at k given by (35), this part of the radiation grows spatially while at the same time, maintaining a fairly uniform amplitude. Thus it has a structure like a shelf which creeps outward from the soliton, in both directions, and the front of the shelf moving with the group velocity of the resonance. Of course, there will also be a much smaller amount of high-k radiation out in front, as well as some slower moving low-k radiation in the rear. Using stationary phase on (14) to estimate the amplitude behind the front, one finds

$$\mu \approx \frac{\eta\gamma_m l(k_0+i)^2 e\sqrt{\pi}}{8\pi k_0 cosh(\frac{\pi}{2}k_0)} \cdot e^{-2\pi it/l} \cdot$$

$$\left[e^{ik_0\theta}H(\theta)H(k_0 - \frac{\theta}{8\eta^2 t}) + e^{-ik_0\theta}H(-\theta)H(k_0 + \frac{\theta}{8\eta^2 t})\right] \tag{37}$$

where k_0 is the positive root of (35) and $H(x)$ is the Heaviside function. We emphasize that (37) is only an approximation in that the front of the shelf does have a structure whose width varys as $\sqrt{8\eta^2 tk_0}$. However, we shall apply (37) only when the width of the front is wide compared to a soliton's width.

By (8) and considering the phases, the right-going part of the first-order solution [see (2)] is

$$q_{1+} = e^{i(\alpha+k_0\theta)}\mathcal{A}H(k_0 - \frac{\theta}{8\eta^2 t}) \tag{38}$$

and the left-going part is

$$q_{1-} = e^{i(\alpha-k_0\theta)}\mathcal{A}H(k_0 + \frac{\theta}{8\eta^2 t}) \tag{39}$$

where the amplitude is

$$\mathcal{A} \approx \frac{\eta\gamma_m le\sqrt{\pi}(k_0+i)^2}{8\pi k_0 cosh(\frac{\pi}{2}k_0)} \cdot e^{-2\pi it/l} \tag{40}$$

Consider the right-going radiation. As it approaches the next soliton, to that soliton it will appear to be an almost plane wave of wavevector k_0 with an adiabatically varying amplitude. Of course, as the radiation passes thru the soliton, it undergoes a phase shift in accordance to (A.4). From (8),(14) and (A.4), in the vicinity of the next soliton, we have

$$\mu = \mathcal{B}e^{ik_0\theta_+}\left[1 - \frac{2ik_0e^{-\theta_+}}{(k_0+i)^2cosh\theta_+} + \frac{1}{(k_0+i)^2cosh^2\theta_+}\right]\cdot H(k_0 - \frac{\theta_0}{8\eta^2\tau})$$

$$+\frac{\mathcal{B}^*e^{-ik_0\theta_+}}{(k_0-i)^2cosh^2\theta_+}H(k_0 - \frac{\theta_0}{8\eta^2 t}) \tag{41}$$

where θ_0 is θ measured from the first soliton, θ_+ is θ measured from the next one,

$$\mathcal{B} = (\frac{k_0+i}{k_0-i})^2 e^{ik_0(\theta_0-\theta_+)}e^{i(\alpha_0-\alpha_+)}\mathcal{A}, \tag{42}$$

α_0 is the phase of the first soliton and α_+ is the phase of the next one.

Let us now ask what will be the dominate terms in (12) when a single soliton is perturbed by the above radiation field. First, the terms linear in μ would have oscillations like $e^{-2\pi it/l}e^{ik_0\theta}$.

Since k_0 is large, the integral over θ would reduce the amplitude by a factor of order $1/k_0$, and then the forcing would also be rapidly oscillatory in τ. Thus time-integrated quantities would have an amplitude on the order of $l\mathcal{B}/(2\pi k_0)$ and oscillatory.

Now consider the second-order (in μ) terms in (12). Some of these terms will go as $\mathcal{B}^*\mathcal{B}$ and therefore lack the oscillatory structure present in the first order. These terms are also phase independent and are the last two terms in (12). Evaluating only them, we have

$$\epsilon\mathcal{R} \approx -\frac{4A\epsilon^2\mathcal{B}^*\mathcal{B}}{(k_0^2+1)^2cosh\theta_+}\cdot H^2(k_0 - \frac{\theta_0}{8\eta^2 t})\cdot$$

$$\left[\ (k_0^2+1)^2 - \frac{k_0^2+3}{cosh^2\theta_+} + \frac{3}{cosh^4\theta_+} + \frac{2ik_0tanh\theta_+}{cosh^2\theta_+}\ \right] \tag{43}$$

To determine how this perturbation affects this next soliton, we evaluate the inner products with F as in (15) - (18). One finds that (43) has no effect on the evolution of the eigenvalue parameters η and ξ, but the position and phase are affected. In particular, (26)-(27) are replaced by

$$\partial_t\bar{x} + 4\xi = 4\epsilon C_1(\eta^2+4\xi^2) - \frac{2\epsilon^2k_0\mathcal{B}^*\mathcal{B}}{3\eta(k_0^2+1)^2}H^2(k_0 - \frac{\Delta\theta}{8\eta^2 t}) \tag{44}$$

$$\partial_\tau\bar{\alpha} - 4(\eta^2+\xi^2) = 2(A^2-4\eta^2) + 16\epsilon\xi C_1(\eta^2-\xi^2) + \frac{4}{\eta}\epsilon^2\mathcal{B}^*\mathcal{B}\frac{k_0^4+k_0^2-2/3}{(k_0^2+1)^2}H^2(k_0 - \frac{\Delta\theta}{8\eta^2 t} \tag{45}$$

where $\Delta\theta$ is the θ difference between the adjacent solitons.

For the standard case, $\mathcal{B}^*\mathcal{B} = \mathcal{A}^*\mathcal{A} \approx 10^{-3}$ whence (44)-(45) gives

$$\partial_t \bar{x}_+ \approx -2 \times 10^{-5} H^2 (k_0 - \frac{\Delta\theta}{8\eta^2 t}) \tag{46}$$

$$\partial_t \bar{\alpha}_+ \approx +4 \times 10^{-3} H^2 (k_0 - \frac{\Delta\theta}{8\eta^2 t}) \tag{47}$$

This is a small effect, and for a distance of $t = 20, \bar{x}_+$ would change only by 4×10^{-4}, an effect which would be undectectable. To shift by one soliton width in the standard case, one would require a 200,000km cable.

Note that the interaction is attractive. The second soliton has been shifted toward the first one by the integrated amount in (46). Thus the interaction is an unstabilizing one. Consider an array of equal spaced solitons and of equal amplitudes. In this case the Raman-compensating generated radiation will reach the next soliton at the exact same time as that from the soliton on the other side. Thus they nullify with each radiation packet generating a shift which exactly cancels the other. Now slightly increase the position of one. It's damping-generated radiation will now reach the one in front of it just a little earlier than that from the one on the other side. Thus until this second packet of radiation is received, the attactive nature will cause the shortest separation distance to shorten even more. As this continues, the distance shortens more and more, destabilizing the array.

APPENDIX

The operator L was discussed in Ref. 12 and has the following eigenstates

$$L|\psi, k> = (k^2+1)|\psi, k> \tag{A.1}$$

$$L|\bar{\psi}, k> = -(k^2+1)|\bar{\psi}, k> \tag{A.2}$$

$$L|\phi_0> = 0 = L|\phi_e> \tag{A.3}$$

where

$$|\psi, k> = e^{ik\theta}[1 - \frac{2ike^{-\theta}}{(k+i)^2 cosh\theta}] \begin{pmatrix} 0 \\ 1 \end{pmatrix} + \frac{e^{ik\theta}}{(k+i)^2 cosh^2\theta} \begin{pmatrix} 1 \\ 1 \end{pmatrix} \tag{A.4}$$

$$|\bar{\psi}, k> = \sigma_1 |\psi, k> \tag{A.5}$$

$$|\phi_e> = \frac{1}{cosh\theta} \begin{pmatrix} 1 \\ -1 \end{pmatrix}, |\phi_0> = \frac{sinh\theta}{cosh^2\theta} \begin{pmatrix} 1 \\ 1 \end{pmatrix} \tag{A.6}$$

The closure of these states require two more states.

$$L|\chi>=-|\phi_e> \tag{A.7}$$

$$L|\theta\phi_e>=-|\phi_0> \tag{A.8}$$

where

$$|\chi>=\frac{\theta tanh\theta-1}{cosh\theta}\begin{pmatrix}1\\1\end{pmatrix} \tag{A.9}$$

$$|\theta\phi_e>=\theta|\phi_e> \tag{A.10}$$

Although L is self-adjoint, instead of using (A.1) and (A.2) as solutions of the adjoint problem, it is best to use the states

$$<\phi,k|=e^{-ik\theta}[1-\frac{2ike^{\theta}}{(k+i)^2cosh\theta]}\widetilde{\begin{pmatrix}1\\0\end{pmatrix}}+\frac{e^{-ik\theta}}{(k+i)^2cosh^2\theta}\widetilde{\begin{pmatrix}1\\1\end{pmatrix}} \tag{A.11}$$

$$<\bar{\phi},k|=<\phi,k|\sigma_1 \tag{A.12}$$

where $\sim$ indicates the matrix transpose. Then

$$<\phi,k|L^A=-(k^2+1)<\phi,k| \tag{A.13}$$

$$<\bar{\phi},k|L^A=+(k^2+1)<\bar{\phi},k| \tag{A.14}$$

For the adjoint bound states and closure states, we use the matrix transpose of (A.6), (A.9) and (A.10).

The only nonzero inner products are

$$<\phi,k'|\sigma_3|\bar{\psi},k>=2\pi a^2\delta(k-k') \tag{A.15}$$

$$<\bar{\phi},k'|\sigma_3|\psi,k>=-2\pi a^2\delta(k-k') \tag{A.16}$$

$$<\theta\phi_e|\sigma_3|\phi_0>=2=<\phi_0|\sigma_3|\theta\phi^e> \tag{A.17}$$

$$<\phi_e|\sigma_3|\chi>=-2=<\chi|\sigma_3|\phi_e> \tag{A.18}$$

where

$$a=\frac{k-i}{k+i} \tag{A.19}$$

and the inner product is defined as

$$< u|\sigma_3|v >= \int_{-\infty}^{\infty} d\theta \widetilde{u(\theta)}\sigma_3 v(\theta) \tag{A.20}$$

with no complex conjugations involved.

ACKNOWLEDGEMENTS

This work has been supported by the National Science Foundation through Grant No. DMS-8803471 and by the Air Force Office of Scientific Research through URI Grant No. AFOSR-89-0510.

References

[1] L.F. Mollenauer, J.P. Gordan and M.N. Islam, IEEE J. Quan. Elec., QE-22, 157-173 (1986).

[2] D.J. Kaup, SIAM J. Appl. Math. 31, 121-133 (1976).

[3] J.P. Keener and D.W. McLaughlin, Phys. Rev. A 16, 777 (1977).

[4] D.W. McLaughlin and A.C. Scott, Phys. Rev. A 18, 1652 (1978).

[5] V.I. Karpman, Pis'ma Zh. Eksp. Teor. Fiz. 25, 296 (1977); [JETP Letts, 25, 271 (1977)].

[6] V.I. Karpman and E.M. Maslov, Zh. Eksp. Teor. Fiz. 73, 537 (1977); [JETP 46, 281 (1977)].

[7] D.J. Kaup and A.C. Newell, Proc. Royal Soc. (London) Ser. A, 361, 413 (1978).

[8] Y. Kodama and K. Nozaki, Optics Letters 12, 1038 - 1040 (1987).

[9] K.J. Blow, N.J. Doran, and David Wood, Journal of Optical Society of America B 5, 1301 - 1304 (1988).

[10] Keith J. Blow and David Wood, IEEE Journal of Quantum Electronics 25, 2665 - 2673 (1989).

[11] Yuri S. Kivshar and Borsi A. Malomed, Rev. Mod. Phys. 61, 763 (1989).

[12] D.J. Kaup, Phys. Rev. A 42, 5689 (1990).

[13] A. Hasegawa and Y. Kodama, A Guiding Center Soliton in Optical Fibers. (preprint).

[14] A. Hasegawa and Y. Kodama, The Guiding Center Soliton. (preprint).

[15] K. Smith and L.F. Mollenauer, Optics Letters 14, 1284-1286 (1989).

[16] D.J. Kaup (submitted to Phys. Rev. A).

[17] K.J. Blow, N.J. Doran, and David Wood, Optics Letters 12, 1011 - 1013 (1987).

[18] K.J. Blow, N.J. Doran, and David Wood, Journal of Optical Society of America 5, 1301-1304 (1988).

[19] Boris A. Malomed, Optics communications 61, 192-194 (1987).

[20] Yuji Kodama and Akira Hasegawa, Optics Letters 7, 339-341 (1982).

[21] F.M. Mitschke and L.F. Mollenauer, Optics Letters 11, 659 - 661 (1986).

[22] J.P. Gordan, Optics Letters 11, 662 - 664 (1986).

[23] Keith J. Blow and David Wood, IEEE Journal of Quantum Electronics 25, 2665-2673 (1989).

[24] See Ref. 12 for a discussion of choosing this relationship.

[25] L.G. Cohen and W.L. Mammel, Electronics Letters 18, 1023-1024 (1982).

Similarity Solutions of Equations of Nonlinear Optics

P. Winternitz*

Centre de recherches mathématiques, Université de Montréal
CP 6128 - A, Montréal (Québec) H3C 3J7, Canada.

Abstract. Symmetry reduction is used to obtain exact analytic solutions for two different nonlinear optical systems, namely the stimulated Raman scattering system and the pumped Maxwell-Bloch system. In both cases the solution is expressed in terms of the fifth Painlevé transcendent. The asymptotic behaviour is studied and is shown to agree with experimental data.

1. INTRODUCTION

The purpose of this contribution is to review some recent work on the application of group theory and singularity analysis to obtain exact analytic solutions of some nonlinear partial differential equations, having their origin in nonlinear optics. The two phenomena to be analyzed are the stimulated Raman scattering system (SRS) and the pumped Maxwell-Bloch system (PMB). Work on the first system was done in collaboration with D.Levi and C.R. Menyuk [1], on the second with A.V. Mikhailov [2].

The equations describing these high intensity optical phenomena that have become particularly important since the advent of optical lasers, have several features in common. They both have soliton and multisoliton solutions and can be integrated by inverse scattering techniques [3], or modifications thereof [4, 5, 2]. Both systems are invariant under nontrivial Lie point symmetry groups and these symmetries can be used to generate similarity solutions. These solutions are physically relevant and describe the observed asymptotic behavior in realistic situations. The solutions are expressed in

* Work supported in part by the Natural Sciences and Engineering Research Council of Canada and by the Fonds FCAR du Québec.

terms of Painlevé transcendents, as can be expected in the case of integrable systems.

As usual, similarity methods are complementary to those making use of integrability : they provide a different class of solutions.

2. STIMULATED RAMAN SCATTERING EQUATIONS

We write the SRS equation in a symmetric form, namely

$$v_{1x} = ia_1 v_2^* v_3^*, \quad v_{2x} = ia_2 v_3^* v_1^*, \quad v_{3t} = ia_3 v_1^* v_2^*, \tag{1}$$

The a_i are real coupling constants that can be normalized to ± 1 and we have set $v_1 = iA_1^*, v_2 = A_2, v_3 = X$, where A_1, A_2 and X are the complex pump, Stokes and material excitation wave envelopes, respectively. The stars denote complex conjugation, the subscripts x and t denote partial derivatives. Eqs. (1) are actually a special degenerate case of the full three wave resonant interaction equations [6,7,8].

The symmetry group G of system (1), i.e. the group of local point transformations, leaving the solution set of Eq. (1) invariant, can be otained using a standard algorithm [9], realized as a MACSYMA program [10]. The result is an infinite dimensional Lie group, with a Lie algebra spanned by the following vectors fields:

$$\begin{aligned}
P &= \partial_x, \\
D &= x\partial_x - \frac{1}{2}(M_1\partial_{M_1} + M_2\partial_{M_2} + 2M_3\partial_{M_3}), \\
V &= -\partial_{\alpha_2} + \partial_{\alpha_3}, \\
U(h) &= h(t)(-\partial_{\alpha_1} + \partial_{\alpha_2}), \\
T(f) &= f(t)\partial_t - \frac{1}{2}f'(t)(M_1\partial_{M_1} + M_2\partial_{M_2}).
\end{aligned} \tag{2}$$

The algebra is infinite-dimensional, because of the two arbitrary functions $f(t)$ and $h(t)$. We have replaced the complex waves v_i by their moduli and phases, putting $v_k = M_k \exp(i\alpha_k), M_k \geq 0, 0 \leq \alpha_k < 2\pi$. The algebra L of Eq. (2) is a direct sum of three algebras

$$L = \{P, D\} \oplus \{V\} \oplus \{T(f), U(h)\}, \tag{3}$$

where $\{T(f), U(h)\}$ is a $\hat{U}(1)$ Kac-Moody-Virasoro algebra.

In order to perform symmetry reduction in a systematic manner, we must first classify the subalgebras of L into conjugacy classes. This can be done using methods described elsewhere [12]. Actually, we only need the one-dimensional subalgebras and they are summed up in Ref.1. The algebra $\{P+\epsilon T(1)+aV\}$ with $\epsilon=\pm 1, a\epsilon\mathbb{R}$ leads to travelling wave solutions, that can be either periodic (expressed in terms of Jacobi elliptic functions), or solitary waves. The algebra $\{P+\kappa U(1)+aV\}(\kappa=0,\pm 1, a\epsilon\mathbb{R})$ leads to so-called phase wave solutions and $T(1)+\kappa V$ provides plane wave solutions.

We are interested in similarity solutions, which are invariant under the one-dimensional group generated by $\{D+\epsilon T(t)+aV, \epsilon=\pm 1\}$. Invariance requires that the solution have the following form:

$$v_1=t^{-\frac{1+\epsilon}{2}}\rho_1 e^{i\phi 1},\quad v_2=t^{-\frac{1+\epsilon}{2}}e^{-i\epsilon a l n t}\rho_2 e^{i\phi_2},\quad v_3=\frac{1}{x}e^{ialnx}\rho_3 e^{i\phi 3}, \tag{4}$$

$$\xi=xt^{-\epsilon},\quad \rho_i=\rho_i(\xi),\quad \phi_i=\phi_i(\xi),\quad i=1,2,3$$
$$\rho_i\geq 0,\quad 0\leq\phi_i<2\pi.$$

We substitute (4) into the SRS equations (1), introduce a phase $\phi=\phi_1+\phi_2+\phi_3+aln\,\xi$, separate out the real and imaginary parts of the equations and obtain a system of 6 real ordinary differential equations (ODE)

$$\begin{aligned}
\dot\rho_1&=-\frac{1}{\xi}\rho_2\rho_3\sin\phi & \rho_1\dot\phi_1&=-\frac{1}{\xi}\rho_2\rho_3\cos\phi\\
\dot\rho_2&=\frac{1}{\xi}\rho_3\rho_1\sin\phi & \rho_2\dot\phi_2&=\frac{1}{\xi}\rho_3\rho_1\cos\phi\\
\dot\rho_3&=-\epsilon\rho_1\rho_2\sin\phi & \rho_3\dot\phi_3&=-\epsilon\rho_1\rho_2\cos\phi.
\end{aligned}\tag{5}$$

Eqns (5) allow two first integrals

$$I_1=\rho_1^2+\rho_2^2,\quad I_2=\rho_1\rho_2\rho_3\cos\phi-\frac{a}{2}\rho_1^2, \tag{6}$$

which we use to decouple the equations. In terms of ρ_1 we have

$$\begin{aligned}
\rho_2=(I_1-\rho_1^2)^{1/2},\quad \rho_3\sin\phi&=-\xi\dot\rho_1(I_1-\rho_1^2)^{-1/2}\\
\rho_3\cos\phi&=-\frac{1}{\rho_1}(I_2+\frac{a}{2}\rho_1^2)(I_1-\rho_1^2)^{-1/2}
\end{aligned}\tag{7}$$

The amplitude ρ_1 then satisfies a second order nonlinear ODE namely

$$\xi\ddot{\rho}_1 + \dot{\rho}_1 = -\xi\frac{\rho_1\dot{\rho}_1^2}{I_1 - \rho_1^2} - \frac{(I_2 + \frac{a}{2}\rho_1^2)^2}{\xi\rho_1(I_1 - \rho_1^2)} + (I_1\epsilon - \frac{a^2}{4\xi})\rho_1 - \epsilon\rho_1^3 + \frac{I_2^2}{\xi\rho_1^3} \tag{8}$$

Eq. (8) was subjected to the Painlevé test [13], performed using a MACSYMA program [14]. It passes the test and since it belongs to the class of equations, studied by Painlevé and Gambier, it can be reduced to one of their standard forms [15]. The appropriate transformation is

$$\rho_1(\xi) = \sqrt{I_1}\left[\frac{W(\xi)}{W(\xi) - 1}\right]^{1/2} \tag{9}$$

and $W(\xi)$ satisfies the equation

$$\ddot{W} = \left(\frac{1}{2W} + \frac{1}{W-1}\right)\dot{W}^2 - \frac{1}{\xi}\dot{W} + \frac{(W-1)^2}{\xi^2}(\alpha W + \frac{\beta}{W}) + \frac{\gamma}{\xi}W + \frac{\delta W(W+1)}{W-1}, \tag{10}$$

where the constants are

$$\alpha = \frac{1}{2}\left(\frac{2I_2}{I_1} + a\right)^2, \beta = \frac{2I_2^2}{I_1^2}, \gamma = 2\epsilon I_1, \delta = 0 \tag{11}$$

Eq. (10) is one of the 6 irreducible Painlevé equations, namely the one defining the Painlevé transcendent $P_V(\xi; \alpha, \beta, \gamma, \delta)$. For $\gamma = \delta = 0$ this transcendent can be expressed in terms of elementary functions. We are dealing with a less special care, namely $\gamma \neq 0, \delta = 0$, when $P_V(\xi; \alpha, \beta, \gamma, 0)$ can be expressed in terms of the third Painlevé transcendent P_{III} [16, 17].

The P_V trascendent provides an exact physically interesting solution that reproduces the asymptotic behavior found earlier in approximate numerical calculations [18, 19, 20] and in experimental studies [21, 22, 23].

Ideally, we would wish to know how an initial pulse, determined by ρ_1 and $\dot{\rho}_1$ for some $\xi = \xi_0$, will develop as a function of ξ for $\xi \to \infty$. This is closely related to the connection problem for the Painlevé transcendents, relating their behaviour at two different singular points [24, ..., 28]. Here we shall address a less ambitious problem, namely what are the possible asymptotic behaviors of the solutions of Eq. (8).

The technique that we apply is that of the Boutroux transformation, [29] that makes no direct use of the integrability of the equation and is hence also applicable to situations with dissipation. We are interested in solutions of Eq. (8) that satisfy $\rho_1 \to 0$ for $\xi \to \infty$ and hence expect an asymptotic expansion of the form

$$\rho_1 = A\xi^{-k} + o(\xi^{-s}), \quad s > k > 0. \tag{12}$$

The balance of leading powers requires $k = 1/4$. This suggests the transformation

$$\rho_1(\xi) = \xi^{-1/4}u(z), \quad z = 2\sqrt{\xi} \tag{13}$$

where the new variable z was chosen so as to obtain an equation with constant coefficients for the leading term in the expansion

$$u(z) = u_0(z) + \frac{1}{z}u_1(z) + \frac{1}{z^2}u_2(z) + \ldots \tag{14}$$

For $\epsilon = -1$ we do indeed obtain an expansion for which all $u_i(z)$ are finite and periodic, with period $T = \pi/\sqrt{I_1}$. For $I_2 \neq 0$ (and $I_1 > 0$ which is always satisfied), we obtain

$$u_0 = I_2^{1/2} I_1^{-1/4}[\cosh\delta + \sinh\delta \sin 2\sqrt{I_1}(z - z_0)]^{1/2}, \tag{15}$$

where δ and z_0 are constants.

Returning to the wave amplitudes v_i (4) we have, for $I_2 \neq 0$:

$$\begin{aligned} v_1 &= (xt)^{-1/4} e^{i\phi_1(xt)} I_2^{1/2} I_1^{-1/4}\{\cosh\delta + \\ &\qquad \sinh\delta \sin[4\sqrt{I_1}(\sqrt{xt} - \xi_0)]\}^{1/2}[1 + o(\frac{1}{\sqrt{xt}})] \\ v_2 &= e^{i(a\ell nt + \phi_2(xt))} I_1^{1/2}[1 + o(\frac{1}{\sqrt{xt}})] \\ v_3 &= t^{1/4} x^{-3/4} e^{i(a\ell nx + \phi_3(xt))} I_2^{1/2} I_1^{-1/4} \\ &\qquad \{\cosh\delta - \sinh\delta \sin[4\sqrt{I_1}(\sqrt{xt} - \xi_0)]\}^{1/2}[1 + o(\frac{1}{\sqrt{xt}})] \end{aligned} \tag{16}$$

The phases $\phi_j(xt)$ are obtained from (5) by quadratures in terms of ρ_1 of Eq. (9) with $W = P_V(\xi, \alpha, \beta, \gamma, 0)$.

For $I_2 = 0, a \neq 0$ and $I_2 = a = 0$ the results are somewhat different and are presented in Ref.1.

Looking at Eq. (16) we see that the asymptotic behaviour is the experimentally observed one. Namely, the pump intensity $A_1 = -iv_1^*$ undergoes a slow decrease with distance (given by the factor $(xt)^{-1/4}$ for t fixed, $x \to \infty$), while oscillating. The Stokes intensity $A_2 = v_2$ approaches a constant value $\sqrt{I_1}$, large with respect to the pump intensity. The material excitation envelope $X = v_3$ also oscillates towards $X \to 0$ as $x \to \infty$.

3. THE PUMPED MAXWELL BLOCH SYSTEM

The pumped Maxwell-Bloch sytem (PMB) describes the propagation of a radiation pulse with complex envelope E in a two level atomic system with population N and polarization ρ. The equations are

$$E_\xi = \rho, \quad N_t + \frac{1}{2}(\rho^* E + \rho E^*) = 4c, \quad \rho_t = NE, \quad \xi = x - t. \tag{17}$$

The constant c represents the pumping of atoms, the asterisk denotes complex conjugation. The PMB system is integrable both for $c = 0$ [30] and $c \neq 0$ [5], i.e. it is the compatibility condition required for a pair of linear operators to commute. For $c \neq 0$, i.e. nonzero pumping, one of the operators involves a derivative with respect to the spectral parameter λ and the corresponding inverse scattering transform is not isospectral. The dressing method [31] has been extended to cope with such situations [5, 32]. More specifically, soliton solutions on various backgrounds have been obtained for the PMB system [2].

The PMB system is also invariant under dilations and this can be used to obtain a similarity solution [2]. Interestingly, this solution is obtained in terms of the fifth Painlevé transcendent P_V.

We eliminate ρ and N from equations (17) and obtain a partial differential equation for $E(\xi, t)$:

$$EE_{\xi tt} - E_{\xi t}E_t + \frac{1}{2}E^2(EE_\xi^* + E^* E_\xi) - 4cE^2 = 0. \tag{18}$$

Dilationally invariant solutions will satisfy

$$E = \frac{F(\eta)}{t}, \qquad \eta = t\xi^{1/2} \tag{19}$$

For simplicity, let us restrict ourselves to real solutions ($F = F^*$). The function $F(\eta)$ then satisfies a third order ODE:

$$\eta^2 F\dddot{F} - \eta^2 \dot{F}\ddot{F} + \eta F\ddot{F} + F^3\dot{F} - 8c\eta F^2 = 0. \tag{20}$$

A first integral of the form

$$A(\eta, F, \dot{F})\ddot{F}^2 + B(\eta, F, \dot{F})\ddot{F} + C(\eta, F, \dot{F}) = K \tag{21}$$

exists. It can be used to replace (20) by the second order equation

$$\ddot{F}^2 - 16cF\ddot{F} + \frac{1}{\eta^2}F^2\dot{F}^2 - \frac{8c}{\eta^2}F^4 + \left(64c^2 - \frac{K}{\eta^2}\right)F^2 = 0. \tag{22}$$

This equation belongs to a class analyzed by Bureau [33] in his study of ODEs with fixed critical points. Modifying Bureau's results slightly, we obtain a contact transformation relating eq. (22) to the ODE for the P_V transcendent. The transformation is

$$F(\eta) = i\frac{W - 1 - 2zW_z}{W(W-1)} \qquad \eta = 2\sqrt{z} \tag{23a}$$

$$W(z) = \frac{-iFF_\eta + \eta F_{\eta\eta}}{-iFF_\eta + \eta(F_{\eta\eta} - 8cF)}, \quad z = \frac{1}{4}\eta^2 \tag{23b}$$

$$F_{\eta\eta} - 8cF \neq 0.$$

If $F(\eta)$ satisfies Eq. (22) then $W(z)$ satisfies Eq. (10) with

$$\alpha = \frac{K}{64c}, \quad \beta = -\frac{1}{8}, \quad \gamma = -4c, \quad \delta = 0. \tag{24}$$

Thus we obtain a solution for E in terms of the Painlevé transcendent $P_V(z;\alpha,\beta,\gamma,0)$. Its asymptotic behavior for $z \to \infty$ is known and can also be explicitly obtained via the Boutroux transformation [29].

Acknowledgements

The authors research was partly supported by research grants from NSERC of Canada and FCAR du Québec.

References

1. D. LEVI, C.R. MENYUK AND P. WINTERNITZ, Phys. Rev. A (to appear).
2. A.V. MIKHAILOV AND P. WINTERNITZ, (to be published).
3. M.J. ABLOWITZ AND H. SEGUR, "Solitons and the Inverse Scattering Transform," SIAM, Philadelphia, 1981.
4. D.J. KAUP AND C.R. MENYUK, Phys. Rev. **A42** (1990), 1712.
5. S.P. BURTSEV, V.E. ZAKHAROV AND A.V. MIKHAILOV, Teor. Mat. Fiz. **70** (1987), 323.
6. H. CORNILLE, J. Math. Phys. **20** (1979), 1653.
7. D.J. KAUP, Physica **D1** (1980), 5.
8. L. MARTINA AND P. WINTERNITZ, Ann. Phys. **196** (1989), 231.
9. P.J. OLVER , "Applications of Lie Groups to Differential Equations," Springer, New York, 1974.
10. B. CHAMPAGNE AND P. WINTERNITZ, Preprint CRM-1278 (1985), Montréal.
11. B. CHAMPAGNE, W. HEREMAN AND P. WINTERNITZ, Comp. Phys. Commun. (to appear).
12. P. WINTERNITZ, in "Partially Integrable Evolution Equations in Physics," Kluwer, Dordrecht, 1990, pp. 515–567.
13. M.J. ABLOWITZ, A. RAMANI AND H. SEGUR, J. Math. Phys. **21** (1980), 715.
14. D. RAND AND P. WINTERNITZ, Comp. Phys. Commun. **42** (1986), 359.
15. E.L. INCE , "Ordinary Differential Equations," Dover, NY, 1956.
16. F.J. BUREAU, Bull. Classe des Sciences Acad. Roy. Belgique - 5e série **70** (1984), 290.
17. V.I. GROMAK, Diff. Uravn. **11** (1975), 373; Diff. Uravn. **20** (1984), 1674.
18. C.R. MENYUK AND G. HILFER, Opt. Lett. **12** (1989), 227.
19. ——————, J. Opt. Soc. Am. **B7** (1990), 739.
20. C.R. MENYUK, Phys. Rev. Lett. **62** (1989), 2937.
21. R.L. CARMAN, F. SHIMIZU, C.S. WANG AND N. BLOEMBERGEN, Phys. Rev. **A2** (1970), 60.
22. R.L. CARMAN, M.E. MACK, F. SHIMIZU AND N. BLOEMBERGEN, Phys. Rev. Lett. **23** (1969), 1327.
23. R.L. CARMAN AND M.E. MACK, Phys. Rev. **A5** (1972), 341.
24. B.M. McCOY AND S. TANG, Physica **D18** (1986), 190; Physica **D19** (1986), 42; Physica **D20** (1986), 187.

25. A.R. ITS AND V. YU NOVOKSHENOV, "The Isomonodromic Deformation Method in the Theory of Painlevé Equations," Springer, Berlin, 1986.

26. A.V. KITAEV, Math. USSR Sbornik **62** (1989), 421.

27. A.S. ABDULLAEV, Sov. Math. Dokl. **31** (1985), 45.

28. D. LEVI AND P. WINTERNITZ, (Editors), "Painlevé Transcendents, their Asymptotics and Physical Applications (to appear)," Plenum, New York, 1991.

29. M.P. BOUTROUX, Ann. Ec. Norm. Sup. **30** (**3**) (1913), 255; Ann. Ec. Norm. Sup. **31** (**3**) (1914), 99.

30. M.J. ABLOWITZ, D.J. KAUP AND A.C. NEWELL, J. Math. Phys. **15** (1974), 1852.

31. V.E. ZAKHAROV AND A.V. MIKHAILOV, Zh. Eksp. Teor. Fiz. **74** (1978), 1953.

32. V.A. BELINSKII AND V.E. ZAKHAROV, Zh. Eksp. Teor. Fiz. **75** (1978), 1953.

33. F.J. BUREAU, Ann. Math. Pura Appl.IV **41** (1972), 163.

Heisenberg Ferromagnet, Generalized Coherent States and Nonlinear Behaviour

V. G. Makhankov.

Joint Institute for Nuclear Research,
P.O.Box 79, Moscow (Dubna) USSR.

Various version of the so-called Heisenberg ferromagnet model

$$\hat{H}_{ea} = -J \sum_j \left[\hat{\vec{S}}_j \hat{\vec{S}}_{j+1} + \delta\, \hat{S}^z_j \hat{S}^z_{j+1} \right] \tag{1}$$

are well-known to be treated via certain techniques. Among those there are mean field method, trial function method, that of correlation functions and others; for example, in the case of $s=1/2$ spins situated along a chain the model can be solved exactly. Nowadays hamiltonian (1) is of a great interest both from the application point of view and from that of mathematical physics. Indeed, we shall see in what follows that the system governed by (1) displays many interesting and even intriguing features.

In treating (1) we take $s=1$ and apply the trial function method. The first choice is very easy to understand: model (1) then is neither integrable nor can be studied by means of the Holstein-Primakoff representation. Therefore one might expect some pecularities in behaviour of the system. The second choice probably is not so transparent, but referring to [1] we are able to say that taking as trial functions the so-called generalized coherent states (GCS) (in Perelomov's sense) extends the mean field method and gives a powerful tool to handle spin and pseudo-spin hamiltonians. These states are very much appropriate to do so and admit of further generalization to the so-called quantum or q-deformed algebras.

As far as we are dealing with $s=1$ model (i.e. with three dimensional representation of the SU(2) group) we can apply GCS constructed either on the $SU(2)/U(1) \simeq \mathbb{CP}^1$ coset space or on the $SU(3)/SU(2) \otimes U(1) \simeq \mathbb{CP}^2$ one. These are respectively

$$|\,\psi\,\rangle = \frac{1}{1+|\psi|^2} \left\{ |0\rangle + \sqrt{2}\psi|1\rangle + \psi^2|2\rangle \right\} \tag{2}$$

$$|\,\zeta\,\rangle = (1+|\zeta_1|^2+|\zeta_2|^2)^{-1/2} \left\{ |0\rangle + \zeta_1|1\rangle + \zeta_2|2\rangle \right\} \tag{3}$$

and one can easily check that when

$$\zeta_1 = \sqrt{2}\psi, \qquad \zeta_2 = \frac{1}{2}\zeta_1^2 \tag{4}$$

formula (3) goes to (2). It means that the section (4) in $\mathbb{CP}^2$ is none but $\mathbb{S}^2$.

On the other hand it is well-known [1] that having been averaged over (2) hamiltonian (1) becomes the so-called phenomenological Landau-Lifshitz hamiltonian which governs orientation dynamics of the classical magnetization vector lying on the $\mathbb{S}^2$.

Dimension of the total spin phase space in a lattice site is easy to calculate 2(2s+1)-2=4 to give the dimension of the $\mathbb{CP}^2$.

Classical behaviour of system (1) with s=1 is descibed by the SU(2)/U(1) GCS and Landau-Lifshitz hamiltonian, and trajectories in spin phase space, $\mathbb{CP}^2$, lie on its subspace $\mathbb{CP}^1$ defined by (4). It means that the magnitude

$$\Delta = \sup_{x\in[-L,L]} |\zeta_2 - \zeta_1^2/2| \tag{5}$$

describes a deviation in the behaviour of the system from the classical one.

Averaging eq.(1) over GCS(3) we have the hamiltonian

$$H_{ea} = \int \left\{ \frac{\alpha_0^2}{2} \left[\langle \hat{S}^+\rangle_x \langle \hat{S}^-\rangle_x + (\langle \hat{S}^z\rangle_x)^2 \right] - \right.$$
$$\left. - \langle \hat{S}^+\rangle\langle \hat{S}^-\rangle - (\langle \hat{S}^z\rangle)^2 - \delta(\langle \hat{S}^z\rangle)^2 \right\} dx$$

giving the equations of motion which up to the lowest nonlinear terms are

$$i\dot{\zeta}_1 - \zeta_{1xx} + 2\delta\zeta_1 - 4\zeta_1\zeta_2 + 2(1-\delta)|\zeta_1|^2\zeta_2 = 0,$$
$$i\dot{\zeta}_2 + 4(1+\delta)\zeta_2 - 2\zeta_1^2 = 0 \tag{6}$$

Vacuum (ground state) solution is $\zeta_1=\zeta_2=0$ so will $\Delta=0$. Linearized version of eqs.(6) describes trajectories which span all the four-dimensional spin phase space and $\Delta\neq 0$. Nonlinear localized stationary solutions to (6) are shown to be soliton-like

$$\begin{aligned} \zeta_1(x,t) &= \eta_1(x)\, e^{i\omega_1 t} \\ \zeta_2(x,t) &= \eta_2(x)\, e^{i\omega_2 t} \end{aligned} \tag{7}$$

where $\omega_2 = 2\omega_1$ and

$$\eta_1 = \frac{b}{\cosh(\sqrt{2\delta - \omega_1}\, x)}, \qquad \eta_2 = \frac{\eta_1^2}{2 - \omega_1 + 2\delta} \tag{8}$$

$$b^2 = (2 - \omega_1/\delta) \tag{9}$$

with $\Delta=0$ so lie in the SU(2)/U(1) subspace (4) of the $\mathbb{CP}^2$. Then one can suppose that this 2-dimensional subspace of $\mathbb{CP}^2$ defined with(4) is a "classical" attractor in the spin phase space and classical behaviour (description) of the system is only realized asymptotically in time.

Computer experiments carried out in Dubna on stability of (7)-(9) and dynamics of some initial packets (7),(8) showed as it was the case: solutions (7)-(9) proved to be stable and the derivative $d/t|\Delta|$ is negative.

More details will be published elsewhere.

The author is indebted a lot to his colleagues A.Makhankov, A.Maksudov, K.Muminov and K.Abdulloev with whom some of the above results have been obtained.

1.V.Makhankov.Soliton Phenomenology,Kluwer Acad.Publ.,Dordrecht, 1990.

Integrable Supersymmetric Models and Phase Transitions in One Dimension

O.K. Pashaev.

Dipartimento di Fisica, Università di Lecce,
73100, Lecce, Italy.
Permanent address: Joint Institute for Nuclear Research, Dubna, 141980, USSR.

Supersymmetry concepts originally introduced in the high-energy physics have recently penetrated in the nuclear physics, condensed matter and statistical mechanics [1]. In the last cases anticommuting variables are first introduced for pure combinatorical reasons, to replace the $N\to 0$ replica trick. But one might expect the supersymmetry in condensed matter physics to arise as a relation between fermionic and bosonic excitations. Recently there have been made an attempt to describe the High Temperature Superconductivity (HTSC) and strong correlated electronic systems in general, using symmetry between fermionic and bosonic excitations (hole and spin excitations) [2]. The order parameter in this case can appear in a very sophisticated form as the Grassman constant or the Clifford number. Related to this integrable generalizations of the Ginzburg-Landau model are possible in the form of the graded vector or matrix Nonlinear Schrödinger Equations (NLSE) with global symmetry in the space of order parameters [3]. The close analogies between antiferromagnetism and superfluidity in the context of classical integrable models have exact meaning in the form of the gauge-equivalence of the Heisenberg model and NLSE.

Let us introduce a matrix-valued function $S(x,t)$ on the super algebra $SU(N|M)$ and diagonalized it in the form

$$S = g^{-1}\Sigma g \tag{1}$$

where $g(x,t) \in SU(N|M)$. S is invariant under the local transformations $g\to \exp[ih(x,t)]g$, where $[\Sigma,h]=0$, and belong to the coset space $SU(N|M)/H$. H is the subgroup generated by h. For quadratic constraints on S we have two types of symmetry breaking:

$$\text{I.}\quad H=S(U(N)\times U(M)\times U(1)), \qquad \text{II.}\quad H=S(U(N-p)\times U(p|M)\times U(1)) \quad . \tag{2}$$

From the linear problem defined by the matrix operators

$$U = i\lambda S \quad , \qquad V=i\lambda^2 S+ \frac{\lambda}{2}\,[S,S_x] \tag{3}$$

we obtain equations of motion

$$i\,S_t = [S,S_{xx}]$$

of the isotropic supersymmetric $SU(N|M)/H$ Heisenberg model. Then we decompose the super algebra $SU(N|M) = L^{(0)} \oplus L^{(1)}$, where $[L^{(i)}, L^{(j)}\} \subset L^{(i+j)\bmod 2}$ and $L^{(0)}$ is the algebra generated subgroup H. It is easy to see that now we have two types of $\mathbb{Z}_2$ graduation: first-grassman graduation, second-symmetric space graduation. Matrix $g(x,t)$ from eq. (1) defines a current $I_\mu = g_\mu g^{-1}$ ($\mu=0,1$), which can be restricted in such a way, thanks to local symmetry, that $I_1 \subset L^{(1)}$. Parametrize I_1 in the form

$$I_1 \equiv i \begin{pmatrix} 0 & Z \\ Z^+ & 0 \end{pmatrix} ,$$

where $Z = (\underbrace{\Phi}_{p} \mid \underbrace{\Psi}_{M})\}N-p$ is $(N-p) \times (M+p)$ matrix first p columns of which form grassman-even matrix $\Phi(x,t)$ and rest M columns form grassman-odd matrix $\Psi(x,t)$. In the case when both grassman and symmetric space graduation coincides the even part of Z, the matrix Φ, disappears and we have only grassman odd matrix Ψ.

Performing the gauge transformations of linear problem (3) generated by g from eq. (1) we obtain the graded matrix NLSE [4]:

$$\begin{aligned} & i\,\Phi_t + \Phi_{xx} + 2\Phi\,\Phi^+\Phi + 2\,\Psi\,\Psi^+\Phi = 0 \\ & i\,\Psi_t + \Psi_{xx} + 2\Psi\,\Psi^+\Phi + 2\,\Phi\,\Phi^+\Psi = 0 \end{aligned} \tag{4}$$

with the Hamiltonian

$$H = \operatorname{str} \int_{-\infty}^{\infty} \left[Z_x^+ Z_x - (Z^+Z)^2 \right] dx \ .$$

More details for these relations for the case of $SU(2|1)$ and $OSPU(1,1|1)$ supergroups have been done in [5,6,10]. As it has been shown in the last case the constraints on S are cubic essentially and lead to an additional terms in equations of motion of higher order nonlinearity.

To obtain an anisotropic supersymmetric Heisenberg model we can use the following trick [7],[11]. Let us consider $U(n|m)$ NLSE:

$$i\,Q_t + Q_{xx} + 2(\bar{Q}\,Q)Q = 0 \tag{5}$$

where $Q^T = (Q_1, \ldots, Q_n; Q_{n+1}, \ldots, Q_{n+m})$ is a spinor field from the graded vector space $V(n|m)$ with n-Bose and m-Fermi dimensions, and $(\bar{Q}Q) = \sum_{a=1}^{n} |Q^{(a)}|^2 + \sum_{a=n+1}^{n+m} \bar{Q}^{(a)} Q^{(a)}$. Using the Bäcklund transformations

$$Q_x^{(a)} - Q_x^{(a)\prime} = (Q^{(a)} + Q^{(a)\prime})\sqrt{\Delta - ((\bar{Q}-\bar{Q}')\cdot(Q-Q'))} \quad ,$$

$$Q_t^{(a)} - Q_t^{(a)'} = i\,(Q_x^{(a)} + Q_x^{(a)'})\,\sqrt{\Delta - (\overline{(Q-Q')}\cdot(Q-Q'))} + \qquad (6)$$

$$+\; i\,((\overline{Q}Q) + (\overline{Q'}Q'))\,(Q^{(a)} - Q^{(a)'})$$

for new variables $\Psi^{(a)} \equiv \frac{i}{4\mu}\,(Q^{(a)} - Q^{(a)'})$,

$(a=1, \ldots, n+m)$, $\Delta=16\mu^2$, from the right invariant coset space $(U(n|m)\times U(1))\backslash U(n+1|m)$ we obtain equations of motion in the form of modified graded NLSE:

$$i\,\Psi_t^{(a)} + \Psi_{xx}^{(a)} + \frac{\Delta}{2}\,(\overline{\Psi}\Psi)\,\Psi + \frac{1}{2}\,\frac{(\overline{\Psi}_x\Psi_x)\,\Psi^{(a)} + (\overline{\Psi}\Psi)_x\,\Psi_x^{(a)}}{1-(\overline{\Psi}\Psi)} = 0, \qquad (7)$$

where

$$(\overline{\Psi}\Psi) = \sum_{a=}^{n} |\,\varphi^{(a)}|^2 + \sum_{a=1}^{m} \overline{\chi}^{(a)}\chi^{(a)}$$

is $U(n|m)$ invariant form. Bäcklund transformations parameter Δ in this approach can be interpreted as one axis anisotropy parameter of the Heisenberg model. New reductions of Zakharov-Shabat problem for NLSE related to the "easy plane" anisotropy $(\Delta<0)$ are presented in Ref. [7].

If we relate the classical integrable models considered above to the quantum one, the later can be interpreted as a bosonic model with reduced number of order parameter components. This result follows from the identity

$$\frac{[\det\,(-\,\Delta + m^2 + i\sigma)]^{M}}{[\det\,(-\,\Delta + m^2 + i\sigma)]^{N/2}} = [\det\,(-\Delta + m^2 + i\sigma)]^{-(N-2M)/2}$$

in the functional integral representation of Green functions after integration over bosonic and fermionic variables, where σ is auxiliary grassman even collective field [8]. Thus we see that M fermionic (grassman odd) degrees of freedom are equivalent to 2M negative bosonic one. If N=2M the supersymmetric models arise and the $N\to 0$ limit is reproduced. For N<2M the theory is equivalent to bosonic one with N<0 components. But in this case as it was shown by Balian and Toulouse [9] a phase transition at finite temperature in one-dimension can occur and the theorem of Lieb and Mattis suffer violations.

At the classical level reduction of the component number N is related to non-positivity of bilinear form

$$(\overline{\Psi}\Psi) = \sum_{a=1}^{N} |\,\varphi^{(a)}|^2 + \sum_{a=1}^{M} \overline{\chi}^{(a)}\chi^{(a)} = \sum_{a=1}^{N} |\,\varphi^{(a)}|^2 - \sum_{a=1}^{M} \overline{\xi}^{(a)}\,\xi^{(a)}$$

where $\xi^{(a)} \equiv \overline{\chi}^{(a)}$. Integrable models are mostly one-dimensional systems. So this result shows that the phase transition in supersymmetric or in general in the grassman extended integrable models can take place. Supersymmetry breaking and appearance of Goldstone modes (bosonic and fermionic) are related to the existence of nonlinear excitations of corresponding classical models such like solitons.

Acknowledgements

I would like to thank Prof. M. Boiti and F. Pempinelli for hospitality in University of Lecce and INFN for financial support.

Many discussions with Prof.V.G.Makhankov were also fruitful.

References

[1] V.A. Kostelecky, D.K. Campbell (eds.) "Supersymmetry in Physics", Physica D, v. 15, 1985.

[2] Z. Zou, P.W. Anderson, Phys. Rev. B37, 1988, 627.
P.B. Weigmann, Phys. Rev. Lett. 69, 1988, 821.

[3] V.G. Makhankov, O.K. Pashaev, Theor. Math. Phys., vol.53, 1982, 55.

[4] O.K. Pashaev, Prepr. JINR p17-89-146, Dubna 1989
and in "Group representations in physics" Proc. of Workshop, Tambov-89, Nauka, Moscow 1990.

[5] V.G. Makhankov, R. Myrzakulov and O.K. Pashaev, Lett. Math. Phys., 16, 1988, 83.

[6] V.G. Makhankov, O.K. Pashaev, Phys. Lett. A141, 1989, 285 and in "Nonlinear World" Proc. of Workshop Kiev-1989.

[7] O.K. Pashaev "Bäcklund transformations, reduction problem and anisotropic right-invariant Heisenberg models" (to be published).

[8] G. Parisi, N. Sourlas, J. Physique-Lettres 41, 1980, L103.
A.Y. McKane, Phys. Lett. 76A, 1980, 22.

[9] R. Balian, G. Toulouse, Ann. Phys. 83, 1974, 28.

[10]V.G.Makhankov and O.K.Pashaev in "Solitons and Applications" Proc of IV International workshop (Dubna,August,1989) eds. V.G.Makhankov,V.K.Fedyanin and O.K.Pashaev, World Scientific Pub.Co. 1990, p.31.

[11]O.K.Pashaev in "NEEDS'90" Proc.of VI Workhop (Dubna,July,1990) eds.V.G. Makhankov and O.K.Pashaev, Springer 1991.

Denaturation of DNA in a Toda Lattice Model

P. Christiansen[1], A.C. Scott[2],
P.S. Lomdahl[3] and V. Muto[4]

[1,2]Laboratory of Applied Mathematical Physics,
The Technical University of Denmark,
DK-2800 Lyngby, Denmark.
[3]Center for Nonlinear Studies and Theoretical Division,
Los Alamos National Laboratory,
Los Alamos, 87545 New Mexico, USA.
[4]Department of Applied Mathematics,
Faculty of Sciences, University of Pais Vasco,
Apartado 644, E-48080-Bilbao, Spain.

Abstract: In deoxyribonucleic (DNA) the two polynucleotide strands are linked together through hydrogen bonds. The phosphodiester bridges in the backbone are described by a Toda potential and the hydrogen bonds by a Lennard-Jones potential. This anharmonic model of the vibration and wave propagation in the molecules predicts a significant increase in the lifetime of the open states of the hydrogen bonds at physiological temperatures. Thus anharmonicity may play a role in DNA denaturation. The paper summarizes results for which a full account is given in Refs. 1–3.

Introduction

In recent years the possibility that anharmonic exitations could play a role in the dynamics of DNA has been considered by several authors listed in Refs. 2 and 3. It has been suggested that the phosphodiester bridges in the polynucleotide strands can be modelled by a Toda potential [1]. Thus solitons are generated thermally at biological temperatures. The solitons can easily be counted because of the integrability of the unperturbed Toda system [1,2].

In the present paper we consider the transversal coupling of the two strands through hydrogen bonds in the two base pairs modelled by a Lennard-Jones potential [3].

The Model

Instead of the double helix of DNA we consider two parallel strands, each having the form of a straight line. Each of the two chains is a spring and mass system. Each mass represents a single base of the base pair. We assume a homogeneous DNA molecule. Therefore each particle has the mass M ($M = 1.28 \times 10^{-24}$ kg), all the Toda potentials ($V_{T,I}$ and $V_{T,II}$) connecting the masses longitudinally along strands are identical, and all the Lennard-Jones potentials (V_{LJ}) connecting the masses transversally are identical. As indicated in Fig. 1 there are four degrees of

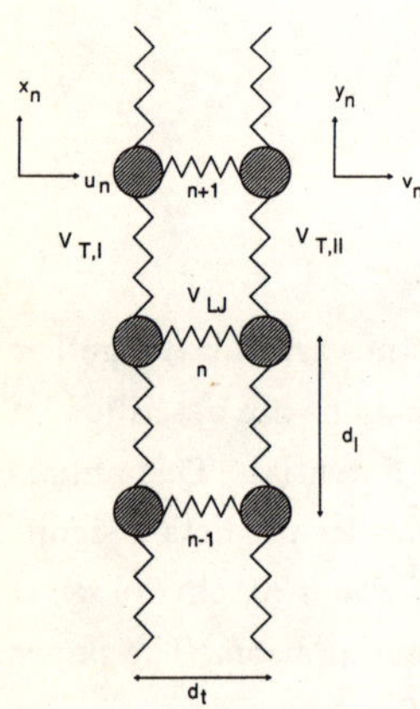

Figure 1. DNA double helix. Two identical Toda chains are connected by Lennard-Jones potentials representing the hydrogen bonds between the two strands. From [3]

freedom, x_n, u_n, y_n, v_n, for each base pair. $x_n(t)$ and $y_n(t)$ denote longitudinal displacements of masses $n = 1,2,\cdots,N$ on strands I and II. $u_n(t)$ and $v_n(t)$ denote corresponding transversal displacements. Being interested in ring-shaped DNA-molecules, we assume periodicity of the displacements, namely $x_{n+N} = x_n$ etc., where N is the number of base pairs in the DNA.

The Toda potential for the first strand is given by

$$V_{T,I}\,(\lambda_n' - d_\ell) = \frac{a}{b} \exp[-b(\lambda_n' - d_\ell)] + a(\lambda_n' - d_\ell)\ , \tag{1}$$

where λ_n' denotes the distance between the n'th and the (n+1)'th base in strand I given by

$$\lambda_n' = \sqrt{(d_\ell + x_{n+1} - x_n)^2 + (u_{n+1} - u_n)^2}\ . \tag{2}$$

Here d_ℓ is the equilibrium distance between adjacent bases in the same strand measured along the helix axis. According to [4] d_ℓ = 3.4 Å. The Toda potential makes a reasonable approximation to the more commonly used van der Waals potential for the phosphodiester bridge and provides a sound speed velocity along the strands in accordance with measurements if we choose the parameters $a = 2.56 \times 10^{-10}$ N and $b = 6.18 \times 10^{10}$ m^{-1} [3]. An expression similar to (1) is valid for the potential $V_{T,II}(\lambda''_n - d_\ell)$ on the second strand. Here the distance λ''_n is given by

$$\lambda''_n = \sqrt{(d_\ell + y_{n+1} - y_n)^2 + (v_{n+1} - v_n)^2} \; . \tag{3}$$

The Lennard-Jones potential for the transversal springs modelling the hydrogen bonds is given by

$$V_{LJ}(\tau_n - d_t + d_h) = 4\varepsilon \times \left[\left(\frac{\tau}{\tau_n - d_t + d_h}\right)^{12} - \left(\frac{\tau}{\tau_n - d_t + d_h}\right)^{6}\right] , \tag{4}$$

where τ_n denotes the distance between two bases of the two strands and is given by

$$\tau_n = \sqrt{(d_t + v_n - u_n)^2 + (y_n - x_n)^2} \; . \tag{5}$$

Here d_t is the equilibrium distance between the bases in a pair, namely the diameter of the helix (d_t = 20 Å [4]), and d_h is the equilibrium length of the hydrogen bond ($d_h = 2^{1/6}\,\rho$ [5]). Approximate values of the parameters ε and ρ in (4) are $\varepsilon = \varepsilon_{LJ} = 0.352 \times 10^{-19}$ Nm and $\rho = 4.01 \times 10^{-10}$ m [6].

Hamiltonian and Thermalization of the Problem

The equations of motion are obtained from the Hamiltonian of the system

$$H = \sum_{n=1}^{N} \left\{\frac{1}{2} M(\dot{x}_n^2 + \dot{u}_n^2) + \frac{1}{2} M(\dot{y}_n^2 + \dot{v}_n^2) + \right.$$

$$\left. V_{LJ}(\tau_n - d_t + d_h) + V_{T,I}(\lambda'_n - d_\ell) + V_{T,II}(\lambda''_n - d_\ell) \right. , \tag{6}$$

where dot denotes differentiation with respect to time, t, and $V_{T,I}$, $V_{T,II}$, and V_{LJ} are the anharmonic potentials given in the previous section. Introducing $\Delta_n^{(i)}$ such that $\Delta_n^{(1)} = x_n$,

$\Delta_n^{(2)} = u_n$, $\Delta_n^{(3)} = y_n$, $\Delta_n^{(4)} = v_n$ we obtain the equation of motion from the Hamiltonian (6) as

$$M\ddot{\Delta}_n^{(i)} = - \frac{\partial H}{\partial \Delta_n^{(i)}} \qquad i = 1,2,3,3, \text{ and } n = 1,2,\cdots,N \ . \tag{7}$$

In order to describe the interaction of the system with a thermal reservoir at a finite temperature, T, a damping force and a noise force are added to the equations of motion for the displacements (7) leading to the Langevin equations [7]

$$M\ddot{\Delta}_n^{(i)} = - \frac{\partial H}{\partial \Delta_n^{(i)}} - M\Gamma\dot{\Delta}_n^{(i)} + \eta_n^{(i)}(t) \ . \tag{8}$$

Here Γ is the damping coefficient and $\eta^{(i)}(t)$, $i = 1,2,3,4$ are random forces acting on the bases. The noises $\eta_n^{(i)}(t)$ are assumed to be Gaussian processes with zero mean and correlation functions of the form

$$\langle \eta_n^{(i)}(t), \eta_{n'}^{(j)}(t') \rangle = 2M\Gamma k_B T \delta_{nn'} \delta_{ij} \delta(t - t') \ . \tag{9}$$

Here k_B is the Boltzmann constant; $\delta_{nn'}$ and δ_{ij} are Kronecker deltas, and $\delta(t-t')$ is Dirac's delta function.

The numerical solutions of Eqs. (8) are carried out over time intervals which are sufficiently long to reach thermal equilibrium. This implies that the mean kinetic energy is

$$\langle \sum_{n=1}^{N} \frac{1}{2} M(\dot{x}_n^2 + \dot{u}_n^2) \rangle = Nk_B T = \langle \sum_{n=1}^{N} \frac{1}{2} M(\dot{y}_n^2 + \dot{v}_n^2) \rangle \tag{10}$$

for each particle on either strand. Here $\langle \; \rangle$ denotes time average.

Numerical Results and Conclusions

The perturbed system (8) has been integrated using a Runge-Kutta method. The calculations have been initialized choosing initial conditions corresponding to a double chain of N masses at rest in their equilibrium positions. The calculations have been run until thermal equilibrium has been reached and then continued for several hundred normalized time units.

In order to compute the number of solitons (N_s) in the double chain, we have made the average of the values of the number of solitons in each of the two chains. The latter values have been computed using the spectral analyzer of the unperturbed Toda lattice, described in detail in Ref. 2. Our typical results are shown in Fig. 2, where the ratio N_s/N is plotted versus the temperature T in a logarithmic scale, for different choices of the

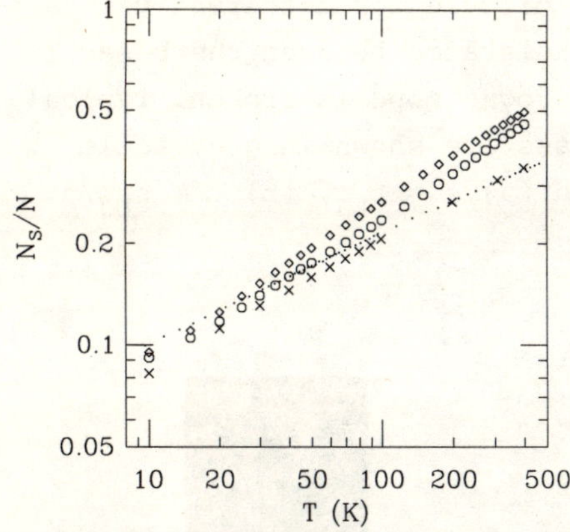

Figure 2. Logarithmic plot to base 10 of ratio N_s/N versus temperature T. The dotted line and the values denoted by crosses are the ones obtained in [1] and [2] for the single-strand model by thermalization of the unperturbed Toda lattice via initial conditions and by the thermalized perturbed Toda lattice respectively. The values denoted by the octagons and diamonds are calculated for the perturbed 2-strand system with DNA parameters, damping coefficient $\Gamma = \sqrt{ab/M}$, number of masses N = 32, and the Lennard-Jones parameter $\varepsilon = \varepsilon_{LJ}$ and $\varepsilon = \varepsilon_{LJ}/10$. From [3]

Lennard-Jones parameter ε. In particular, the perturbed system modelling the double chain immersed in a thermal bath for 64 masses, N = 32 on each of the identical chains, and with a normalized damping coefficient of unity are computed for $\varepsilon = \varepsilon_{LJ}$ (octagons) and $\varepsilon = \varepsilon_{LJ}/10$ (diamonds) and compared to results for the single-strand case (dots and crosses). Like in the single-strand case there is a $T^{1/3}$ dependence on the number of solitons, at least at low temperatures. At room temperature (T = 300) $N_s = 0.31$ N in the single-strand case, and $N_s = 0.39$ N or 0.42 N for $\varepsilon = \varepsilon_{LJ}/10$ or $\varepsilon = \varepsilon_{LJ}$. Thus the anharmonicity is capable of creating a number of solitons which amounts to approximately one third of the number of masses. Increasing the Lennard-Jones potential for the hydrogen bonds gives a slightly higher soliton number.

Next we consider the denaturation of the DNA molecule, the so-called melting phenomenon, which happens when the two strands of the DNA helix come apart because the hydrogen bonds between its paired bases are disrupted. This can be accomplished by heating a solution of DNA or by adding acid or alkali to ionize its bases. The melting temperature depends markedly on its base composition. For the homogeneous DNA molecule we compute the hydrogen bond stretching in the molecule for different values of ε and T. When the transversal distance between the bases in a pair is bigger than 4 Å the hydrogen bond is broken. Typical transversal distances between bases are shown in grey-scale plots in Fig. 3 for T = 310 K and (a) $\varepsilon = \varepsilon_{LJ}$, (b) $\varepsilon = \varepsilon_{LJ}/5$.

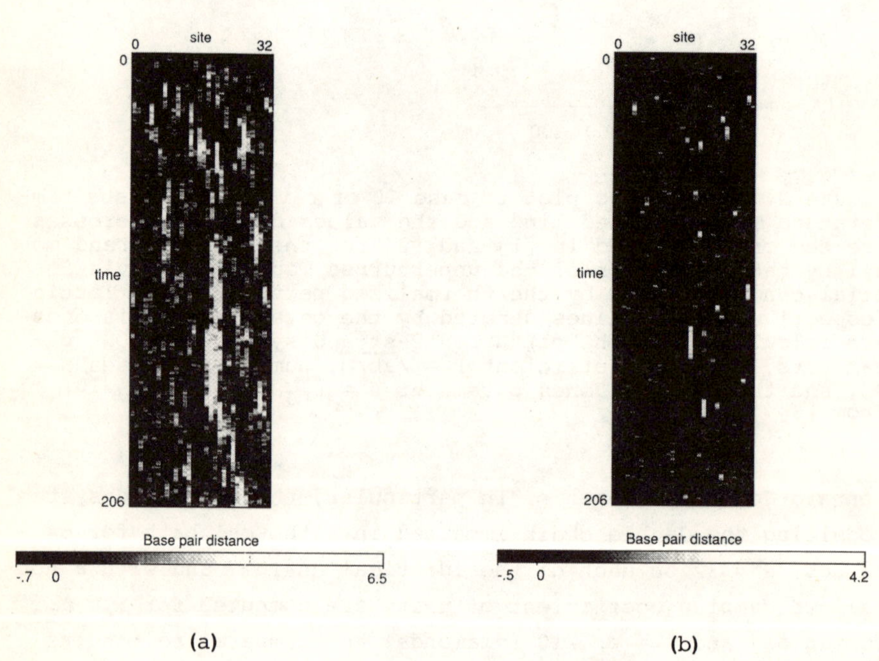

Figure 3. Grey-scale plots of transversal distance between bases versus site number n and time t. T = 310 K, and (a) $\varepsilon = \varepsilon_{LJ}$, (b) $\varepsilon = \varepsilon_{LJ}/5$. From [3]

The system has been integrated up to 1024 normalized time units, corresponding to 206.0 picoseconds. In the grey-scale plots the white areas denote the maximum values of the distance between bases, while the black ones correspond to minimum values of the hydrogen bond stretching. It is noted that at T = 310 K many more sites have open hydrogen bonds in extended times for $\varepsilon = \varepsilon_{LJ}/5$ than for $\varepsilon = \varepsilon_{LJ}$.

Finally, Fig. 4 shows the averaged lifetime of the open states as a function of temperature, for $\varepsilon = \varepsilon_{LJ}/10$. The open states

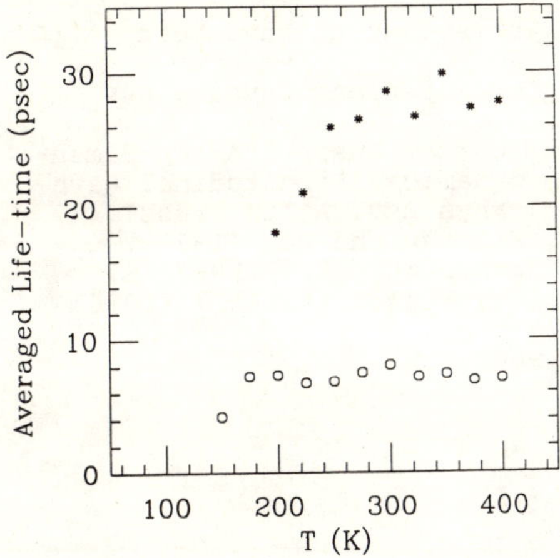

Figure 4. Averaged lifetime of open states versus temperature T. Bin 3 - 17 psec: open symbols. Bin > 18 psec: stars. From [3]

are divided into three bins. The first bin contains open states with a lifetime shorter than 3 psec. We have not considered these states in the calculations of the averaged lifetimes. The second bin contains open states with lifetimes between 3 psec and 17 psec. The averaged values of this group are the ones shown as open symbols. Finally, the third bin contains the open states with a lifetime longer than 18 psec. These averaged values are plotted by stars. It is clear from Fig. 4 that for temperatures larger than 250 K, the presence of open states which last for more than 20 psec becomes significant.

Thus we conclude that anharmonicity may play a role in the DNA denaturation.

Acknowledgement

Part of this work was done under the auspices of the US Department of Energy. One of the authors (PLC) expresses thanks for the warm hospitality of the Center for Nonlinear Studies, Los Alamos National Laboratory. Financial support from the Danish Technical Research Council, NATO Scientific Affairs Division, Grant No. 35-0948/88 and EC Science Program (Contract SCI/0229-C(AM)) is acknowledged.

References

[1] V. Muto, A.C. Scott and P.L. Christiansen, Phys.Lett. A136 (1989) 33.
[2] V. Muto, A.C. Scott and P.L. Christiansen, Physica D44 (1990) 75.
[3] V. Muto, P.S. Lomdahl and P.L. Christiansen, "A two-dimensional discrete model for DNA dynamics: Londitudinal wave propagation and denaturation", Phys.Rev. A (in press).
[4] L. Stryer, Biochemistry, W.H. Freeman and Co. (1981).
[5] J. Halding and P.S. Lomdahl, Phys.Lett. A127 (1987) 37.
[6] P. Perez and N. Theodorakopoulos, Phys.Lett. A117 (1986) 405; Phys.Lett. A124 (1987) 267.
[7] R. Kubo, Rep.Prog.Phys. 29 (1966) 255.

Part III

Dissipative Systems

A Simple Method to Obtain First Integrals of Dynamical Systems

R. Conte[1] *and M. Musette*[2]

[1]Service de physique du solide et de résonance magnétique
Centre d'études nucléaires de Saclay, F-91191 Gif-sur-Yvette Cedex
[2]Theoretische Natuurkunde, Vrije Universiteit Brussel
Pleinlaan 2, B-1050 Brussel

Abstract. We give values of parameters of the Lorenz model for which a first integral probably exists.

Given a dynamical system , e.g. the Lorenz model [1]

$$x' = \sigma(y - x), \quad y' = -y - xz + rx, \quad z' = xy - bz, \qquad \sigma \neq 0 \tag{1}$$

we first represent it by an equivalent ordinary differential equation (ODE) [2] in u

$$x = u, \quad y = u + \frac{u'}{\sigma}, \quad z = r - 1 - \frac{\sigma + 1}{\sigma}\frac{u'}{u} - \frac{u''}{\sigma u} \tag{2}$$

$$\begin{aligned} E \equiv & uu''' - u'u'' + u^3u' + (b + \sigma + 1)uu'' + (\sigma + 1)(buu' - u'^2) \\ & + \sigma u^4 + b(1 - r)\sigma u^2 = 0 \end{aligned} \tag{3}$$

Our goal is to find partially integrable cases, characterized by a value (b, σ, r) and an associated first integral

$$K = F(x, y, z, t) = \mathcal{F}(u, u', u'', t) = \text{const} \tag{4}$$

Up to now, only six cases are known [3,4]. In the representation u, u', u'', they are:

$$(2\sigma, \sigma, r) : K_1 = (-2\sigma(r - 1) + u^2 + 2(\sigma + 1)\frac{u'}{u} + 2\frac{u''}{u})e^{2\sigma t} \tag{5.1}$$

$$(0, \frac{1}{3}, r) : K_2 = (-\frac{3}{4}u^4 + 3u'^2 - 3uu'')e^{\frac{4}{3}t} \tag{5.2}$$

$$\begin{aligned} (1, \sigma, 0) : K_3 = & (1 + u^2 + 2(\sigma + 1)\frac{u'}{\sigma u} + 2\frac{uu'}{\sigma} + \frac{u'^2}{\sigma^2}(1 + \frac{(\sigma + 1)^2}{u^2}) \\ & + 2\frac{u''}{\sigma u} + \frac{u''^2}{\sigma^2 u^2} + 2(\sigma + 1)\frac{u'u''}{\sigma^2 u^2})e^{2t} \end{aligned} \tag{5.3}$$

$$\begin{aligned} (4, 1, r) : K_4 = & (-4(r - 1)^2 + 2(r - 1)u^2 - \frac{1}{4}u^4 + 8(r - 1)\frac{u'}{u} \\ & - 2uu' + u'^2 + 4(r - 1)\frac{u''}{u} - uu'')e^{4t} \end{aligned} \tag{5.4}$$

$$\begin{aligned} (1, 1, r) : K_5 = & ((r - 1)^2 - (r - 1)u^2 - 4(r - 1)\frac{u'}{u} + 2uu' + u'^2 \\ & + 4\frac{u'^2}{u^2} - 2(r - 1)\frac{u''}{u} + \frac{u''^2}{u^2} + 4\frac{u'u''}{u^2})e^{2t} \end{aligned} \tag{5.5}$$

$$\begin{aligned} (6\sigma - 2, \sigma, 2\sigma - 1) : K_6 = & (\sigma^{-1}(\sigma - 1)(3\sigma - 1)u^2 - \frac{1}{4}\sigma^{-1}u^4 - \sigma^{-1}(3\sigma - 1)uu' \\ & + \sigma^{-1}u'^2 - \sigma^{-1}uu'')e^{4\sigma t} \end{aligned} \tag{5.6}$$

We look for other partially integrable cases by the following method. We first consider particular solutions $u(t)$ of (3) of the form:

$$u = a_0 v + a_1, \quad a_0 \neq 0 \tag{6}$$

where v satisfies a Riccati equation

$$v_t + v^2 + \frac{S(t)}{2} = 0 \tag{7}$$

i.e. our unknowns are $S(t)$ and coefficients $a_i(t, S(t))$.

Substituting in Eq. (3) u by (6) and equating to zero the coefficients of the polynomial in v, we obtain five values of (b, σ, r) for which a solution for S exists, which may depend on an arbitrary constant t_0 (see Table I):

$$a_0^2 = -4, \quad a_1 = -a_0 \frac{2b - 3\sigma + 1}{6} \tag{8}$$

As S depends on t, our Ansatz generalizes the procedure introduced by H.Airault and P. Kaliappan [5] for finding particular solutions of a second order nonlinear equation.

Since these particular analytic solutions depend on at most two integration constants, assumption (6) cannot provide a first integral. Let us then consider the most general transformation

$$T(\lambda, \mu, \xi): \ (v, t) \rightarrow (V, z): \ v = \lambda(t) V + \mu(t), \quad z = \xi(t) \tag{9}$$

which applies (7) onto

$$V' + V^2 + \frac{\tilde{S}}{2} = 0 \tag{10}$$

Conditions for that are

$$\lambda - \xi' = 0, \quad \lambda' + 2\lambda\mu = 0 \tag{11}$$

which are solved as

$$\lambda = \xi', \quad \mu = -\frac{\xi''}{2\xi'}, \quad \xi(t) = \text{arbitrary}, \quad \tilde{S} = \xi'^{-2}(S - \{\xi; t\}) \tag{12}$$

where

$$\{\xi; t\} = (\frac{\xi''}{\xi'})' - \frac{1}{2}(\frac{\xi''}{\xi'})^2 \tag{13}$$

denotes the Schwarzian derivative of ξ with respect to t.

Let us choose the gauge so that, in expression $u = a_0 \lambda V + (a_0 \mu + a_1)$, the term independent of V vanishes and, when $\mu = 0$, transformation T is the identity:

$$\mu = \frac{2b - 3\sigma + 1}{6}, \quad \lambda = e^{-2\mu(t - t_0)}, \quad \xi = -\frac{1}{2\mu}(e^{-2\mu(t - t_0)} - 1) \tag{14}$$

With this choice of μ, function $\tilde{S}$ becomes a constant for every of the five solutions. We substitute $u = \pm 2i\lambda V(\xi)$ into (3) to obtain:

$$\begin{aligned} E(V) \equiv & \lambda(-VV''' + V'V'' + 4V^3V') \\ & + \frac{1}{3}(7b - 18\sigma + 2)VV'' - 2(b - 2\sigma)V'^2 - \frac{4}{3}(2b - 6\sigma + 1)V^4 \\ & - \frac{b}{9\lambda^2}((2b - 3\sigma + 1)^2 - 3(\sigma + 1)(2b - 3\sigma + 1) + 9\sigma(1 - r))V^2 = 0 \end{aligned} \quad (15)$$

In four out of the five above mentioned cases, this reduced ODE in V is simple enough to provide easily a first integral, as indicated in Table I. In the fifth case, i.e. $(b, \sigma, r) = (1 - 3\sigma, \sigma, 12\sigma - 3), \sigma \neq \frac{1}{5}$, we have not succeeded in finding a first integral; the associated Riccati subequation

$$u_t + \frac{1}{a_0}u^2 + (1 - 3\sigma)u = 0 \quad (16)$$

does not depend on any arbitrary constant. In this fifth case, we conjecture the existence of a first integral which would be outside the restricted class considered by Kuś, i.e. polynomials of degree at most four in (x, y, z). Our conjecture is consistent with the one already made [6] for $b = 1 - 3\sigma, r$ arbitrary, because in both cases the condition

$$(b - 2\sigma)(b + 3\sigma - 1) = 0 \quad (17)$$

must be satisfied.

The present method is applicable to any dynamical system admitting an *equivalent* ODE.

References.

[1] E. N. Lorenz, Deterministic Nonperiodic Flow, J. Atmos. Sci. **20**, 130-141 (1963).
[2] T. Sen and M. Tabor, Lie symmetries of the Lorenz model, Physica D **44**, 313-339 (1990).
[3] H. Segur, Solitons and the Inverse Scattering Transform, in *Topics in Ocean Physics* , 235-277, by A.R. Osborne and P. Malanotte Rizzoli (ed.), North-Holland Publ. Co. (1982), Proceedings Varenna 7-19 July 1980.
[4] M. Kuś, Integrals of motion for the Lorenz system, J. Phys. **A16**, L689-L691 (1983).
[5] H. Airault et P. Kaliappan, Solutions particulières de l'équation $u'' - \lambda u' = u(u - 1)(u - a)$, C. R. Acad. Sc. Paris **299**, 323-326 (1984).
[6] M. Tabor and J. Weiss, Analytic structure of the Lorenz system, Phys. Rev. **A24**, 2157-2167 (1981).

Table I

(b,σ,r)	S	$\tilde{S}$	-2μ	$E(V)$	First integral
$(2\sigma,\sigma,r_0)$ $r_0=-\frac{(\sigma-2)(2\sigma-1)}{9\sigma}$	$-\frac{(\sigma+1)^2}{18}$	0	$-\frac{\sigma+1}{3}$	$\lambda(-VV'''+V'V''+4V^3V')$ $+\frac{2}{3}(2\sigma-1)(-VV''+2V^4)$	$\lambda^{-\frac{2(2\sigma-1)}{\sigma+1}}(\frac{V''}{V}-2V^2)=\frac{1}{2}K_1$
$(-2,-1,r)$	$1-r$	$1-r$	0	$-VV'''+V'V''+4V^3V'$ $-2(-VV''+2V^4)-2(1-r)V^2$	$(\frac{V''}{V}-2V^2+r-1)e^{-2\xi}=\frac{1}{2}K_1$
$(0,\frac{1}{3},r)$	0	0	0	$-VV'''+V'V''+4V^3V'$ $+\frac{4}{3}(-VV''+V'^2+V^4)$	$(-VV''+V'^2+V^4)e^{\frac{4}{3}\xi}=-\frac{1}{12}K_2$
$(1,\frac{1}{2},0)$	$e^{-(t-t_0)}-\frac{1}{8}$	1	$-\frac{1}{2}$	$-VV'''+V'V''+4V^3V'$	$V'^2-V^4-\frac{1}{2}K_1V^2=-\frac{1}{8}K_6$
$(1-3\sigma,\sigma,12\sigma-3)$	$-\frac{(3\sigma-1)^2}{2}$	0	$3\sigma-1$	$\lambda(-VV'''+V'V''+4V^3V')$ $+2(4\sigma-1)(-VV''+2V^4)$ $+(5\sigma-1)(-VV''+2V'^2)$	$\sigma=\frac{1}{5}:\lambda(\frac{V''}{V}-2V^2)=\frac{1}{2}K_1$ $\sigma\neq\frac{1}{5}:V'+V^2=0$

Transition to Turbulence in 1-D Rayleigh-Bénard Convection

F.Daviaud, M.Bonetti and M.Dubois

SPSRM -CEN Saclay, F-91191 Gif-sur-Yvette Cedex, France

The study of one-dimensional -1D- hydrodynamical systems has recently received much interest. In these systems, a steady periodic cellular state, which depends on one space variable, may destabilize and become turbulent when a control parameter is varied. The spatio-temporal chaos observed thus stands between the deterministic chaos of confined systems and the developped turbulence of extended ones. Much theoretical and numerical work was devoted to the study of 1D phase equations[1], coupled map lattices [2,3] and cellular automata [3]. The results show that the transition to turbulence can occur via Spatio-Temporal Intermittency (STI), a fluctuating mixture of organized laminar domains and incoherent turbulent regions. Many experimental work was also devoted to 1D periodic systems. This is the case of Rayleigh-Bénard convection [4,5], directional solidification [6,7], parametric instability [8] and in the periodic front of the meniscus of a fluid between two rotating cylinders[9]. We present in this paper experimental results on the transition to turbulence in thermal convection and compare them with some numerical ones.

1. Experimental setup

The experimental conditions have been described in details elsewhere [5], but we recall that Rayleigh-Bénard convection was studied in cells with narrow gaps filled with silicon oil of Prandtl number 7. The transverse aspect ratios were respectively $\Gamma_y = 0.29$ and $\Gamma_y = 0.43$ in the annular and the rectangular channels in order to get 1D geometries. The convective structures were visualized by shadowgraphic imaging and their spatio-temporal evolution was recorded using a video camera. The image was digitized along a circle of 1024 pixels (annular cell) or an horizontal line of 512 pixels (rectangular cell) giving a resolution of 15 to 25 pixels per wavelength. The acquisition frequency was varied between 1 image/10 sec. and 10 images/ 1 sec. and time series of several $1000T_0$, with $T_0 \sim 1$ sec. the basic period of oscillation, were needed to perform statistics.

2. Transition to turbulence

When Ra exceeds a critical value $Ra_c = f(\Gamma_y)$, perfect stationnary periodic patterns are observed (domain labeled by S in Fig.1). Above a given ε value, ($\varepsilon = Ra/Ra_c - 1$), which depends on Γ_y and on the present wavelength λ_0, the pattern becomes time dependent with the appearance of thermal oscillators (domain D in Fig.1)[10]. At a larger value of ε, new events leading to a spatial symmetry breaking appear through the propagation of defects (or solitary waves). These defects are made of 2 or 3 wavelengths larger than the others which propagate and can be reflected by lateral walls or interact and give birth to great cells [5,10]. Above a certain threshold which depends on the geometry, turbulent patches are observed and the convective pattern enters in the regime of STI.

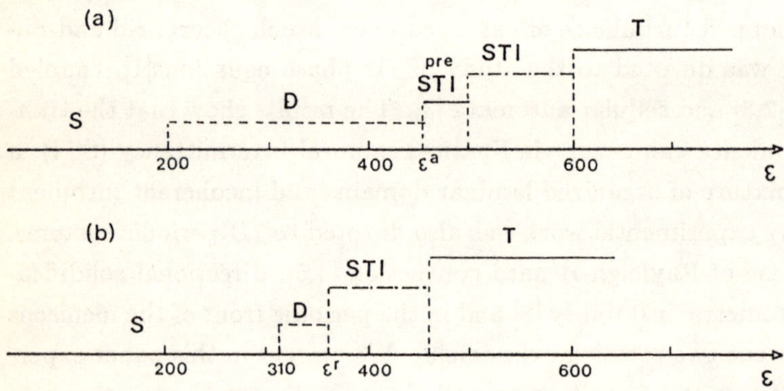

Figure 1. *Phase diagram of the convective states as a function of* ε *(a) annular cell (b) rectangular cell*

In the annulus and at the onset of STI, these patches are localized in space and time and can propagate through the container with a constant velocity or locally break the spatial pattern of rolls (cf.Fig.2). In that case, a turbulent zone can spread by contaminating the next cells or relax to a laminar state. Above a certain threshold ($\varepsilon = \varepsilon_s$), turbulent domains never relax (regime of sustained intermittency). The behavior looks qualitatively the same in the rectangular geometry, but with a stronger confinement of the turbulent regions and the presence of sustained STI just above the onset.

3. Statistical measurements

As turbulent patches are regions without spatial coherence and with a chaotic time evolution, a spatial criterion was taken to define if locally the behavior is laminar or turbulent. If a local wavelength is such as $\lambda_0 - \Delta\lambda < \lambda < \lambda_0 + \Delta\lambda$, with λ_0 the basic wavelength of the pattern and $\Delta\lambda$ a tolerance factor, the behavior is said to be laminar, otherwise it is turbulent. Calculations were also performed with a criterion based on the local time-evolution and they give results similar to those given by the spatial criterion.

A global characterization is given by the evolution of the turbulent fraction F_t: F_t is calculated as the averaged total length occupied by the turbulent cells divided by the length of the container. The results (cf. Fig.3a) show clearly that in the rectangular geometry the variation of F_t with ε reproduces a well-defined transition: $F_t \sim (\varepsilon - \varepsilon_T)^\beta$ with $\beta \simeq 0.3$ and $\varepsilon_T \simeq 360$. In the annular container the ε dependance of F_t is not so well defined and the transition looks imperfect (cf. Fig.3b).

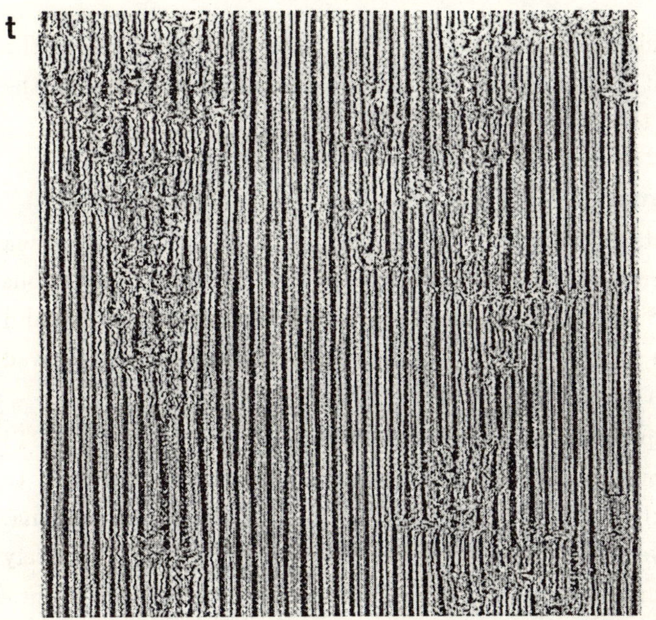

Figure 2. *Space-time evolution of the convective pattern obtained in the annular geometry for* $\varepsilon = 450$

Nevertheless, in the two cases, a power law decay is observed at the threshold for the distribution of the number N(L) of laminar domains of length L and the number N(τ) of laminar domains lasting τ. Above the threshold, the distributions become exponential, defining a characteristic

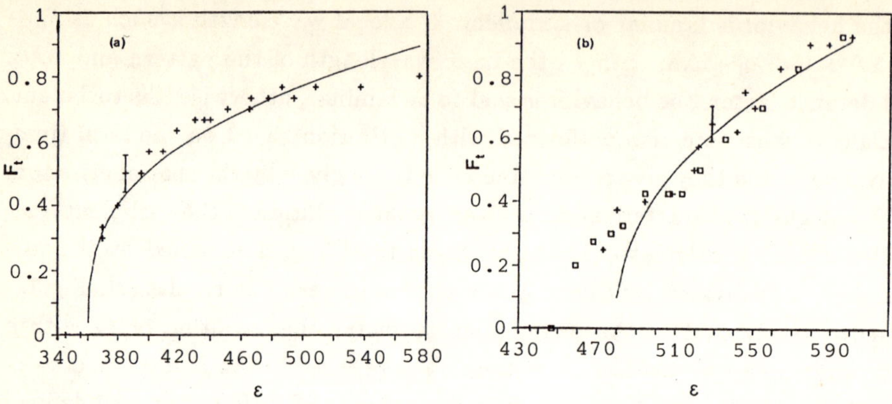

Figure 3. *Mean turbulent fraction F_t as a function of ε (a) rectangular cell, (b) annular cell*

length and time which decay as $(\varepsilon - \varepsilon_s)^{0.5}$. Note that $\varepsilon_s = \varepsilon_T$ in the rectangular case but that they are different in the annulus, owing to the imperfect nature of the transition.

4. Concluding remarks

The experiments reveal the same qualitative route to turbulence via STI as in the numerical simulations of some partial differential equations like the Kuramoto-Sivashinsky equation [1], coupled map lattices [2,3] and directed percolation [3] and most of the characteristic properties observed in the critical region around the threshold of STI are similar.

The relationship between DP and the transition to turbulence via STI had been pointed out by Pomeau [11] who has placed the transition to turbulence for weakly confined systems in the field of critical phenomena. The agreement between the experiments and DP is good, both qualitatively -aspect of spatiotemporal diagrams with a tree-like structure- and quantitatively -second order phase transition and value of the critical exponents-. Nevertheless, the experimental data show that a turbulent region may originate from a laminar one around the threshold of STI by opposition with

DP where the laminar state is absorbing. We have thus performed calculations with a simple model of cellular automaton, defined on a lattice of N sites with 2 possible states: laminar (L) or turbulent (T)and in which the laminar state is not absorbing. By taking totalistic rules, [12], the model is then governed by two parameters: the usual control parameter p plus the probability $\tilde{p} = p(LLL \rightarrow T)$ which stands for the spontaneous apparition of turbulents spots.

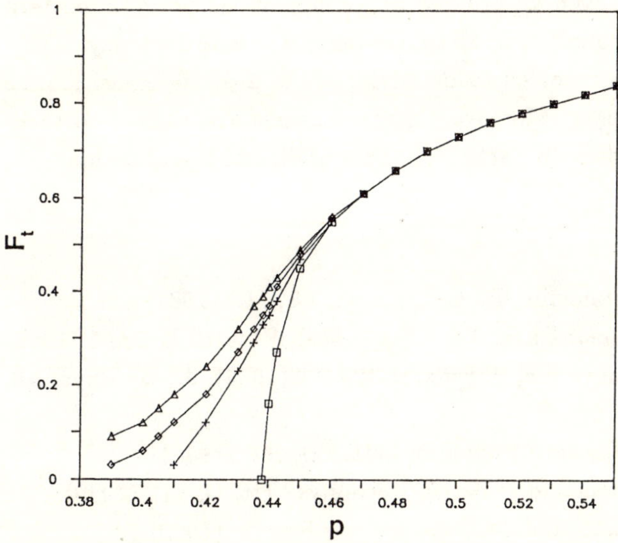

Figure 4. *F_t as a function of p in directed percolation (squares:$\tilde{p} = 0$, crosses:$\tilde{p} = 0.0005$, diamonds:$\tilde{p} = 0.001$, triangles:$\tilde{p} = 0.002$)*

When $\tilde{p} = 0$, a second order phase transition is observed for $p = p_c$ with well defined threshold and critical exponents. The numerical results display important differencies when $\tilde{p}$ is varied from zero to small but non zero values. First of all, the spatiotemporal representations of the automaton exhibits features very similar to the experimental results. The statistical properties are also changed by the presence of $\tilde{p}$, as in the evolution of the turbulent fraction (see Fig.4) and the transition disappears in a strict sense or becomes imperfect. However an algebraic decay is still present around $p = p_c$ if $\tilde{p}$ is small enough in the histograms giving the distributions of lengths or lifetimes of laminar domains, but in a range of scales which shortens when $\tilde{p}$ increases. Finally, these numerical results show that the rectangular geometry seems to be well represented by pure DP, whereas the annular case could be compared to DP with a quasi-absorbing state.

A phenomenological study of a 1D chain of rolls in the frame of the complex Ginzburg-Landau equations has also been performed by P.Coullet and J.Lega [13,14]. By giving a normal form description of the destabilization of a stationnary periodic pattern, they show it may give rise to STI-like behaviours. The first numerical results display spatiotemporal diagrams and statistical results qualitatively similar to the experiments.

In conclusion, 1D Rayleigh-Bénard convection has allowed a complete study of the transition from an ordered and coherent 1D periodic pattern to a turbulent one through typical spatio-temporal intermittency. This route to turbulence has revealed to be analogous to a phase transition, as conjectured by Y.Pomeau. Moreover, many features are similar to these given by numerical studies, in particular these of directed percolation.

References

[1] H.Chaté and P.Manneville, Phys.Rev.Lett. **58**, 112 (1987)
[2] K.Kaneko, Prog.Theor.Phys. **74**, 1033 (1985), Physica D **34**, 1 (1989)
[3] H.Chaté and P.Manneville, Physica D **32**, 409 (1988), J.Stat.Phys. **56**, 357 (1988)
[4] S.Ciliberto and P.Bigazzi, Phys.Rev.Lett. **60**, 286 (1988)
[5] F.Daviaud, M.Dubois and P.Bergé, Europhys.Lett. **9**, 441 (1989) F.Daviaud, M.Bonetti and M.Dubois, Phys.Rev.A (1990)
[6] A.J.Simon, J.Bechhoeffer, A.Libchaber and P.Oswald, Phys.Rev.A **40**, 2042 (1989)
[7] G.Faivre, S.de Cheveigné, C.Guthmannn and P.Kurowski, Europhys.Lett.**9**, 779 (1989)
[8] S.Douady, S.Fauve and O.Thual, Europhys.Lett. **10**, 309 (1989)
[9] M.Rabaud, S.Michalland and Y.Couder, Pys.Rev.Lett. **64**, 184 (1990)
[10] M.Dubois, R.Dasilva, F.Daviaud, P.Bergé and A.Petrov, Europhys.Lett.**8**, 135 (1989)
[11] Y.Pomeau, Physica D **23**, 3 (1986)
[12] W.Kinzel, Zeitschrift fur Physic B **58**, 229 (1985)
[13] P.Coullet and G.Iooss, Phys.Rev.Lett. **64**, 866 (1990)
[14] J.Lega, to appear in Europ.J.Mech.B/Fluids (1991)

Modelling of Low-Dimensional, Incompressible, Viscous, Rotating Fluid Flow

E.A. Christensen[1], J.N. Sørensen[2], M. Brøns[3] and P.L. Christiansen[1]

[1] Laboratory of Applied Mathematical Physics, Bldg.303,
[2] Department of Fluid Mechanics, Bldg.404,
[3] Mathematical Institute, Bldg.303,

The Technical University of Denmark, DK-2800 Lyngby, Denmark.

ABSTRACT

This presentation introduces a low–dimensional model for an incompressible, viscous, rotating fluid flow in a cylindrical vessel. The low–dimensional model is formed by projecting the transport equations on some subspace, spanned by known solutions to the discretized Navier–Stokes equations. Using the software package PATH, a program that analyses finite non–linear ODE–systems, such as our low–dimensional model, we find the bifurcation path with the Reynolds number as modelparameter. Thus, the transition from a stationary to a periodic solution in physical space is recognized as a super–critical Hopf–bifurcation in low–dimensional space.
Further aspects are to determine the bifurcations in space for all aspect ratios, and even further to give the dynamical concepts of a given fluid flow system.

1. THE MODEL

Consider an axisymmetric, incompressible, viscous, rotating fluid flow in a cylindrical vessel with height H and radius R. The flow motion is created by letting the lid rotate with a constant angular velocity Ω. Depending of the aspect ratio $\lambda = H/R$, and the Reynolds number Re different phenoma are seen. Setting for instance $\lambda = 2$ for small Reynolds numbers less than 1500 we have regular vortex flow, at about 1500 we get a bubble breakdown. At 1800, 2 bubbles appear and finally at about Re = 2500 transition from a stationary to a periodic solution occurs, see [1].

2. GOVERNING EQUATIONS IN PHYSICAL SPACE

Employing cylindrical coordinates ($y=r,\theta,x=z$) with corresponding velocity components (u,v,w), the dimensionless, conservative and axisymmetric Navier–Stokes equations expressed in circulation Γ, vorticity ω and streamfunction ψ take the form:
Transport equations for circulation and vorticity:

$$\frac{\partial \Gamma}{\partial t} = \frac{1}{Re}\left\{ \frac{\partial^2 \Gamma}{\partial x^2} - \frac{1}{r}\frac{\partial \Gamma}{\partial y} + \frac{\partial^2 \Gamma}{\partial y^2} \right\} - \frac{\partial (w\Gamma)}{\partial x} - \frac{\partial (u\Gamma)}{\partial y} - \frac{u\Gamma}{r}, \tag{1a}$$

$$\frac{\partial \omega}{\partial t} = \frac{1}{Re}\left\{ \frac{\partial^2 \omega}{\partial x^2} + \frac{1}{r}\frac{\partial \omega}{\partial y} + \frac{\partial^2 \omega}{\partial y^2} - \frac{\omega}{r^2} \right\} - \frac{\partial (w\omega)}{\partial x} - \frac{\partial (u\omega)}{\partial y} + \frac{1}{r^3}\frac{\partial (\Gamma^2)}{\partial x}, \tag{1b}$$

Poisson equation:

$$\omega r = \frac{\partial^2 \psi}{\partial x^2} - \frac{1}{r}\frac{\partial \psi}{\partial y} + \frac{\partial^2 \psi}{\partial y^2}, \tag{2}$$

where Re denotes the Reynolds number and the streamfunction is defined by the equations:

$$\frac{\partial \psi}{\partial x} = ru, \qquad \frac{\partial \psi}{\partial y} = -rw. \tag{3}$$

The flow is completely defined by Dirichlet boundary conditions for the velocities, see [4].

3. REDUCTION TO LOW–DIMENSIONAL SPACE

Replacing the spatial derivates in the transport equations (1a)–(1b) with central, finite difference approximations, we get a finite ODE–system of the order two times the number of interior netpoints:

$$\frac{\partial \Gamma}{\partial t} = F_1(\overset{*}{\Gamma}, \overset{*}{\omega}, \overset{*}{u}, \overset{*}{w}), \tag{4a}$$

$$\frac{\partial \omega}{\partial t} = F_2(\overset{*}{\Gamma}, \overset{*}{\omega}, \overset{*}{u}, \overset{*}{w}). \tag{4b}$$

The asterisk * indicates that the vectors also contain boundary points. The Poisson equation in a fourth–order Hermitian formulation and the definition equations (3) give a linear connection between the velocities u, w and the vorticity ω in all interior nodepoints:

$$u = K_u \, \omega, \tag{5a}$$

$$w = K_w \, \omega, \tag{5b}$$

where K_u, K_w are constant matrices.

Since the circulation Γ is constant on the boundaries, the system becomes closed by replacing the vorticity ω by Taylor expansions in the interior region. Thus we can remove the marks *, say

$$x = (\Gamma^t, \omega^t)^t, \tag{6a}$$

$$X = (\Gamma^t, \omega^t, u^t, w^t)^t, \tag{6b}$$

where superscript ^{t} is the transponing mark. Utilizing eqs. (4a)–(4b) we get the closed, finite ODE–system:

$$\frac{\partial x}{\partial t} = \frac{1}{Re} b_1 + A_0 X + \frac{1}{Re} A_1 X + Q (X , X), \tag{7}$$

where b_1, A_0, A_1 are constants and Q is a constant bilinear matrix.

Assume that the solution vector x = x(t) is almost limited to some subspace $\mathbb{R}^n$, where n is much less than the order of the unreduced ODE–system, and is spanned by an orthogonal base φ, thereby X by some Ψ determined by φ, eq. (5) and eq. (6). Then x(t), X(t) are approximately described by the series:

$$x(t) = \sum_{k=1}^{n} a_k(t) \, \varphi_k, \quad X(t) = \sum_{k=1}^{n} a_k(t) \, \Psi_k. \tag{8}$$

Here we span $\mathbb{R}^n$ by a stationary and points on a periodic solution to the discretized eqs. (4a) and (4b). Then the orthogonal base is developed by the Gram–Schmidts orthogonalization procedure. Substituting eq. (8) into eq. (7) and dotting with respect to the base vectors φ we finally get the low-dimensional model of the described fluid dynamical system. Collecting the amplitude–functions $a_k(t)$ in the vector–function a(t), we get the system:

$$\frac{\partial a(t)}{\partial t} = \frac{1}{Re} b_{low1} + A_{low0} \, a + \frac{1}{Re} A_{low1} \, a + Q_{low} (a,a), \tag{9}$$

where

$$b_{low1}(k) = < \varphi_k, b_1 > / < \varphi_k, \varphi_k >, \tag{10a}$$

$$A_{low0}(k,i) = < \varphi_k, A_0 \, \Psi_i > / < \varphi_k, \varphi_k >, \tag{10d}$$

$$A_{low1}(k,i) = < \varphi_k, A_1 \, \Psi_i > / < \varphi_k, \varphi_k >, \tag{10c}$$

$$Q_{low} (k,i,j) = < \varphi_k, Q (\Psi_i, \Psi_j) > / < \varphi_k, \varphi_k >. \tag{10d}$$

The low–dimensional model is a closed ODE–system of order n and it can be examined by the software package PATH.

4. RESULTS

Following is presented the preliminary results of our investigations.

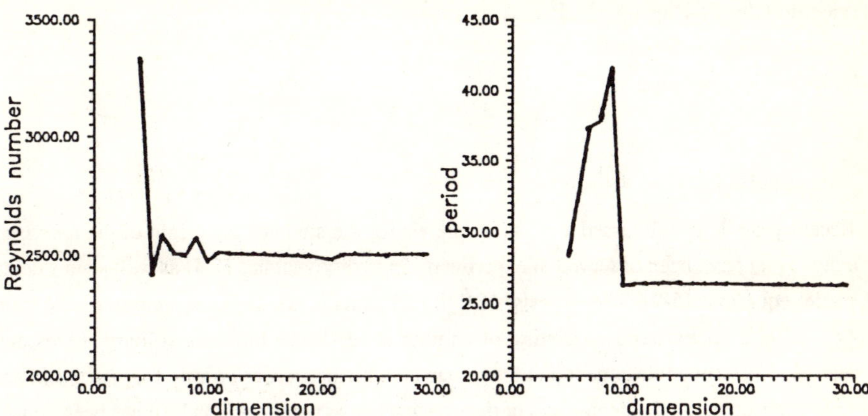

Figure 1.

Supercritical Hopf–bifurcations against the dimensions

Figure 2.

Periods against the dimensions

5. CONCLUSIONS

By projecting the transport equations (1a) and (1b) on a subspace, spanned by known solutions, the transition is found to be a simple bifurcation problem. The critical Reynolds number, where the fluid flow becomes instationary is exactly the Reynolds number where a complex conjugated pair of eigenvalues crosses the imaginary axis.
We did construct the solution space by known solutions, but probably it is possible to construct the space of only stationary solutions, just before the critical value. Remember the constants are independent of the Reynolds number Re.
Notice that the application of known u, w velocities instead of solving the linear equations directly, save us for inverting a matrix of size 100000.

REFERENCES

[1] *M.P. Escudier*, Observations of the flow produced in a cylindrical container by a rotating endwall, Experiments in Fluids, 2, pp 189–196, (1984).

[2] *M. Brøns*, Bifurcations and Instabilities in the Greitzer Model for Compressor System Surge, Math. Engng Ind., Vol. 1, No. 1, (1988).

[3] *O. Daube & J.N. Sørensen*, Simulation numerique de l'ecoulement periodique axisymetrique dans une cavite cylindrique, C.R. Acad. Sci. Paris, 308, 2, pp 443–469, (1989).

[4] *J.N. Sørensen & O. Daube*, Direct simulation of flow structures initiated by a rotating cover in a cylindrical vessel, Adv. in Turb., pp 383–390, Springer (1989).

[5] *J.N. Sørensen & Ta Phuoc Loc*, High–Order Axisymmetric Navier–Stokes Code Description and Evaluation of Boundary Conditions, Int. J. for Num. Meth. in Fluids, Vol. 9, 1517–1537, (1989).

[6] *C.K. Petersen*, PATH – User's Guide, Dept. of Applied Math. and Nonlinear Studies, University of Leeds, (1988).

Spatial Coherent Structures in Dissipative Systems

G. Dewel[1] and P. Borckmans[2]

Service de Chimie-Physique, CP 231,
Université Libre de Bruxelles,
1050 Bruxelles, Belgium.

1. Introduction:

Recent years have witnessed a growing interest for the study of the origin of the *localized structures* that have been observed in experiments in hydrodynamics [1-4] as well as in studies of model equations [5-9]. They consist of finite domains in which the system is in one state embedded in a background consisting of another state. These *intrinsic* solitary waves can appear in a uniform environment and their characteristic properties (width, frequency, ...) are entirely determined by the parameters of the system. The term *autosoliton* [10] has been coined to characterize these patterns.

In this short note we discuss that such localized structures can also exist in reaction-diffusion systems exhibiting uniform oscillations [11]. Their relevance to the problem of the existence of *homogeneous target patterns* [12] that have recently been observed to appear spontaneously in thin layers of oscillating media, such as the Belousov-Zhabotinskii (B-Z) reaction, is also discussed.

2. The target patterns and the " λ - ω " model:

Target patterns consist of concentric concentration waves that propagate outward from randomly distributed centers in a thin layer of reactants. There exists a controversy [12] regarding the origin of these waves. Is some centrally positioned foreign catalytic particle necessary for the generation of such patterns? A particle, with radius ranging between 70 and 100 μm, has indeed been observed in many targets but by no means in all of them. Experimental evidence has moreover recently been collected [13] showing the existence of homogeneous target waves the center of which is free from particles up to a visual resolution of a few micrometers. These waves, that propagate in a motionless fluid and thus arise solely from the coupling between nonlinear chemistry and diffusion, present particular characteristic properties [14].

Because the mechanism of the B-Z reaction is very complex, we will consider a simple reaction-diffusion model exhibiting the uniform concentration oscillations characteristic of the existence of a stable limit cycle in phase space appearing through an inverted Hopf bifurcation.

---

* Research Associates with the National Fund for Scientific Research (Belgium)

Our toy model is defined by the following equations describing the kinetics of the concentrations of two reacting species:

$$\frac{\partial C_1}{\partial t} = \lambda(R)\, C_1 - \omega(R)\, C_2 + D_1 \frac{\partial^2 C_1}{\partial x^2} \tag{1}$$

$$\frac{\partial C_2}{\partial t} = \omega(R)\, C_1 + \lambda(R)\, C_2 + D_2 \frac{\partial^2 C_2}{\partial x^2} \tag{2}$$

λ and ω are real-valued functions of $R^2 = C_1^2 + C_2^2$. We assume that $\lambda(R_+) = 0$ and $\left.\frac{\partial \lambda}{\partial R}\right|_{R+} < 0$ where R_+ is the amplitude of the stable limit cycle with frequency $\Omega_+ = \omega(R_+)$.

For most chemical systems it is reasonable to make the approximation $D_1 = D_2 = D$. In this case, it is convenient to work with polar coordinates in concentration space: $C_1 = R \cos\bar{\theta}$ and $C_2 = R \sin\bar{\theta}$. Then equations (1) and (2) read

$$\frac{\partial R}{\partial t} = R\, \lambda(R) + D \left[\frac{\partial^2 R}{\partial x^2} - R \left(\frac{\partial \bar{\theta}}{\partial x} \right)^2 \right] \tag{3}$$

$$\frac{\partial \bar{\theta}}{\partial t} = \omega(R) + \frac{D}{R^2} \frac{\partial}{\partial x} \left[R^2 \frac{\partial \bar{\theta}}{\partial x} \right] \tag{4}$$

For the sake of concreteness it is useful to consider the following expressions for λ and ω:

$$\lambda(R) = \mu + \beta_r R^2 - \gamma_r R^4$$
$$\omega(R) = \omega_o + \beta_i R^2 \qquad ; \beta_r > 0 , \gamma_r > 0 \tag{5}$$

The case $\beta_r < 0$, $\gamma_r = 0$ was considered in [15].

With these expressions, (3) and (4) are identical to the amplitude and phase equations of the complex Ginzburg-Landau equation with a fifth order nonlinearity that describes the dynamics of a system in the vicinity of a subcritical Hopf bifurcation at $\mu = 0$. β_r and γ_r describe the saturation of the amplitude due to the generation of harmonics, whereas β_i represents nonlinear dispersion.

When $- \beta_r / 4\gamma_r < \mu < 0$, this model exhibits bistability between the basic state $R = 0$ and a uniform limit cycle of amplitude $R_+(\mu)$. Besides these uniform solutions, this model also admits a one-parameter family of plane waves (PW) with wavenumber $k_* = \frac{\partial \bar{\theta}}{\partial x}$, amplitude R_* and frequency Ω_*, such that

$$\lambda(R_*) = Dk_*^2 \tag{6}$$

$$\Omega_* = \omega(R_*) \tag{7}$$

These PW furnish building blocks for the construction of the target waves described above. Indeed, as r --> $\pm \infty$ along a line through their center, the concentric waves asymptotically approach a pair of outgoing PW. In this picture an homogeneous target corresponds to a heteroclinic trajectory joining the two nonlinear fixed points: k_* and $-k_*$.

Therefore we look for solutions of (3) and (4) in the form

$$R(x,t) \equiv R(x) \quad ; \quad \bar{\theta}(x,t) = \Omega_* t + \theta(x) \tag{8}$$

Substituting (8) into (3) and (4), we get

$$\frac{1}{2}\frac{\partial V}{\partial R} = D\left[\frac{\partial^2 R}{\partial x^2} - R\left(\frac{\partial \theta}{\partial x}\right)^2\right] \tag{9}$$

$$\omega(R_*) - \omega(R) = \frac{D}{R^2}\frac{\partial}{\partial x}\left[R^2\frac{\partial \theta}{\partial x}\right] \tag{10}$$

where

$$V(R) = -\mu R^2 - \frac{\beta_r}{2}R^4 + \frac{\gamma_r}{3}R^6$$
$$\omega(R_*) - \omega(R) = \beta_i\left[R^2 - R_*^2\right] \tag{11}$$

In the dispersionless limit, $\beta_i = 0$, (9) and (10) are decoupled and the model displays a gradient structure. Useful analogies with the theory of equilibrium phase transitions can then be developed. In particular, any droplet-like structure corresponding to a pulse or a hole (antipulse) of one state embedded in the other is unstable (critical nucleus).

When the phase is taken into account (complex field) one can obtain a configuration where two uniformly oscillating regions in phase opposition are in contact. The amplitude goes to zero in the boundary region whereas the phase suffers a step-function jump. This structure is stable over the range of values of μ for which the oscillating state, $R = R_+(\mu)$, is dominant. This stability originates from the repulsion between the kink and antikink bounding this hole.

In the next section we construct antipulses in the weak dispersion limit ($\beta_i << 1$) when the deviations from the potential case are small.

3. Antipulses in the weak dispersion limit:

In the presence of dispersion ($\beta_i \neq 0$) such an antipulse plays the role of a pacemaker that generates PW of wavenumber $|k_*| = \beta_i R_+^2 / \delta D$ (where δ^{-1} is the characteristic width of the (anti-)kink).

The amplitude hole thus selects one wavenumber among the one-parameter family of PW. However there is still a phase change of π between $x= \infty$ and $x= -\infty$. This structure may be thought of as a defect of the PW system with waves emanating alternatively on the two sides of the center (one-dimensional version of a spiral wave).

To obtain a structure symmetric about its midpoint, one must consider two adjacent antipulses in such a way that the phase shift induced by the first is reversed by the second. We thus studied the interaction between two antipulses $A_L(x+\sigma)$ and $A_R(x-\sigma)$ respectively centered at $x=-\sigma$ and $x=\sigma$. To get approximate analytical results, we furthermore supposed that the distance between pulses is large with respect to their width so that their interaction is small. Our smallness parameter is thus $\varepsilon = e^{-\delta\sigma}$. We took advantage of this weak coupling to look for a solution in the form ($T = \varepsilon t$)

$$R(x,t) = A_L(x + \sigma(T)) + A_R(x - \sigma(T)) + \varepsilon w(x,T) \quad (12)$$

The solvability condition resulting from the expansion of (3) for the existence of a stationary solution σ_s yields the following relation at the leading order

$$a_1 k_*^4 - a_2 k_*^2 e^{-\delta\sigma_s} + a_3 e^{-2\delta\sigma_s} = 0 \quad (13)$$

where a_1, a_2 and a_3 are positive constants depending on the parameters in the problem.

When k_* is sufficiently small ($\beta_i << 1$), it may be shown that this equation exhibits two roots corresponding to localized structures of which only the wider is stable.

In the presence of a weak dispersion two antipulses can thus form a stable bound state leading to a symmetric localized structure that synchronously emits waves propagating outward from the center. So this model provides a one dimensional idealization of the *intrinsic* target patterns.

A similar calculation can be performed in the weak dissipation limit [16]. There a bound state between two dark solitons is selected by the dissipation terms. Similar structures have been observed in experiments in optical fibers [17] that present a positive group velocity dispersion.

References:

[01] J. Wu, R. Keolian, I. Rudnick, Phys. Rev. Lett. **52** 1421 (1984)

[02] E. Moses, J. Fineberg, V. Steinberg, Phys. Rev. **A35** 2757 (1987)

[03] J.J. Niemela, G. Ahlers, D.S. Cannel, Phys. Rev. Lett.**64** 1365 (1990)

[04] P. Kolodner, D. Bensimon, C.M. Surko, Phys. Rev. Lett. **60** 1723 (1988)

[05] O. Thual, S. Fauve, J. Phys. (France) **49** 1829 (1988)

[06] S. Fauve, O. Thual, Phys. Rev. Lett. **64** 282 (1990)

[07] V. Hakim, P. Jakobsen, Y. Pomeau, Europhys. Lett. **11** 19 (1990)

[08] W. Van Saarloos, P.C. Hohenberg, Phys. Rev. Lett. **64** 749 (1990)

[09] C. Elphick, E. Meron, Phys. Rev. **A40** 3226 (1989)

[10] B.S. Kerner, V.V. Osipov, Sov. Phys. Usp. **32** 101 (1989)

[11] G. Nicolis, I. Prigogine, *Self Organization in Nonequilibrium Systems* (Wiley, New York, 1977)

[12] J. Ross, S.C. Müller, C. Vidal, Science **240** 460 (1988)

[13] C. Vidal, A. Pagola, J. Phys. Chem. **93** 2711 (1989)

[14] Y. Kuramoto, *Chemical Oscillations, Waves, and Turbulence* (Springer, Berlin, 1984)

[15] N. Kopell, L.N. Howard, Studies Appl. Math. **64** 1 (1981)

[16] G. Dewel, P. Borckmans, to be published

[17] A. Hasegawa, *Optical Solitons in Fibers*, Springer Tracts in Modern Physics **116**, (Springer, Berlin, second edition,1990)

Hierarchies of (1+1)-Dimensional Multispeed Discrete Boltzmann Model Equations

H. Cornille

SPhT CEN-Saclay,

91191 Gif-sur-Yvette, France.

We study multispeed discrete Boltzmann models [1] (discrete velocities and densities) in R^d , with coordinates $x_1,...x_d$, satisfying all conservation laws and leading to well-defined temperatures. The $(1+1)$- dimensional restrictions of the system of Pde satisfied by the densities of these models can be classified. One class is characterized , along a chosen coordinate, by the number and location of the independent densities and by the number and physical meaning of the independent collision terms. We obtain different classes associated with velocity speeds 1(densities M_i),$\sqrt{2}$ (densities N_i) or 2 (M_i),$\sqrt{2}$ (N_i)or 1(M_i),$\sqrt{d}$(N_i) or 2(M_i),$\sqrt{d}$ or 0(density R),1,$\sqrt{2}$ or $0,2,\sqrt{2}$. Each class represents a d-hierarchy of a well-known $d=2$ or 3 discrete model. The simplest exact solutions are the similarity shock waves with the variable $\eta = x - \varsigma t$ and we can hope to obtain other $(1+1)$-dimensional solutions which are sums of similarity waves [2]. We restrict our classification of models to those which satisfy; (i) at least two speeds, (ii) microscopic conservation relations, (iii) integer values or zero for the $x_1,..\ ,x_d$ coordinates, (iv) well known physical models [3] for $d=2$ and $d=3$ (v) same mass for the particles of a single gas [4]... Along the $x=x_1$ axis, we have at most six independent densities N_1, N_3, N_2 with $x=-1,0,1$ and M_4, M_2, M_1 with $x=-1,0,1$ or $-2,0,2$. For the collision terms we can have collisions for particles with the same speed $N_iN_j - N_kN_l$, $M_iM_j - M_kM_l$ or collisions with mixed speeds $N_iM_j - N_kM_l$. $M_iR - N_kN_l$, $N_iR - M_kM_l$.

For the shock waves we define the two upstream and downstream states Ma_0, Ma_s with densities n_{0i}, m_{0i} and $s_{0i} = n_{0i} + n_i, p_{0i} = m_{0i} + m_i$ associated with N_i, M_i and $r_0, r_0 + r$ if the density R is included. Two methods exist: (i) The Rankine-Hugoniot (R-H) relations (equivalent to the linear conservation laws), (ii) The determination of the similarity shock waves $N_i = n_{0i} + n_i/\Delta, \Delta = 1 + e^{\eta\gamma}, M_i = m_{0i} + m_i/\Delta, R = r_0 + r/\Delta$. We define $p_\pm = \partial_t \pm \partial_x$, $q_\pm = \partial_t \pm 2\partial_x$ and scaled parameters $\overline{n}_{02} = n_{02}/n_{01}, \overline{m}_{0i} = m_{0i}/n_{01}, \overline{n}_2 = n_2/n_1, \overline{m}_i = m_i/n_1$. Eight different classes of Pde are written down.

1. CLASS I: 5 DENSITIES $(M_1, M_4), M_2, (N_2, M_1)$ with x coordinates $-1,0,1$ and **2 COLLISION TERMS** for two subclasses (i),(ii):

$$C = \overline{c}(N_2M_4 - N_1M_1), \quad dD = 2d_c(M_1M_4 - M_2^2), \quad d_c > 0 \tag{1.1}$$

$$Pde: p_-N_1 = -p_+N_2 = C, M_{2t} = D, p_+M_1 = -(d-1)D + d_*C, p_-M_4 = -(d-1)D - d_*C$$

$$R-H: p_-N_1 + p_+N_2 = p_+M_1 + p_-M_4 + 2(d-1)M_{2t} = p_+M_1 - p_-M_4 - 2d_*p_-N_1 = 0 \tag{1.2}$$

We define the macroscopic quantities: mass M, momentum J and energy E.

$$M = M_1 + M_4 + 2(d-1)M_2 + d_*(N_1 + N_2), \quad J = M_1 - M_4 + d_*(N_2 - N_1)$$

$$E=(M_1+M_4)/2+(d-1)M_2+d_*d_{**}(N_1+N_2),\quad (i)d_{**}=d/2,\quad (ii)d_{**}=1 \quad (1.3)$$

Subclass(i): $(2^d+2d)v_i, d\geq 2$, $\underline{8}, \underline{14}, 24$, speeds$1, \sqrt{d}$, $d_*=2^{d-1}$, $\overline{c}=\sqrt{(d+3)}/2$

$$N_i:(\pm1,\pm1,\cdots,\pm1)\quad M_i:(\pm1,0,0,\cdots,0),(0,\pm1,0,\cdots,0),\cdots,(0,0,\cdots,\pm1)\quad (1.4(i))$$

$8v_i, d=2, (\pm1,\pm1), (\pm1,0), (0,\pm1)$, $14v_i, (\pm1,\pm1,\pm1), (\pm1,0,0), (0,\pm1,0), (0,0,\pm1)$

Subclass(ii): $2(3d-2)v_i$, $\underline{8}v_i, \underline{14}v_i, 20, 36, d\geq 2$, speeds$1, \sqrt{2}, d_*=2(d-1), 2\overline{c}=\sqrt{5}$

$$N_i:(\pm1,\pm1,0,\cdots,0),(\pm1,0,\pm1,\cdots,0),\cdots,M_i:(\pm1,0,\cdots),(0,\pm1,0,\cdots),\cdots\quad (1.4(ii))$$

$8v_i, d=2$, same as (i) $\quad 14v_i, d=3:(\pm1,\pm1,0),(\pm1,0,\pm1),(\pm1,0,0),(0,\pm1,0),(0,0,\pm1)$

An analytical positive similarity solution has been found: $s_{01}=s_{02}=p_{01}=p_{02}=0, p_{04}\neq 0$. Two parameters are arbitrary: n_{01}which is a scaling parameter and ς, $1/\sqrt{1+(d-1)^2}\leq\varsigma\leq 1$ We deduce $\overline{m}_{01}=\overline{m}_{04}\overline{n}_{02}, \overline{m}_{02}=\overline{m}_{04}\sqrt{\overline{n}_{02}}$ and

$$\overline{n}_{02}=(1+\varsigma)/(1-\varsigma), m_{04}=d_*\sqrt{1-\varsigma^2}/(\varsigma(d-1)-\sqrt{1-\varsigma^2}), p_{04}=m_{04}2(d-1)\varsigma/\sqrt{1-\varsigma^2}$$

$$1-1/\Delta=N_1/n_{01}=N_2/n_{02}=M_1/n_{01}=M_2/m_{02},\quad M_4=m_{04}+(p_{04}-m_{04})/\Delta$$

$$\gamma=-\overline{c}p_{04}/(1-\varsigma)\leq 0,\quad d_c=d\varsigma\overline{c}/2\sqrt{1-\varsigma^2}\ >0 \quad (1.5)$$

2. CLASS II: 6 DENSITIES $(N_1,M_4),(M_2,R),(N_2,M_1)$ with x coordinates -1,0,1 and **4 COLLISION TERMS** C,D given in (1.1),A_1, A_2 for three speeds $0,1,\sqrt{2}$

$$A_1=\overline{a}a_c(M_4M_2-N_1R),\ A_2=\overline{a}a_c(M_1M_2-N_2R),\ \overline{a}\sqrt{2}=1 \quad (2.1)$$

It is class I (ii) with a new $v_i=0$ and new collision terms for $0,\sqrt{2}\to 1,1$.

$$\text{Pde}:\ p_-N_1=C+A_1,\ p_+N_2=-C+A_2,\ M_{2t}=-A_1-A_2+D$$

$$p_-M_4=-(d-1)(D+2C+2A_1), p_+M_1=-(d-1)(D-2C+2A_2), R_t=2(d-1)(A_1+A_2)$$

$$\text{R-H}:\quad p_+M_1+p_-M_4+2(R_t+(d-1)M_{2t})=2(d-1)(p_-N_1+p_+N_2)=0$$

$$p_+M_1-p_-M_4+2(d-1)(p_+N_2-p_-N_1)=0 \quad (2.2)$$

$$M=M_1+M_4+2(d-1)(M_2+N_1+N_2)+R,\quad J=M_1-M_4+2(d-1)(N_2-N_1)$$

$$2E=M_1+M_4+(d-1)(2M_2+2N_1+4N_2) \quad (2.3)$$

$3(2d-1)v_i, \underline{9}v_i, \underline{16}v_i, 21,.., d\geq 2, 9v_i, 15v_i$, same as $8v_i, 14v_i$ classI(ii) plus$(0,0,0)$

An analytical solution has been found with $s_{01}=s_{02}=p_{01}=p_{02}=r+r_0=0, p_{04}\neq 0$. We have 2 arbitrary parameters : n_{01} which is a scaling parameter and $0<\varsigma<1$. We deduce $\overline{m}_{01}=\overline{m}_{04}\overline{n}_{02}, \overline{m}_{02}=\overline{m}_{04}\sqrt{\overline{n}_{02}}$,$\overline{r}_0=\overline{m}_{04}^2\sqrt{\overline{n}_{02}}$ and

$$\overline{m}_{04}^3+(d-1)(\overline{m}_{04}^2+4(\varsigma^2-1)/\varsigma^2)=0,\quad \overline{n}_{02}=(1+\varsigma)/(1-\varsigma)(1+\overline{m}_{04}/(d-1))$$

$$\sqrt{n_{02}} = \overline{\varsigma m}_{04}/2(1-\varsigma), \quad m_4/n_{01} = 9d-1)(4+3\overline{m}_{04}/(d-1))/(1+\overline{m}_{04}/(d-1))$$

$$p_{04} = m_{04} + m_4, \quad 1 - 1/\Delta = N_1/n_{01} = N_2/n_{02} = M_1/m_{01} = M_2/m_{02} = R/r_0$$

$$M_4 = m_{04} + m_4/\Delta, \quad \gamma = -p_{04}\sqrt{5}/2(1-\varsigma), \quad d_c = d\sqrt{5}(1 + 2(d-1)/\overline{m}_{04})$$

$$a_c = \sqrt{5}\varsigma\overline{m}_{01}/2\sqrt{5}(1-\varsigma)(d-1) \tag{2.4}$$

3. CLASS III: 5 DENSITIES M_4, N_1, M_2, N_2, M_1 with x coordinates -2,-1,0,1,2 and **3 COLLISION TERMS** for two subclasses (i),(ii) (like classI except $M_i : (\pm 2, 0, \cdots, 0)$...)

$$C_i = \overline{c}(N_2 M_{2i} - N_1 M_i), i = 1,2, \quad dD = 4d_c(M_4 M_1 - M_2^2), \quad d_c > 0 \tag{3.1}$$

$$\text{Pde:} \quad p_- N_1 = -p_+ N_2 = C_1 + C_2, \quad M_{2t} = D + d_*(C_2 - C_1)$$

$$q_+ M_1 = (d-1)(-D + 2d_* C_1), \quad q_- M_4 = (d-1)(-D - 2d_* C_2) \tag{3.2}$$

$$\text{R-H} : p_- N_1 + p_+ N_2 = q_+ M_1 + q_- M_4 + 2(d-1)M_{2t} = q_+ M_1 - q_- M_4 - 2d_*(d-1)p_- N_1 = 0$$

$$M = M_1 + M_4 + 2(d-1)(M_2 + d_*(N_1 + N_2)), \; J = 2(M_1 - M_4) + 2(d-1)d_*(N_2 - N_1)$$

$$E = 2(M_1 + M_4 + 2(d-1)M_2) + (d-1)d_* d_{**}(N_1 + N_2), (i) \; 2d_{**} = d, (ii) \; d_{**} = 1 \tag{3.3}$$

$$\text{Subclass (i)} : \; (2^d + 2d)v_i, \; \underline{8}v_i, \; \underline{1}4v_i, 24, 42, ..., \; d \geq 2 \quad \text{speeds} \quad 2, \sqrt{d}$$

$$(d-1)d_* = 2^{d-2}, \; d\overline{c} = (d-1)\sqrt{8+d}, \quad 8v_i, d = 2, \; (\pm 1, \pm 1), (\pm 2, 0), (0, \pm 2)$$

$$14v_i, d = 3, \quad (\pm 1, \pm 1, \pm 1), (\pm 2, 0, 0), (0, \pm 2, 0), (0, 0, \pm 2) \tag{3.4(i)}$$

$$\text{Subclass (ii)} : 2(3d-2)v_i, \; \underline{8}v_i, \underline{14}v_i, 20, 26, .., \quad d \geq 2 \quad \text{speeds } 2, \sqrt{2}, \; d_* = 1 \tag{3.4(ii)}$$

$$2\overline{c} = \sqrt{10}, 8v_i \text{ as (i)}, \; 14v_i : \; (\pm 1, \pm 1, 0), (\pm 1, 0, \pm 1), (\pm 2, 0, 0), (0, \pm 2, 0), (0, 0, \pm 2)$$

4.CLASS IV: 6 DENSITIES $M_4, N_1, (M_2, R), N_2, M_1$ with x coordinates $-2, -1, 0, 1, 2$ and **6 COLLISION TERMS** for three speeds $0, 2, \sqrt{2}$:

$$2C_i = \sqrt{10}(N_2 M_{2i} - N_1 M_i), i = 1, 2, \quad dD = 4d_c(M_1 M_4 - M_2^2), \quad d_c > 0$$

$$dA_1 = 2a_c(N_2^2 - M_1 R), \; dA_4 = 2a_c(N_1^2 - M_4 R), \; A_2 = a_c(N_1 N_2 - M_2 R), \; d_c > 0 \tag{4.1}$$

That class is the same as ClassIII(ii) with an additional velocity zero and the new collision terms A_i correspond to $0, 2 \to \sqrt{2}, \sqrt{2}$.

$$\text{Pde} : p_- N_1 = C_1 + C_2 - A_2 - A_4, p_+ N_2 = -C_1 - C_2 - A_1 - A_2, M_{2t} = -C_1 + C_2 + A_2 + D$$

$$q_+ M_1 = (d-1)(2C_1 + A_1 - D), q_- M_4 = (d-1)(-2C_2 + A_4 - D), R_t = (d-1)(A_1 + A_4 + 2A_2)$$

$$\text{R-H} \quad (d-1)(p_- N_1 + p_+ N_2) + R_t = q_+ M_1 + q_- M_4 + 2(d-1)M_{2t} - R_t = 0$$

$$q_+ M_1 - q_- M_4 - (d-1)(p_- N_1 - p_+ N_2) = 0 \tag{4,2}$$

$$M = 2(d-1)(N_1 + N_2 + M_2) + M_1 + M_4 + R, \quad \hat{J} = M_1 - M_4 + (d-1)(N_2 - N_1)$$

$$\hat{J} = J/2, \quad 2E = 4(M_1 + M_4 + 2(d-1)M_2) + 4(d-1)(N_1 + N_2) \tag{4.3}$$

$$3(2d-1)v_i, \underline{9}v_i, \underline{15}v_i, 21, .., d \geq 2, \; 9v_i, 15v_i \text{ like } 8v_i, 14v_i \text{ ClassIII(ii) plus } (0,0,0) \tag{4.4}$$

5.CLASS V: 6 DENSITIES (M_4, N_1),(M_2, N_3),(M_1, N_2) with x coordinates -1,0,1 and **6 COLLISION TERMS**. The new independent density is N_3 with $x = x_1 = 0$.

$$D = d_c\overline{d}(M_1M_4 - M_2^2), \overline{d} = 2/d, \quad \overline{b} = 2(d-2)(1/(d-1) + \sqrt{2}/d), B = b_c\overline{b}(N_1N_2 - N_3^2)$$

$$\overline{c}_3 = 2(d-2)/\sqrt{3}, \quad C_{13} = \overline{c}_3(M_4N_3 - M_2N_1), C_{23} = \overline{c}_3(M_1N_3 - M_2N_2) \tag{5.1}$$

$$2C_{12} = \sqrt{5}(N_2M_4 - N_1M_1), \overline{c} = (d-2)\sqrt{5}/2(d-1), C_3 = \overline{c}(N_1M_1 + N_2M_4 - 2N_3M_2)$$

This class, like classI, has speeds $\sqrt{2}$ for the N_i and 1 for the M_i. TheM_i $(\pm1, 0, \cdots, 0)$ ... are those of classI. The N_i contain terms of classI $(\pm1, \pm1, 0, \cdots, 0)$,$(\pm1, 0, \pm1, 0, \cdots, 0)$ ... associated to N_1, N_2 . New terms exist$(0, \pm1, \pm1, 0, \cdots, 0)$,$(0, \pm1, 0, \pm1, \cdots, 0)$... associated to N_3 .The B,D,C_i terms represent respectively collisions of two $\sqrt{2}$ speeds,two speeds 1 and mixed collisions with speeds 1,$\sqrt{2}$. The new C_i collision terms come from terms like N_3M_i,i=1,2,4.The physical model is the $18v_i, d = 3$ model.

$$\text{Pde}: \quad p_-N_1 = -B + C_{12} - C_3 + C_{13}, \quad p_+N_2 = -B - C_{12} - C_3 + C_{23}$$

$$(d-2)N_{3t} = 2B + 2C_3 - C_{13} - C_{23}, \quad M_{2t} = 2C_3 + C_{13} + C_{23} + D \tag{5.2}$$

$$p_+M_1 = (d-1)(2C_{12} - 2C_3 - 2C_{23} - D), \quad p_-M_4 = (d-1)(-2C_{12} - 2C_3 - 2C_{13} - D)$$

$$\text{R-H}: p_-N_1 + p_+N_2 + (d-2)N_{3t} = p_+M_1 + p_-M_4 + 2(d-1)M_{2t} = 0$$

$$p_+M_1 - p_-M_4 + 2(d-1)(p_+N_2 - p_-N_1) = 0$$

$$M = M_1 + M_4 + 2(d-1)(M_2 + N_1 + N_2 + (d-2)N_3), J = M_1 - M_4 + 2(d-1)(N_2 - N_1)$$

$$2E = M_1 + M_4 + 2(d-1)(M_2 + 2N_1 + 2N_2) \tag{5.3}$$

$$2d^2v_i, d \geq 3, \quad \underline{18}v_i, \quad 32, 50, .. \quad 18v_i : (N_1, N_2) = (\pm1, \pm1, 0), (\pm1, 0, \pm1)$$

$$N_3 : (0, \pm1, \pm1, 0), \quad (M_1, M_4) : (\pm1, 0, 0), \quad M_2 : (0, \pm1, 0), (0, 0, \pm1) \tag{5.4}$$

6. CLASS VI: 7 DENSITIES $(M_4, N_1), (M_2, N_3, R), (M_1, N_2)$ with x coordinates -1, 0, 1 and **9 COLLISION TERMS** which are the six previous ones of classV,(5.1) plus three new for $1, 1 \to 0, \sqrt{2}$. Here the speeds are $0, 1, \sqrt{2}$.

$$2A_1 = \sqrt{2}(M_2M_4 - N_1R) \quad 2A_2 = \sqrt{2}(M_2M_1 - N_2R) \quad 2A_3 = \sqrt{2}(M_2^2 - N_3R) \tag{6.1}$$

$$\text{Pde}: \quad p_-N_1 = -B + C_{12} - C_3 + C_{13} + A_1 \quad p_+N_2 = -B - C_{12} - C_3 + C_{23} + A_2$$

$$(d-2)N_{3t} = 2B + 2C_3 - C_{13} - C_{23} + A_3(d-2), R_t = 2(d-1)(A_1 + A_2 + (d-2)A_3)$$

$$M_{2t} = 2C_3 + C_{13} + C_{23} + D - A_1 - A_2 - 2(d-2)A_3$$

$$p_+M_1 = (d-1)(2C_{12}-2C_3-2C_{23}-D-2A_2), p_-M_4 = (d-1)(-2C_{12}-2C_3-2C_{13}-D-2A_1)$$

$$\text{R-H}: \quad 2(d-1)(p_-N_1 + p_+N_2 + (d-2)N_{3t}) - R_t = 0 \tag{6.2}$$

$$p_+M_1 + p_-M_4 + 2(d-1)M_{2t} + 2R_t = p_+M_1 - p_-M_4 + 2(d-1)(p_+N_2 - p_-N_1) = 0$$

$$M = 2(d-1)(N_1+N_2+(d-2)N_3+M_2)+M_1+M_4+R, J = 2(d-1)(N_2-N_1)+M_1-M_4$$

$$2E = 2(d-1)(2N_1 + 2N_2 + 2(d-2)N_3 + M_2) + M_1 + M_4 \tag{6.3}$$

$$(2d^2+1)v_i, \quad d \geq 3, \quad \underline{19v_i}, 33, 51, \ldots \tag{6.4}$$

$$19v_i: \ (\pm1,\pm1,0),(\pm1,0,\pm1), \ (0,\pm1,\pm1), \ (\pm1,0,0),(0,\pm1,0),(0,0,\pm1), \ (0,0,0)$$

7.CLASS VII: 6 DENSITIES $M_4, N_1, (M_2, N_3), N_1, M_4$ with x coordinates -2,-1,0,1,2 and **4 COLLISION TERMS.** The v_i are those of classV with M_i: $(\pm2, 0, \cdots, 0)$,...

$$\overline{d} = 4/d \quad D = d_c\overline{d}(M_1M_4 - M_2^2), \quad B = \overline{b}b_c(N_1N_2 - N_3^2)$$

$$\overline{c} = \sqrt{10}/2 \quad C_1 = \overline{c}(M_2N_2 - M_1N_1), \quad C_2 = \overline{c}(M_4N_2 - M_2N_1) \tag{7.1}$$

For this class the M_i,N_i have speeds $2, \sqrt{2}$. The B,D,C_i terms correspond respectively to collisions of two $\sqrt{2}$ speeds, two speeds 2 and mixed $\sqrt{2}, 2$ speeds.

$$\text{Pde}: \quad p_-N_1 = -B + C_1 + C_2, \ p_+N_2 = -B - C_1 - C_2, \ d_*N_{3t} = 2B$$

$$M_{2t} = -C_1 + C_2 + D, \quad q_+M_1 = (d-1)(2C_1 - D), \quad q_-M_4 = (d-1)(-2C_2 - D)$$

$$\text{R-H}: \quad p_-N_1 + p_+N_2 + d_*N_3 = q_+M_1 + q_-M_4 + 2(d-1)M_{2t} = 0$$

$$q_+M_1 - q_-M_4 + (d-1)(p_+N_2 - p_-N_1) = 0 \tag{7.2}$$

$$M = 2(d-1)(N_1 + N_2 + d_*N_3 + M_2) + M_1 + M_4, J = 2(d-1)(N_2 - N_1) + 2(M_1 - M_4)$$

$$E = 2(d-1)(N_1 + N_2 + d_*N_3 + 2M_2) + 2(M_1 + M_4) \tag{7.3}$$

$$2d^2v_i, \quad d \geq 3, \quad \underline{18v_i}, 32, 50, \cdots \quad d_* = d-2, \quad \overline{b} = 2d_*(1/(d-1) + \sqrt{2}/d) \tag{7.4}$$

$$N_1, N_2: (\pm1, x_2, \cdots, x_j, \cdots, x_d) \quad x_j = 0 \text{ except one term} \pm 1$$

$$N_3: (0, x_2, \cdots, x_j, \cdots, x_d) \quad x_j = 0 \text{ except two terms} \pm 1,$$

$$M_1, M_4: (\pm2, 0, \cdots, 0), M_2: (0, x_2, \cdots, x_j, \cdots, x_d) \quad x_j = 0 \text{ except one term} \pm 2$$

$$18v_i: \quad (\pm1,\pm1,0),(\pm1,0,\pm1),(0,\pm1,\pm1), \quad (\pm2,0,0),(0,\pm2,0),(0,0,\pm2)$$

In our classification of models we have excluded those for which the x_j coordinates are not integer. If we relax this condition we can introduce here another classVIIbis with the same number of densities,collision terms, Pde, R-H relations. The only change is for the velocities associated to N_3 with only one $\pm\sqrt{2}$ instead of the two ±1 for x_j.

$$\text{ClassVIIbis}: \quad 2(4d-3)v_i, d \geq 2, \quad \underline{10v_i, 18v_i}, 26, 34, \cdots, \quad d_* = 1, 3\overline{b} = 2\sqrt{2}$$

N_1, N_2, M_1, M_4, M_4 like ClassVII, $N_3 : (0, x_2, \cdots, x_j, \cdots, x_d), x_j = 0$ except one $\pm\sqrt{2}$

$$\underline{10v_i}: \quad (\pm1,\pm1), \quad (0,\pm\sqrt{2}), \quad (\pm2,0),(0,\pm2) \tag{7.4bis}$$

$$\underline{18v_i}: (\pm1,\pm1,0),(\pm1,0,\pm1), \ (0,0,\pm\sqrt{2}),(0,\pm\sqrt{2}),0), \ (\pm2,0,0),(0,\pm2,0),(0,0,\pm2)$$

8.CLASS VIII: 7 DENSITIES $M_4, N_1, (M_2, N_3, R), N_2, M_4$ with x coordinates -2,-1,0,1,2 and **7 COLLISION TERMS.** The v_i are those of classVII plus zero for R. The first four collision terms are written down in (7.1) with for B the change of $1/d$ instead of $1/(d-1)$ in (7.4). The three new collision terms correspond to the collisions of particles N_iN_j with speeds $\sqrt{2}$ giving particles M_jR with speeds 0,2:

$$dA_4/2 = -M_4R + N_1^2, \ dA_1/2 = -M_1R + N_2^2, \ dA_3/2 = M_2R - N_3^2 \tag{8.1}$$

For simplicity we write down $-dA_2 = 2(M_2R - N_1N_2) = dA_3 + 2(N_1N_2 - N_3^2)$ which, being a linear combination of A_3, B, is not a new independent collision term. For this class with three speeds the R, N_i, M_i correspond respectively to speeds $0, \sqrt{2}, 2$. The associated velocities are $(0,0,\ldots,0)$ for R and written down in (7.4) for N_1, N_2, N_3, M_1, M_2, M_4.

$$\text{Pde} \quad : p_-N_1 = -B + C_1 + C_2 - A_4 - A_2, \quad p_+N_2 = -B - C_1 - C_2 - A_1 - A_2$$

$$N_{3t} = 2B/(d-2) + 2A_3, \quad M_{2t} = -C_1 + C_2 + D + A_2 - (d-2)A_3$$

$$q_+M_1 = (d-1)(2C_1 + D + A_1), \quad q_-M_4 = (d-1)(-2C_2 - D + A_4)$$

$$R_t = (d-1)(A_1 + A_4 + 2A_2 - 2(d-2)A_3) \tag{8.2}$$

$$\text{R-H}: \quad (d-1)(p_-N_1 + p_+N_2 + (d-2)N_{3t}) + R_t = 0$$

$$(d-1)(p_+N_2 - p_-N_1) + q_+M_1 - q_-M_4 = 0, \quad q_+M_1 + q_-M_4 + 2(d-1)M_{2t} = R_t$$

$$M = 2(d-1)(N_1 + N_2 + (d-2)N_3 + M_2) + M_1 + M_4 + R$$

$$E = 2(d-1)(N_1 + N_2 + (d-2)N_3 + 2M_2) + 2M_1 + 2M_4 \tag{8.3}$$

$$J = 2(d-1)(N_2 - N_1) + 2(M_1 - M_4), \ (2d^2+1)v_i, \ d \geq 3, \ \underline{19}v_i, 33, 51, \ldots \tag{8.4}$$

$$19v_i: \ (\pm1,\pm1,0),(\pm1,0,\pm1),(0,\pm1,\pm1),(\pm2,0,0),(0,\pm2,0),(0,0,\pm2),(0,0,0)$$

REFERENCES

[1] Gatignol R Lect. Notes Phys.36, Springer 1975; Platkowski T,Illner R SIAM Rev 30,213,1988; d'Humieres D Proceed.in Phys. ed. Manneville Springer-Verlag 1989,p186

[2] Cornille H Partially Integrable Eqs in Physics ed. Conte R Kluwer Academic Publishers, the Netherlands 1990, p39; Cornille H and Qian YH J.Stat.Phys. 61,683,1990; Cornille H to appear in Phys. Let.A and Discrete Models of Fluid Dynamics, Euromech 267,Figuera da Foz September 1990.

[3] d'Humieres D, Lallemand P and Frisch U Europh.Let. 2,291,1986; Cabannes H J.Mecan. 14,703,1975, Mech.Research Commun. 12,289,1985

[4] On the contrary, for gas mixture, see the recent work by Monaco R, Pandolfi Bianchi M, Platkowski T Acta Mechanica 84,175,1990.

Part IV

Hamiltonian Systems

Universality of the Long Time Tail in Hamiltonian Dynamics

– An Approach to the f^{-1} Noise of Quartz Oscillators –

Y. Aizawa and K. Tanaka

Department of Applied Mathematics,
Waseda University, 169 Shinjuku,
Tokyo, Japan.

Abstract: A universal law of the long time tail in nearly integrable systems is discussed and the onset mechanism of "non-stationary" and "$f^{-\nu}$ fluctuation" is briefly reviewed using a lattice vibration model. The mechanism of f^{-1} fluctuations in quartz crystals is numerically pursued and compared with several experimental observations such as phase noises, phonon number fluctuations and dissipation coefficients.

INTRODUCTION

One of the most striking chaotic phenomena in Hamiltonian dynamics is the universality of long time tails (LTT). The theory of dynamical systems succeeded to understand the basic mechanism leading to chaos; the stability criterion of quasi-periodic motions and the destabilized homoclinic structure. But the fundamental law imbedded in the LTT has not yet been fully elucidated. It is especially important to study the measure-theoretical and geometrical features of phase space to understand the generality of the LTT in Hamiltonian flow [1, 2].

The problem of the LTT should be recast from various angles with long perspectives, not only because it is a universal one which appears in a lot of different systems, but also because it is tightly connected to unsolved problems in statistical mechanics. The most remarkable one of them is the f^{-1} fluctuation which appears in the time course of transport coefficients and in phase noises [3]. Weakly irreversible processes are usually accompanied by f^{-1} fluctuations. Controversial points are followings; Dose the f^{-1} fluctuation occur in equilibrium or non-equilibrium ? Is the ergodicity guaranteed or not ? Furthermore, how can we decouple the characteristic time scales into two regimes—microscopic and macroscopic—? The separability of time scales is necessary for local equilibrium description, but it dose not hold in f^{-1} fluctuations.

Nearly integrable Hamiltonian systems are quite relevant models to explain the origin of f^{-1} spectral behaviors in many examples. In this article we will propose a new approach to phase noises in quartz crystal oscillators [4].

STAGNANT MOTIONS NEAR INVARIANT KAM TORI

Let us consider the nonlinear lattice vibration of one dimensional chain with the following Hamiltonian,

$$H = \frac{1}{2m}\sum_{i=1}^{N} p_i^2 + \frac{k}{2}\sum_{i=0}^{N}(q_{i+1} - q_i)^2 + \frac{\mu}{4}\sum_{i=0}^{N}(q_{i+1} - q_i)^4 \tag{1}$$

under fixed boundary conditions, i.e., $q_0 = q_{N+1} = 0$. The action-angle variable for each normal mode (I_j, θ_j) satisfies,

$$\begin{aligned} \dot{\theta}_j &= \omega_j + \mu f_j \\ \dot{I}_j &= \mu g_j \qquad (j = 1, 2, \cdots, N) \end{aligned} \tag{2}$$

where ω_j is the eigen frequency of the j-th normal mode, f_j and g_j are certain functions of θ_k's and I_k's.

Figure 1 shows an example of induction phenomena observed in this system [5]. Almost all energy is stored in the initially excited mode ($E_i = \omega_i \cdot I_i$) for long induction period T before it spreads over other modes. The "quasi-periodic" behavior during the induction period is considered to be the stagnant motion which was studied in the previous papers [1, 2].

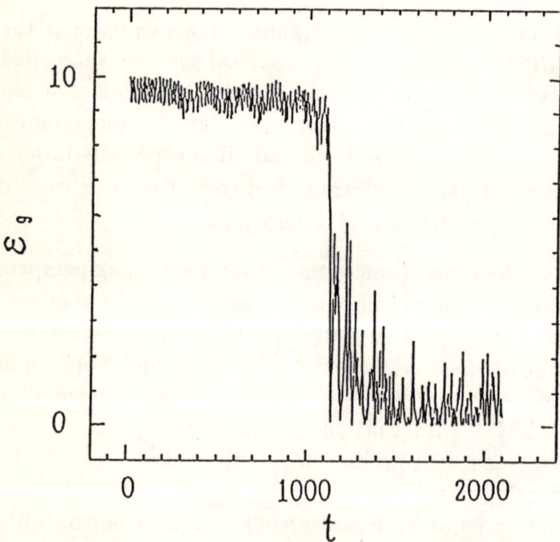

Fig. 1. The energy mixing process for $N = 16$ ($m = 1, k = 1, \mu = 0.8$). The initial energy is concentrated only in the 9-th mode E_9, and after a long induction period the energy is transfered successively over other modes E_7, E_{11}, E_{13}, etc. Owing to the symmetry of the system the energy of even modes is not excited

Then the distribution of the induction time $P(T)$ obeys,

$$P(T) \sim \frac{1}{T \log T} \qquad (T \gg 1) \tag{3}$$

for large T and the power spectrum $S(f)$ for the initially excited mode-energy, $\log(I_i(t)/I_i(0))$, becomes,

$$S(f) \sim f^{-2} \qquad (f \ll 1) \tag{4}$$

in the "quasi-periodic" pre-induction regime. These are surmised to be universal for the Arnold diffusion described by the Nekhoroshev theorem [1, 2, 6]. The essential point is that the stagnant motion is non-stationary, since the mean induction period as well as the mean recurrence time become infinity.

After the induction period the behavior looks stochastic, but it is not completely thermal equilibrium. We can not say anything about the detailed structure of the post-induction regime, but the law of energy equi-partition is surmised to be realized gradually after extremely long period.

In the weakly non-integrable system described by Eq. (1), we can expect that many KAM invariant tori remain stablly in phase space even if the system includes many degrees of freedom just like a single crystal. Then the phase space structure is very complicated due to the entanglement of many Arnold whiskers [7], and the topological diffusion in high dimensional phase space generates very slow motions which reveal $f^{-\nu}$ fluctuations.

PHASE NOISE AND PHONON–NUMBER FLUCTUATION

Here we assume that the dynamical state of quartz crystals is not quasi-periodic but is rather in the stochastic regime of the post-induction, and that many off-resonant modes are accompanied by an effective damping due to the loss in resonator with the free energy dissipation. Then the mode selected by the resonant effect in the resonator obeys Eq. (2), but on the other hand the decay-mode with the resonator loss should obey,

$$\begin{aligned} \dot{\theta}_l &= \omega_l + \gamma R_{1l} + \mu f_l \\ \dot{I}_l &= \gamma R_{2l} + \mu g_l \end{aligned} \tag{5}$$

where γ is a small friction constant, R_{1l} and R_{2l} are certain functons of θ_l and I_l. As the energy of the mode selected by the resonance in the resonator is much larger than that of the decay-mode, we will initially excite only the resonant mode and study the post-induction regime in order to compare with experiments. When the value of the friction constant γ is small enough, the induction phenomanon occurs in the same manner even in this dissipative system though the behavior in the post-induction regime is quite different from the frictionless system.

To realize the stable regime of the post-induction, we adopt a two-mode system which describes the lattice vibration by the discrete time mapping [2],

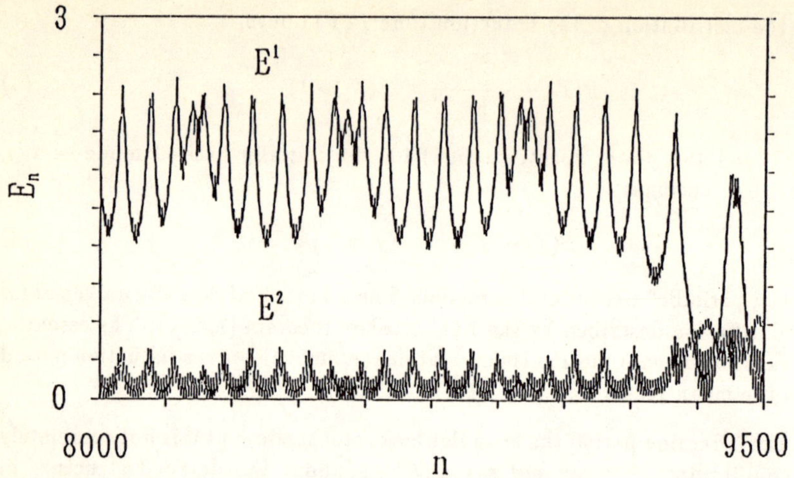

Fig. 2. The induction phenomenon in the discrete time lattice vibration described by Eq. (6) with $\mu = 0.0945$ and $\gamma = 10^{-7}$

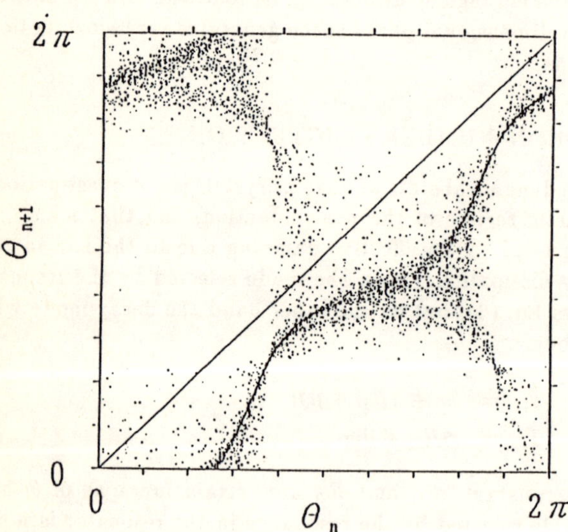

Fig. 3. The return map $\phi : \theta^1_n \mapsto \theta^1_{n+1}\,(mod.\ 2\pi)$ in the post-induction regime corresponding to Fig. 2

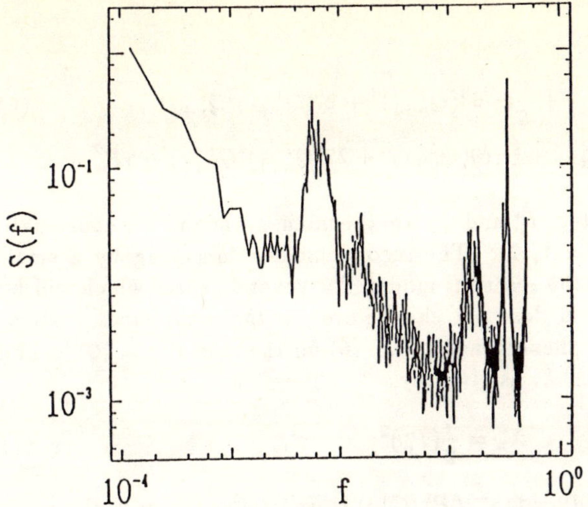

Fig. 4. The power spectrum for the phase noise $\Delta_n^1 (= \theta_{n+1}^1 - \theta_n^1)$ corresponding to Fig. 3, where the value of indeces is $\nu = 1.48$

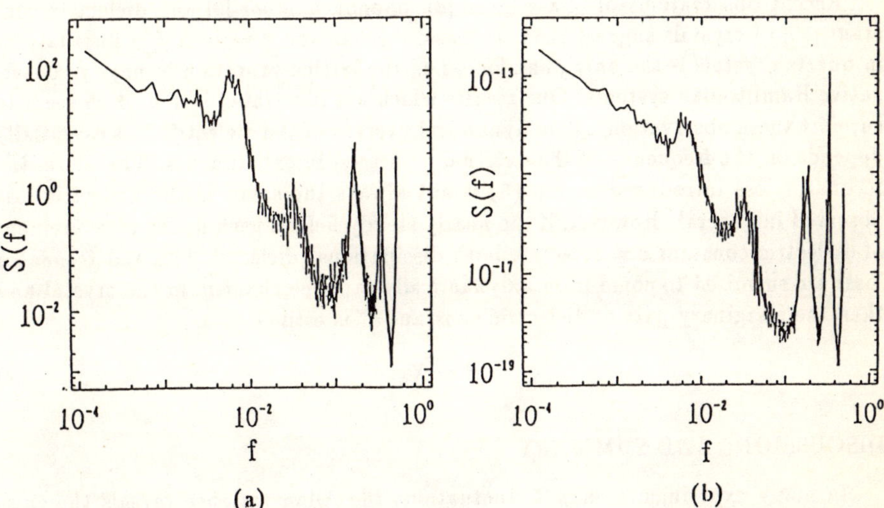

Fig. 5. Fluctuations of the phonon-number ($\propto E_n^1$) and the Rayleigh's dissipation function (X_n). (a) The power spectrum of $\log(E_n^1/E_0^1)$. (b) The power spectrum of X_n. Both power spectra $S(f)$ reveal $f^{-\nu}$ fluctuations; $\nu = 1.31$ and $= 1.23$ respectively

$$
\begin{aligned}
Q_{n+1}^1 &= Q_n^1 + P_n^1 \\
Q_{n+1}^2 &= Q_n^2 + P_n^2 \\
P_{n+1}^1 &= P_n^1 - Q_{n+1}^1 + \frac{\mu}{8}(-63(Q_{n+1}^2)^3 + 27(Q_{n+1}^1)^2 Q_{n+1}^2) \\
P_{n+1}^2 &= P_n^2 - 3Q_{n+1}^2 + \frac{\mu}{8}((9(Q_{n+1}^1)^3 + 27(Q_{n+1}^2)^2 Q_{n+1}^1) - \gamma P_n^2
\end{aligned} \tag{6}
$$

where n stands for the time step, P_n^i and Q_n^i are the momentum and the coordinate of the i-th normal mode (i = 1, 2). The second mode is decaying by a small damping, and the first one is the resonant mode. The general aspect which will be shown in the two-mode system dose not change even in the many-mode system. Figure 2 shows the induction phenomenon in Eq. (6) for the case of $\gamma = 10^{-7}$. The Rayleigh's dissipation function X_n is given by,

$$X_n = \frac{\gamma}{2}(P_n^2)^2 \tag{7}$$

The phase of the first mode ($\theta_n^1 = \tan^{-1}(P_n^1/Q_n^1)$) is well defined even in the post-induction regime as is shown in Fig. 3. The frequency fluctuation of the first mode ($\Delta_n^1 \equiv \theta_{n+1}^1 - \theta_n^1$) reveals the $f^{-\nu}$ ($\nu \simeq 1.48$) spectrum (Fig. 4).

The fluctuations of the phonon-number ($\propto E_n^1 = (P_n^1)^2 + Q_{n+1}^1 Q_n^1$) and the dissipation function X_n also reveal $f^{-\nu}$ spectra (Fig. 5).

Recent observations of phase noise [8], phonon-number [9] and dielectric constant [10] of crystals suggest that the basic mechanism of various f^{-1} fluctuations in quartz crystals is the only one alone, i.e., the lattice vabration in nearly conservative Hamiltonian systems. Our results which are illustrated in Fig. 4~5 seem to support these observations. The dynamical process of the dielectric loss essentially depends on the frequency of the external field used in experiments. Therefore, the resonator loss introduced in Eq. (5) is not always the same as the dielectric loss observed in general. However, if the nearly steady field is used in the measurement of dielectric constant $\varepsilon = \varepsilon_r + i\varepsilon^*$, both dissipations (dielectric loss and resonator loss) are surmised to come from a dynamically same mechanism in the crystal and then the imaginary part of dielectric constant ε^* is estimated as,

$$\varepsilon^* \propto X_n \tag{8}$$

DISCUSSIONS AND SUMMARY

In many experiments on $f^{-\nu}$ fluctuations the Allan variance reveals the existence of non-stationary regimes ($\nu > 1$), where the mean value of the time series itself changes very slowly. This situation suggests that the system under consideration is ceaselessly evolving, and the macroscopic state is always innovated through the evolution mode with a ghost frequency of $f = 0$. The evolution mode plays an essential role only in the asymptotic scaling regime ($f \to 0$). In nearly integrable Hamiltonian systems the same asymptotic regime is universally generated by the orbit in the boundary layer between KAM tori and chaos [1, 2], where the non-stationary stagnant motion reveals a kind of irreversibility. The irreversibility is the direct cause for the LTT in Hamiltonian flows, of which anomalous large deviation properties can be discussed in terms of the multi-ergodic theory [11].

In this paper we have tried to construct a unified physical picture to explain the generality of f^{-1} fluctuations in quartz crystals based on the Hamiltonian dynamics. In our approach a weak dissipation was introduced in order to derive the $f^{-\nu}$ noise of dielectric loss, but the dynamical origin of the dielecric loss is not so clear. Furthermore, it is not obvious whether the energy dissipation is really important or not for the onset of $f^{-\nu}$ phase noises in physical systems. According to recent simulations for many-mode lattice vibrations with continuous time, $f^{-\nu}$ spectral phase noises are widely observed even in non-dissipative cases [12]. In our models the value of the spectral index ν is not universal, but depends on the friction constant γ. Figure 6 shows the Allan variance of the phase nose Δ_n^1 for the case of $\gamma = 10^{-2}$ and $\mu = 0.0945$, where the so-called flicker floor is clearly observed, $\sigma_A^2(\tau) \sim O(\tau^0)$ for $\tau \gg 1$.

It is an especially important fact that fluctuations in nearly integrable Hamiltonian systems are non-stationary in general. Then the recurrence property in the sense of Poincaré-Hopf theorem dose not hold, and as the result the ergodicity also is not guaranteed in a global sense. These remarkable features may lead us to a new interpretation of the second law in thermodynamics as well as to a new paradigm of the evolution beyond the irreversibility based on the time-reversal symmetry breaking. The concept of non-stationarity is much wider than that of the time-asymmetry.

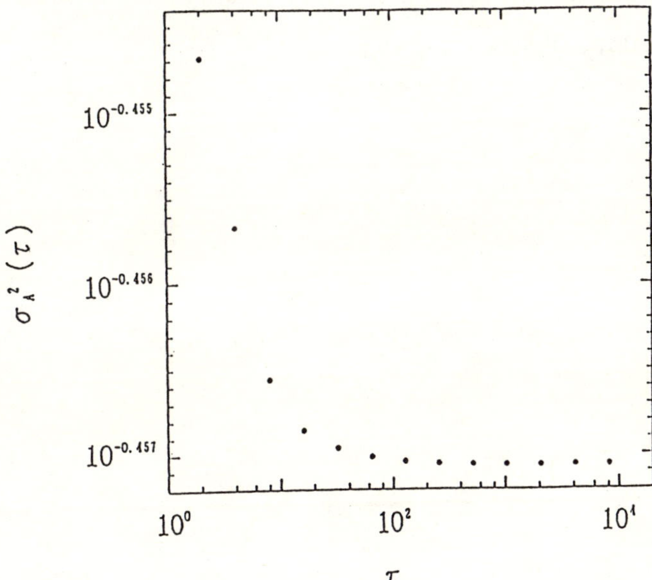

Fig. 6. The Allan variance of the phase noise Δ_n^1 for the case of $\gamma = 10^{-2}$ and $\mu = 0.0945$. Initial conditions are the same as used in Figs. 2–5.

References

[1] Y. Aizawa. Prog. Theor. Phys. , 81 (1989), 249.

[2] Y. Aizawa, Y. Kikuchi, T. Harayama, K. Yamamoto, M. Ota and K. Tanaka. Prog. Theor. Phys. Suppl. No.98 (1989), 36.

[3] M. B. Weissman. Rev. Mod. Phys., 60 (1988), 537.

[4] V. F. Kroupa, editor. *Frequency Stability: Fundamentals and Measurement.* IEEE PRESS, New York, 1983.

[5] N. Saito, N. Ooyama, Y. Aizawa and H. Hirooka. Prog. Theor. Phys. Suppl. No.45 (1970), 209.

[6] N. N. Nekhoroshev. Russian Math. Surveys, 32 (1977), 1.

[7] V. I. Arnold and A. Avez. *Ergodic Problems of Classical Mechanics.* Benjamin, New York, 1968.

[8] Y. Noguchi, Y. Teramachi and T. Musha. Jpn. J. Appl. Phys., 21 (1982), 61.

[9] T. Musha, B. Gábor and M. Shoji. Phys. Rev. Lett., 64 (1990), 2394.

[10] T. Musha, A. Nakajima and H. Akabane. Jpn. J. Appl. Phys. Pt.2, 27 (1988), L311.

[11] Y. Aizawa. Prog. Theor. Phys. Suppl. No.99 (1989), 149.

[12] Y. Aizawa, K. Tanaka, J. Tonotani. , in preparation.

Why some Hénon-Heiles Potentials are Integrable[†]

A.P. Fordy

Department of Applied Mathematical Studies and Centre for Nonlinear Studies,
University of Leeds,
Leeds, LS2 9JT, UK.

Abstract. The known integrable cases of the Hénon-Heiles system are shown to be closely related to the stationary flows of the known (and only) integrable 5th order (single component and polynomial) nonlinear evolution equations. This is further evidence that these are the only integrable cases of the Hénon-Heiles system. Lax pairs are deduced for each of the integrable cases and used to construct the constants of motion.

1. Introduction

The general Hénon-Heiles system and its energy are given by:

$$\ddot{q}_1+c_1q_1 = bq_1^2-aq_2^2 \tag{1.1a}$$

$$\ddot{q}_2+c_2q_2 = -2aq_1q_2 \tag{1.1b}$$

$$E = \frac{1}{2}(\dot{q}_1^2+\dot{q}_2^2+c_1q_1^2+c_2q_2^2)+aq_1q_2^2-\frac{1}{3}bq_1^3 . \tag{1.1c}$$

With parameter values $c_1=c_2=a=b=1$, this system is known to have irregular behaviour at high energies [2]. On the other hand there are three known integrable cases of the Hénon-Heiles system, characterised by certain ratios of the parameters:

$$\text{(i)} \quad a/b = -1, \; c_1=c_2 , \tag{1.2a}$$

$$\text{(ii)} \quad a/b = -1/6, \; c_1, c_2 \text{ arbitrary} , \tag{1.2b}$$

$$\text{(iii)} \quad a/b = -1/16, \; c_2=16c_1 . \tag{1.2c}$$

These parameter ratios were isolated by the Painlevé method [3-5], although second integrals were already known for the first two cases. There was some speculation in [3] that these are not the only integrable cases, but direct Hamiltonian analysis did not isolate any further ones [6,7] and this situation has not changed to date. In the present paper, the variable q_1

---

† The original title of the talk was "The Hénon-Heiles System Revisited". This is the title of [1], where further details can be found.

is shown to satisfy a fourth order ordinary differential equation for all values of a and b (c_1 and c_2 are set to zero without loss of generality). The three integrable cases are shown to correspond to three well known integrable fifth order nonlinear evolution equations.

The known spectral problems of these nonlinear evolution equations are used to derive the two constants of motion for each of the integrable Hénon-Heiles systems.

2. An Equation for q_1

To give the appearance of a stationary partial differential equation let:

$$u = q_1,\ u_x = \dot{q}_1\ ,\ \text{etc.} \tag{2.1}$$

and, without loss of generality set $c_1=c_2=0$. After some manipulation we obtain an autonomous, fourth order equation for u:

$$u_{xxxx}+(8a-2b)uu_{xx}-2(a+b)u_x^2-\frac{20}{3}abu^3 = -4aE\ . \tag{2.2}$$

This equation has the following scale symmetry:

$$x\to s^{-1}x\ ,\ u\to s^2u\ . \tag{2.3a}$$

Fifth order (polynomial and autonomous) nonlinear evolution equations with this scale symmetry are of the form:

$$u_t = u_{xxxxx}+Auu_{xxx}+Bu_xu_{xx}+Cu^2u_x = (u_{xxxx}+Auu_{xx}+\frac{1}{2}(B-A)u_x^2+\frac{1}{3}Cu^3)_x\ . \tag{2.3b}$$

There are three known integrable cases of this class of equations:

(i) $$u_t = (u_{xxxx}+5uu_{xx}+\frac{5}{3}u^3)_x\ , \tag{2.4a}$$

(ii) $$u_t = (u_{xxxx}+10uu_{xx}+5u_x^2+10u^3)_x\ , \tag{2.4b}$$

(iii) $$u_t = (u_{xxxx}+10uu_{xx}+\frac{15}{2}u_x^2+\frac{20}{3}u^3)_x\ , \tag{2.4c}$$

known respectively as the Sawada-Kotera equation, Lax's fifth order KdV flow and the Kaup-Kupershmidt equation (see [8] and references therein). According to [9] and [10] these are the <u>only</u> integrable cases. Comparing these with (2.2), we find:

(i) $$a = \frac{1}{2}\ ,\ b = -\frac{1}{2}\ , \tag{2.5a}$$

(ii) $$a = \frac{1}{2}\ ,\ b = -3\ , \tag{2.5b}$$

(iii) $$a = \frac{1}{4}\ ,\ b = -4\ . \tag{2.5c}$$

<u>Remark</u>

The integrability of (2.3b) is invariant under the scale change $u \to \sigma u$, which means that $a \to \sigma a$, $b \to \sigma b$, so that only the ratios a/b are important.

Thus equation (2.2) corresponds to the stationary flow of an integrable fifth order nonlinear evolution equation of the form (2.3b) only for the ratios (1.2), corresponding to the three integrable cases of the Hénon-Heiles system (1.1).

3. The Spectral Problems

Equations (2.4) have Lax form:

$$L_t = [P,L] \ , \tag{3.1}$$

where L and P are respectively (see [8]):

$$\text{(i)} \quad L = \partial^3 + u\partial \ , \quad P = 9\partial^5 + 15u\partial^3 + 15u_x\partial^2 + (5u^2 + 10u_{xx})\partial \ , \tag{3.2a}$$

$$\text{(ii)} \quad L = \partial^2 + u \ , \quad P = 16\partial^5 + 40u\partial^3 + 60u_x\partial^2 + (50u_{xx} + 30u^2)\partial + 15u_{xxx} + 30uu_x \ , \tag{3.2b}$$

$$\text{(iii)} \quad L = \partial^3 + 2u\partial + u_x \ , \quad P = 9\partial^5 + 30u\partial^3 + 45u_x\partial^2 + (20u^2 + 35u_{xx})\partial + 10u_{xxx} + 20uu_x \ . \tag{3.2c}$$

We can use these to derive a zero curvature representation for each of the integrable Hénon-Heiles systems. We can then derive the constants of motion. The details will only be presented for case (ii).

<u>Case (ii)</u>

We have:

$$\psi_{xx} + u\psi = \lambda\psi \tag{3.3a}$$

$$\psi_t = 16\psi_{xxxxx} + 40u\psi_{xxx} + 60u_x\psi_{xx} + (50u_{xx} + 30u^2)\psi_x + (15u_{xxx} + 30uu_x)\psi \ . \tag{3.3b}$$

<u>Remark</u>

An integro-differential Lax operator for this case was presented in [11]. The relationship between this and (3.3b) is explained in [1].

Using (3.3a), the time evolution can be rewritten:

$$\psi_t = (16\lambda^2 + 8\lambda u + 6u^2 + 2u_{xx})\psi_x - (4\lambda u_x + 6uu_x + u_{xxx})\psi \ . \tag{3.4a}$$

If u is independent of t (stationary case) we can separate t in ψ to give (when written in terms of q_1, q_2):

$$(16\lambda^2 + 8\lambda q_1 - q_2^2)\psi^{(1)} + (q_2\dot{q}_2 - 4\lambda\dot{q}_1)\psi^{(0)} = \mu\psi^{(0)} \ , \tag{3.4b}$$

where $(\psi^{(0)}, \psi^{(1)}) = (\psi, \psi_x)$ and μ is the separation constant. This leads to the following Lax form (written in terms of q_i, $p_i = \dot{q}_i$) for this Hénon-Heiles system:

$$\mathcal{L}\Psi = \mu\Psi \ , \quad \dot{\Psi} = \mathcal{P}\Psi \quad \Rightarrow \quad \dot{\mathcal{L}} = [\mathcal{P},\mathcal{L}] \ , \tag{3.5a}$$

where:

$$\mathcal{L} = \begin{bmatrix} q_2p_2-4\lambda p_1 & 16\lambda^2+8\lambda q_1-q_2^2 \\ 16\lambda^3-8\lambda^2q_1+\lambda(4q_1^2+q_2^2)+p_2^2 & 4\lambda p_1-q_2p_2 \end{bmatrix} , \mathcal{P} = \begin{bmatrix} 0 & 1 \\ \lambda-q_1 & 0 \end{bmatrix} \tag{3.5b}$$

and $\Psi=(\psi^{(0)},\psi^{(1)})^T$. The integrability conditions (3.5a) are just the (canonical) Hamiltonian form of this Hénon-Heiles system. The two constants of motion follow at once:

$$\mathrm{tr}\tfrac{1}{2}\mathcal{L}^2 = 256\lambda^5+32E\lambda^2-8K\lambda \ , \tag{3.6a}$$

where:

$$E = \tfrac{1}{2}(p_1^2+p_2^2)+q_1^3+\tfrac{1}{2}q_1q_2^2 \ , \quad K = q_2p_1p_2-q_1p_2^2+\tfrac{1}{8}q_2^4+\tfrac{1}{2}q_1^2q_2^2 \ . \tag{3.6b}$$

E is just the energy of (1.1c) and K is the second integral, first written down by Greene (see [3] and [4]).

<u>Remark</u>

This K is just the flux (written in the present co-ordinates) corresponding to the conserved density $\frac{1}{2}u^2$ of the KdV hierarchy.

Since (3.3a) is only second order, the case (ii) Hénon-Heiles system has a 2x2 matrix Lax pair. The other two cases possess 3x3 matrix Lax pairs, but their characteristic equations are still fifth degree in λ.

<u>Case (i)</u>

In this case we have:

$$(9\lambda-3\dot{q}_1)\psi^{(2)}-\tfrac{1}{2}(3q_1^2+q_2^2)\psi^{(1)}+6\lambda q_1\psi^{(0)} = \mu\psi^{(0)} \ , \tag{3.7a}$$

where $(\psi^{(0)},\psi^{(1)},\psi^{(2)})=(\psi,\psi_x,\psi_{xx})$. $\mathcal{L}$ is now a 3x3 matrix and:

$$\mathrm{tr}\tfrac{1}{3}\mathcal{L}^3 = 729\lambda^5-162E\lambda^3+K^2\lambda \ , \tag{3.7b}$$

where E and K are the two constants of motion:

$$E = \tfrac{1}{2}(\dot{q}_1^2+\dot{q}_2^2)+\tfrac{1}{2}q_1q_2^2+\tfrac{1}{6}q_1^3 \ , \quad K = 3\dot{q}_1\dot{q}_2+\tfrac{1}{2}q_2(3q_1^2+q_2^2) \ . \tag{3.7c}$$

This system separates in co-ordinates $q_1 \pm q_2$, these variables having individual energies $E_{\pm}$. In terms of these, $K=\frac{1}{2}(E_+-E_-)$.

<u>Case (iii)</u>

In this case ψ satisfies:

$$9\lambda\psi^{(2)}+\tfrac{1}{4}q_2^2\psi^{(1)}+(12\lambda q_1-\tfrac{1}{2}q_2\dot{q}_2)\psi^{(0)} = \mu\psi^{(0)}. \tag{3.8a}$$

The 3x3 matrix $\mathcal{L}$ gives the two constants of motion E and K [6]:

$$\mathrm{tr}\frac{1}{3}\mathcal{L}^3 = 243\lambda^5-162E\lambda^3+\frac{3}{4}K\lambda \ , \tag{3.8b}$$

where

$$E = \frac{1}{2}(\dot{q}_1^2+\dot{q}_2^2)+\frac{4}{3}q_1^3+\frac{1}{4}q_1q_2^2 \ , \ K = 3\dot{q}_2^4+3q_1q_2^2\dot{q}_2^2-q_2^3\dot{q}_1\dot{q}_2-\frac{1}{4}q_1^2q_2^4-\frac{1}{24}q_2^6 \ . \tag{3.8c}$$

4. The Modified Systems

The second of the Hénon-Heiles equations can be written:

$$q_1 = -\frac{q_{2xx}}{2aq_2} \ . \tag{4.1a}$$

Thus, defining v as follows, we have:

$$v = \frac{q_{2x}}{2aq_2} \ \Rightarrow \ u \equiv q_1 = -v_x-2av^2 \ . \tag{4.1b}$$

and v satisfies the fourth order equation:

$$v_{xxxx}+2(b+6a)v_xv_{xx}+4a(b-4a)(v^2v_{xx}+vv_x^2)-16a^3bv^5 = 0 \ . \tag{4.2}$$

The mapping (4.1b) thus takes solutions of (4.2) onto solutions of (2.2) for <u>all</u> values of a and b. For the three integrable cases, v then satisfies:

$$\text{(i)} \quad v_{xxxx}+5v_xv_{xx}-5(v^2v_{xx}+vv_x^2)+v^5 = 0 \ , \tag{4.3a}$$

$$\text{(ii)} \quad v_{xxxx}-10(v^2v_{xx}+vv_x^2)+6v^5 = 0 \ , \tag{4.3b}$$

$$\text{(iii)} \quad v_{xxxx}-5v_xv_{xx}-5(v^2v_{xx}+vv_x^2)+v^5 = 0 \ . \tag{4.3c}$$

<u>Remark</u>

The map (4.1b) is, in fact, a Poisson map [1].

5. Conclusions

It is important to understand that the integrable Hénon-Heiles systems <u>correspond to</u> but are not equivalent to the stationary flows of (2.4). For instance, we can add a term proportional to q_2^{-2} onto the case (ii) potential and still obtain the stationary fifth order KdV equation.

On the other hand, the fifth order KdV equation can be written as a canonical Hamiltonian system [12], with co-ordinates:

$$Q_1 = u \ , \ Q_2 = u_x \ , \ P_1 = -u_{xxx}-10uu_x \ , \ P_2 = u_{xx} \ . \tag{5.1}$$

The transformation from (Q_i,P_i) to the Hénon-Heiles variables (q_i,p_i) is not canonical. The fifth order stationary KdV equation is bi-Hamiltonian [13] if we embed it into a 5-dimensional system. By taking E to be a dynamical variable we can similarly embed the case (ii) Hénon-Heiles equation into a

5-dimensional, bi-Hamiltonian system:

$$\frac{d}{dt}(q_1,q_2,p_1,p_2,E)^T = (p_1,p_2,-3q_1^2-\frac{1}{2}q_2^2\ ,\ \omega,0)\ , \tag{5.2a}$$

where:

$$\omega = \frac{2}{q_2}\left(\frac{1}{2}p_1^2+\frac{1}{2}p_2^2+\frac{1}{2}q_1q_2^2+q_1^3-E\right) + q_1q_2\ , \tag{5.2b}$$

which reduces to the Hénon-Heiles system on the energy surface:

$$E = \frac{1}{2}(p_1^2+p_2^2)+\frac{1}{2}q_1q_2^2+q_1^3\ . \tag{5.2c}$$

This more general equation (5.2a) is of bi-Hamiltonian form:

$$\frac{d}{dt}(q_1,q_2,p_1,p_2,E)^T = \tilde{\pi}_1\nabla\tilde{h}_1 = \tilde{\pi}_2\nabla(-E)\ , \tag{5.3a}$$

where

$$\nabla = (\partial_{q_1},\partial_{q_2},\partial_{p_1},\partial_{p_2},\partial_E)^T\ ,\ \tilde{h}_1 = q_1p_1^2+q_2p_1p_2+2q_1^4+\frac{3}{2}q_1^2q_2^2+\frac{1}{8}q_2^4-2q_1E\ , \tag{5.3b}$$

and:

$$\tilde{\pi}_1 = \begin{pmatrix} 0 & 0 & 0 & \frac{1}{q_2} & 0 \\ 0 & 0 & \frac{1}{q_2} & -2q_1/q_2^2 & 0 \\ 0 & -\frac{1}{q_2} & 0 & p_2/q_2^2 & 0 \\ -\frac{1}{q_2} & 2q_1/q_2^2 & -p_2/q_2^2 & 0 & 0 \\ 0 & 0 & 0 & 0 & 0 \end{pmatrix}\ ,$$

$$\tag{5.3c}$$

$$\tilde{\pi}_2 = \begin{pmatrix} 0 & 0 & -1 & 0 & -p_1 \\ 0 & 0 & 0 & -1 & -p_2 \\ 1 & 0 & 0 & 0 & 3q_1^2+\frac{1}{2}q_2^2 \\ 0 & 1 & 0 & 0 & \omega \\ p_1 & p_2 & -(3q_1^2+\frac{1}{2}q_2^2) & -\omega & 0 \end{pmatrix}\ .$$

The constant $\tilde{h}_1$ reduces to K of (3.6b) on the energy surface (5.2c). With E now a function of q_1,p_1 the Poisson matrix $\tilde{\pi}_2$ is just canonical and $\nabla=(\partial_{q_1},\partial_{q_2},\partial_{p_1},\partial_{p_2})^T$.

This 5 dimensional extension of the Hénon-Heiles system is Poisson related to the 5 dimensional extension of the stationary fifth order KdV flow given in [13], with $\tilde{\pi}_1$ being the image of π_1. This Poisson relation is lost when we restrict to the energy surface (5.2c), as was done in our construction of (2.2) from the Hénon-Heiles system.

The Hamiltonian pair (5.3c) could equally be obtained from the modified system (4.3b) through the Miura map (4.1b), when extended to the 5-dimensional space. A similar construction would only give rise to one Poisson bracket for the other two cases since the Sawada-Kotera and Kaup-Kupershmidt equations are not bi-Hamiltonian. Indeed, they are reductions of the fifth order member of the (two-component) Boussinesq hierarchy, which should give rise to a larger integrable finite dimensional system which has separate reductions to cases (i) and (iii) of the Hénon-Heiles system.

References

[1] A.P.Fordy, The Hénon-Heiles system revisited, Physica D (in press).

[2] M.Hénon and C.Heiles, The applicability of the third integral of motion: some numerical experiments, Astron.J. 69, 73 (1964).

[3] Y.F.Chang, M.Tabor and J.Weiss, Analytic structure of the Hénon-Heiles Hamiltonian in integrable and nonintegrable regimes, J.Math.Phys. 23, 531-8 (1982).

[4] T.Bountis, H.Segur and F.Vivaldi, Integrable Hamiltonian systems and the Painlevé property, Phys.Rev. A25, 1257-64 (1982).

[5] B.Grammaticos, B.Dorizzi and R.Padjen, Painlevé property and integrals of motion for the Hénon-Heiles system, Phys.Letts.A. 89, 111-3 (1982).

[6] L.S.Hall, A theory of exact and approximate configuration invariants, Physica 8D, 90-116 (1983).

[7] A.P.Fordy, Hamiltonian symmetries of the Hénon-Heiles system, Phys.Letts.A. 97, 21-3 (1983).

[8] A.P.Fordy and J.Gibbons, Factorization of operators I. Miura transformations, J.Math.Phys. 21, 2508-10 (1980).

[9] A.Fujimoto and Y.Watanabe, Classification of fifth-order evolution equations with nontrivial symmetries, Math.Japonica 28, 43-65 (1983).

[10] A.V.Mikhailov, A.B.Shabat and V.V.Sokolov, The symmetry approach to the classification of integrable equations, preprint.

[11] A.C.Newell, M.Tabor and Y.B.Zeng, A unified approach to Painlevé expansions, Physica 29D, 1-68 (1987).

[12] O.I.Bogoyavlenski and S.P.Novikov, The relationship between Hamiltonian formalisms of stationary and non-stationary problems, Func.Anal. & Apps. 10, 8-11 (1976).

[13] M.Antonowicz, A.P.Fordy and S.Wojciechowski, Integrable stationary flows: Miura maps and bi-Hamiltonian structures, Phys.Letts.A. 124, 143-50 (1987).

Chaotic Pulsations in Variable Stars with Harmonic Mode Coupling

Frank Verheest[1] *and Willy Hereman*[2]

[1]Instituut voor theoretische mechanika, Rijksuniversiteit Gent
Krijgslaan 281, B–9000 Gent, Belgium
[2]Department of Mathematical and Computer Sciences, Colorado School of Mines
Golden, CO 80401, U. S. A.

Some variable stars show multi–periodic behaviour with, among others, peaks in their power spectra at harmonically spaced frequencies with ratios 1:2:4. Such modes are nonlinearly coupled by two second–harmonic interactions and their amplitude equations are shown by a Painlevé analysis to be nonintegrable in a hamiltonian sense. Chaotic phenomena are thus expected, especially when other modes and dissipation are included. An example of stars to which this might apply is G191–16 among the variable white dwarfs.

1. Introduction

Some variable stars, such as certain white dwarfs, have individual pulsation amplitudes which vary with time. Power spectra of *e.g.* nonradial oscillations in certain *ZZ Ceti* stars contain, among others, prominent peaks at harmonically spaced frequencies with ratios 1:2:3, 1:2:4 or 1:2:3:4. Simultaneous nonlinear interactions between regularly spaced frequencies can theoretically be modelled by very special cases of mode coupling, in that selection rules combine three–mode coupling with degenerate cases of second harmonic generation (SHG), where two modes coalesce at half the frequency of another one. The nonlinear amplitude equations are fundamentally different from the usual three–mode or SHG equations, but can be brought into hamiltonian form for the regime without dissipation. Painlevé analysis shows that they are *not* integrable, in contrast to the simple three–mode or SHG cases. The 1:2:3 frequency spacing is covered elsewhere (*Verheest, Hereman & Serras* 1990), while the case 1:2:4 will be addressed here. Eventually, the system evolves chaotically, depending very sensitively upon initial conditions.

Before applying these findings to real stars, two inherent restrictions have to be discussed. First of all, other stellar modes have been left out, and secondly the analysis was done in a conservative framework. Real stars are dissipative, but the motivation for nevertheless concluding something about real stars is that if a simpler, hamiltonian model already points to chaotic behaviour, then the inclusion of additional modes and/or dissipation cannot improve matters (*Verheest, Hereman & Serras* 1990). Other researches reached similar conclusions concerning the possibility of chaotic pulsations, especially for the case of harmonic ratios 1:2:4 indicative of period–doubling phenomena, notably for the variable white dwarfs PG1351+489 (*Goupil, Auvergne & Baglin* 1988) and G191–16 (*Vauclair et al.* 1989).

2. Basic formalism

Adhering to a model of three interacting modes, where the second mode is the second harmonic of the fundamental and the third mode is the second harmonic of the second, the combined selection rules for the angular frequencies ω_j ($j = 1, 2, 4$) read

$$\omega_2 = 2\omega_1, \qquad \omega_4 = 2\omega_2 = 4\omega_1. \tag{1}$$

Consequently, the third mode is the fourth harmonic of the fundamental, hence the mnemonic use of the index 4. The equations governing the slow time changes in the complex mode amplitudes a_j are different from either the usual three-mode or the simple SHG cases, and also from the 1:2:3 frequency spacing case (*Verheest, Hereman & Serras* 1990), and are of the form (*Verheest* 1976)

$$\dot{a}_1 = 2i\lambda\overline{a}_1 a_2, \qquad \dot{a}_2 = i\lambda a_1^2 + 2i\mu\overline{a}_2 a_4, \qquad \dot{a}_4 = i\mu a_2^2, \tag{2}$$

plus their complex conjugates. These equations are derivable from the Hamiltonian $H = \lambda(a_1^2\overline{a}_2 + \overline{a}_1^2 a_2) + \mu(a_2^2\overline{a}_4 + \overline{a}_2^2 a_4)$, in a description where complex conjugate variables are at the same time canonically conjugate (*Verheest* 1987). Besides the Hamiltonian, there is a second independent first integral $E = a_1\overline{a}_1 + 2a_2\overline{a}_2 + 4a_4\overline{a}_4$, a measure for the global mode energy, but that is not yet enough for complete integrability. Indications about integrability are given through a Painlevé analysis, see *e.g. Menyuk, Chen & Lee* (1983).

3. Painlevé analysis

We formally rewrite the system (2) as a set of ODEs in real variables a_j and A_j (coming from $\overline{a}_j$) by taking $\tau = it$ as a new independent variable, hence

$$\begin{aligned} \dot{a}_1 &= 2\lambda A_1 a_2, & \dot{A}_1 &= -2\lambda a_1 A_2, \\ \dot{a}_2 &= \lambda a_1^2 + 2\mu A_2 a_4, & \dot{A}_2 &= -\lambda A_1^2 - 2\mu a_2 A_4, \\ \dot{a}_4 &= \mu a_2^2, & \dot{A}_4 &= -\mu A_2^2. \end{aligned} \tag{3}$$

For the weak Painlevé test, we expand all variables as

$$a_j = c_j\tau^{p_j} + d_j\tau^{p_j+r} + \cdots, \qquad A_j = C_j\tau^{P_j} + D_j\tau^{P_j+r} + \cdots \tag{4}$$

and try to find the most singular terms in each equation (with $c_j \neq 0$ and $C_j \neq 0$). Using (4) in (3) gives

$$\begin{aligned} p_1 c_1 \tau^{p_1-1} &= 2\lambda C_1 c_2 \tau^{P_1+p_2}, \\ p_2 c_2 \tau^{p_2-1} &= \lambda c_1^2 \tau^{2p_1} + 2\mu C_2 c_4 \tau^{P_2+p_4}, \\ p_4 c_4 \tau^{p_4-1} &= \mu c_2^2 \tau^{2p_2}, \\ P_1 C_1 \tau^{P_1-1} &= -2\lambda c_1 C_2 \tau^{p_1+P_2}, \\ P_2 C_2 \tau^{P_2-1} &= -\lambda C_1^2 \tau^{2P_1} - 2\mu c_2 C_4 \tau^{p_2+P_4}, \\ P_4 C_4 \tau^{P_4-1} &= -\mu C_2^2 \tau^{2P_2}. \end{aligned} \tag{5}$$

All equations, except the second and the fifth, are easy to balance and give

$$\begin{aligned} p_2 &= p - P - 1, & P_2 &= -p + P - 1, \\ p_4 &= 2p - 2P - 1, & P_4 &= -2p + 2P - 1, \end{aligned} \tag{6}$$

if we call $p_1 = p$ and $P_1 = P$. Whether in the second and the fifth equation of (5) all terms are dominant or not (in the latter case $p + P + 2 > 0$ is inferred), we find already useful results in the form

$$c_4 = \frac{\mu c_2^2}{2p - 2P - 1}, \qquad C_4 = \frac{\mu C_2^2}{2p - 2P + 1}, \qquad c_2 C_2 = -\frac{pP}{4\lambda^2}. \tag{7}$$

In the case where $p + P + 2 > 0$, we try to determine the values for c_j and C_j from (5) reduced to its most singular terms. We calculate that

$$c_2 C_2 = \frac{(p - P + 1)(2p - 2P + 1)}{2\mu^2} = \frac{(p - P - 1)(2p - 2P - 1)}{2\mu^2} \tag{8}$$

and have now three expressions for $c_2 C_2$. The two expressions in (8) are only compatible provided $p = P$, in which case (7) and (8) yield

$$c_2 C_2 = -\frac{p^2}{4\lambda^2} = \frac{1}{2\mu^2}. \tag{9}$$

There are thus no acceptable values for p, λ or μ, and we are led to the case where all terms are dominant, implying that $P = -p - 2$. The values for c_j and C_j now have to be found from

$$\begin{array}{rclrcl} pc_1 &=& 2\lambda C_1 c_2, & (p+2)C_1 &=& 2\lambda c_1 C_2, \\ (2p+1)c_2 &=& \lambda c_1^2 + 2\mu C_2 c_4, & (2p+3)C_2 &=& \lambda C_1^2 + 2\mu c_2 C_4, \\ (4p+3)c_4 &=& \mu c_2^2, & (4p+5)C_4 &=& \mu C_2^2. \end{array} \tag{10}$$

Combining some of these equations in a judicious way gives us two expressions for $c_1 C_1$, namely

$$c_1 C_1 = \frac{p}{2\lambda^2}\left\{2p + 1 - \xi\frac{p(p+2)}{4p+3}\right\} = \frac{p+2}{2\lambda^2}\left\{2p + 3 - \xi\frac{p(p+2)}{4p+5}\right\}, \tag{11}$$

having put for brevity $\xi = \mu^2/2\lambda^2$. Both expressions in (11) are only compatible if either $p = -1$, so that all the weights become equal to -1, or

$$p = -1 \pm \sqrt{\frac{1-\xi}{16-\xi}} \equiv -1 \pm q. \tag{12}$$

This requires that $1-\xi$ and $16-\xi$ have the same sign, in other words that $0 < \xi \leq 1$ or that $16 < \xi$. We will return to this case further on, but first address the simpler case where all the weights are -1. From (7), (10) and (11) we obtain

$$\begin{array}{lll} c_2 = -\dfrac{\lambda c_1^2}{1+\xi}, & C_2 = \dfrac{\lambda C_1^2}{1+\xi}, & c_1 C_1 = \dfrac{1+\xi}{2\lambda^2}, \\ c_4 = -\dfrac{\mu\lambda^2 c_1^4}{(1+\xi)^2}, & C_4 = \dfrac{\mu\lambda^2 C_1^4}{(1+\xi)^2}. & \end{array} \tag{13}$$

We note that only one of the constants c_j and C_j can be taken arbitrary, either c_1 or C_1. The determination of the leading terms in (4) is thus complete and we move on to the next step in the Painlevé analysis, the determination of the resonances r. Keeping only the terms linear in d_j and D_j results in

$$\begin{array}{rcl}(1-r)d_1+2\lambda(C_1d_2+c_2D_1) & = & 0,\\ (1-r)d_2+2\lambda c_1d_1+2\mu(C_2d_4+c_4D_2) & = & 0,\\ (1-r)d_4+2\mu c_2d_2 & = & 0,\\ (1-r)D_1-2\lambda(c_1D_2+C_2d_1) & = & 0,\\ (1-r)D_2-2\lambda C_1D_1-2\mu(c_2D_4+C_4d_2) & = & 0,\\ (1-r)D_4-2\mu C_2D_2 & = & 0.\end{array} \tag{14}$$

For this linear and homogeneous system in d_j and D_j to have a non–trivial solution we must equate the determinant of the coefficient matrix to zero. Using (13), the possible values for r are then given by

$$(r+1)r(r-2)(r-3)(r-1-\xi)(r-1+\xi)=0. \tag{15}$$

A resonance $r=-1$ corresponds to an arbitrary shift in the origin of τ, and $r=0$ to an arbitrary constant (c_1 *or* C_1) in the most singular terms. For the system to be integrable, the other resonances have to be non–negative integers and realisable, as we are expanding in ascending powers of τ. This requires that

$$1+\xi\geq 0, \qquad 1-\xi\geq 0, \tag{16}$$

therefore $\xi=1$ or $\xi=0$. A value $\xi=1$ would lead to *two* resonances zero, imposing that one could choose *two* of the c_j and C_j arbitrary, which cannot be done, however. The other possibility, $\xi=0$ or $\mu=0$, corresponds to simple SHG, long known to be integrable. Hence, we must conclude that the double SHG studied here is not integrable. Numerical computations have yielded positive Lyapunov exponents, even when initially all the energy is in the fundamental, indicating chaotic behaviour (see *e.g. Steeb, Louw & Villet* 1987).

We return now to the case where $p=-1\pm q$, so that (7), (10) and (11) give

$$\begin{array}{lll} c_2=-\dfrac{\lambda c_1^2}{2(1+q)}, & C_2=\dfrac{\lambda C_1^2}{2(1-q)}, & c_1C_1=\dfrac{1-q^2}{\lambda^2},\\ c_4=\dfrac{\mu\lambda^2c_1^4}{4(4q-1)(1+q)^2}, & C_4=\dfrac{\mu\lambda^2C_1^4}{4(4q+1)(1-q)^2}. & \end{array} \tag{17}$$

As in the previous case, where all the weights were -1, only one of the constants c_j and C_j can be taken arbitrary and we arrive at the system

$$\begin{array}{rcl}(1-r-q)d_1+2\lambda(C_1d_2+c_2D_1) & = & 0,\\ (1-r-2q)d_2+2\lambda c_1d_1+2\mu(C_2d_4+c_4D_2) & = & 0,\\ (1-r-4q)d_4+2\mu c_2d_2 & = & 0,\\ (1-r+q)D_1-2\lambda(c_1D_2+C_2d_1) & = & 0,\\ (1-r+2q)D_2-2\lambda C_1D_1-2\mu(c_2D_4+C_4d_2) & = & 0,\\ (1-r+4q)D_4-2\mu C_2D_2 & = & 0,\end{array} \tag{18}$$

for the d_j and D_j. The resonances r are obtained from

$$(r+1)r(r-2)(r-3)(r-1-\sqrt{1+60q^2})(r-1+\sqrt{1+60q^2})=0, \tag{19}$$

leading to nonintegrability on similar grounds as in the previous case. Because the system studied is not integrable, over long time periods we expect irregular phenomena, in sharp contrast to the usual periodic three–mode interactions. The presence of other modes would only increase the complexity and hence enforce the nonintegrability of the model, and so would dissipation.

4. Chaotic pulsations

Stars in which the power spectrum *includes* peaks at a fundamental frequency and its second and fourth harmonics (and hence for which our conclusions might be relevant) include certain *ZZ Ceti* stars. These are single, normal (hydrogen) DA white dwarfs with luminosity variations and hence denoted by DAV. There are also pulsating helium white dwarf (DBV) stars (see *Winget* (1988) for a review of these compact pulsators). The most pronounced of the relevant DAV stars is G191–16, with a light curve dominated by a frequency $\nu_0 = 1.12$ mHz and its harmonics at $2\nu_0$, $3\nu_0$ and $4\nu_0$ (*Vauclair et al.* 1989). Another example is the DBV star PG1351+489 with $\nu_0 = 1.028$ mHz (*Goupil, Auvergne & Baglin* 1988). Since both the special cases with frequency ratios 1:2:3 and 1:2:4 have now been shown to be nonintegrable, the general case with 1:2:3:4 spacing cannot be integrable either. Other *ZZ Ceti* stars which include in their spectra the ratios 1:2:3:4 are *VY Hor* (= BPM31594) (with frequencies at 1.620, 3.240, 4.864 and 6.484 mHz) and its northern hemisphere twin *BG CVn* = GD154 (*O'Donoghue* 1986). The conclusions about deterministic low–order chaos can at this stage only be indicative, in view of the few stars studied so far observationally in any serious detail (*Perdang* 1990).

Acknowledgments

It is a pleasure for FV to thank the National Fund for Scientific Research (Belgium) for a research grant.

References

Goupil M J, Auvergne M and Baglin A 1988 *Astron. Astrophys.* **196**, L13–L16

Menyuk C F, Chen H H and Lee Y C 1983 *Phys. Rev. A* **27**, 1597–1611 & *J. Math. Phys.* **24**, 1073–1079

O'Donoghue D 1986 in: *Seismology of the Sun and the Distant Stars* (ed. D O Gough, Reidel, Dordrecht) 467–472

Perdang J 1990 in: *Rapid variability of OB–stars: Nature and diagnostic value* (in press)

Steeb W–H, Louw J A and Villet C M 1987 *Aust. J. Phys.* **40**, 587–592

Vauclair G, Goupil M J, Baglin A, Auvergne M and Chevreton M 1989 *Astron. Astrophys.* **215**, L17–L20

Verheest F 1976 *Plasma Phys.* **18**, 225–234

Verheest F 1987 *J. Phys. A: Math. Gen.* **20**, 103–110

Verheest F, Hereman W and Serras H 1990 *Mon. Not. R. astron. Soc.* **245**, 392–396

Winget D E 1988 in: *Advances in Helio- and Asteroseismology* (eds. J Christensen–Dalsgaard and S Fransen, Reidel, Dordrecht) 305–324

Canonical Forms for Compatible BiHamiltonian Systems

P.J. Olver†

School of Mathematics, University of Minnesota,
Minneapolis, MN 55455,
USA.

In this note, I will review recent results on the canonical forms for compatible biHamiltonian systems of complex-analytic ordinary differential equations based on Turiel's classification, [8], of compatible non-degenerate Hamiltonian pairs. The resulting explicit forms for general biHamiltonian systems in canonical coordinates lead to a complete analysis of their integrability. More details of these results can be found in the author's paper [6].

A system of differential equations is called *biHamiltonian*, [3], [5], if it can be written in Hamiltonian form in two distinct ways:

$$\dot{x} = J_1 \nabla H_1 = J_2 \nabla H_0 . \qquad (1)$$

Here $J_1(x)$, $J_2(x)$ are Hamiltonian operators (matrices), not constant multiples of each other, determining Poisson brackets: $\{F, G\}_\nu = \nabla F^T J_\nu(x) \nabla G$. The Hamiltonian pair J_1, J_2 is *compatible* if $J_1 + J_2$ also determines a Poisson bracket, i.e. the Jacobi identity holds. The pair is *nondegenerate* if one of the Poisson structures is symplectic. According to the fundamental theorem of Magri, [3], any biHamiltonian system associated with a nondegenerate Hamiltonian pair induces a hierarchy of commuting Hamiltonians and flows, and, provided enough of these Hamiltonians are functionally independent, is therefore completely integrable.

Theorem 1. Suppose J_1, J_2 form a compatible Hamiltonian pair, with J_1 symplectic. Given a biHamiltonian system (1), there exists a hierarchy of Hamiltonian functions $H_0, H_1, H_2, H_3, \ldots,$ all in involution with respect to either Poisson bracket, $\{H_j, H_k\}_\nu = 0$, and generating mutually commuting biHamiltonian flows

$$\dot{x} = J_1 \nabla H_k = J_2 \nabla H_{k-1} . \qquad (2)$$

† *Research supported in part by NSF Grant DMS 89-01600.*

We classify Hamiltonian pairs pointwise according to the algebraic invariants of the skew-symmetric matrix pencil $\lambda J_1(x) + \mu J_2(x)$ at each x. According to the Weierstrass theory, cf. [1], the complete algebraic invariants of a non-degenerate matrix pencil are provided by the eigenvalues, elementary divisors and Segre characteristic. (Degenerate pairs of skew-symmetric matrices are handled by the more detailed Kronecker theory.) A pencil is called *elementary* if it has just one complex eigenvalue, and *irreducible* if it has Segre characteristic [(nn)], analogous to a single Jordan block. Every non-degenerate complex matrix pencil is algebraically the direct sum of irreducible matrix pencils. (For simplicity, we restrict our attention to complex-analytic systems in this paper, although the real case offers little additional difficulty.)

The algebraic invariants, i.e. eigenvalues, elementary divisors and Segre characteristic, of a Hamiltonian pair are invariant under the flow of any associated biHamiltonian system. A Hamiltonian pair is *generic* on a domain M if it has constant Segre characteristic, and the number of functionally independent eigenvalues does not change on M. The main classification theorem for nondegenerate biHamiltonian systems is the following:

Theorem. Every generic non-degenerate, compatible Hamiltonian pair can be locally expressed as a Cartesian product of elementary Hamiltonian pairs. Every associated biHamiltonian system decomposes into independent subsystems corresponding to the elementary sub-pairs, each of which consists of an autonomous Hamiltonian system whose dimension is twice the number of irreducible sub-pairs for the given eigenvalue, coupled with a sequence of linear, non-autonomous Hamiltonian systems.

When an eigenvalue is constant, the elementary sub-pair decomposes into a Cartesian product of irreducible sub-pairs; however, this decomposition does *not* hold in the case of non-constant eigenvalues. We will now present the details of the Turiel classification and the structure of associated biHamiltonian systems.

Without loss of generality, we may assume that neither 0 nor ∞ is an eigenvalue, so that the Hamiltonian pair is determined by two compatible symplectic Hamiltonian operators. (Otherwise, replace J_1, J_2 by two other linearly independent members of the corresponding pencil.) Darboux' theorem, [5; Theorem 6.22], implies that we can write the first Hamiltonian operator in canonical form

$$J_1 = \begin{pmatrix} 0 & I \\ -I & 0 \end{pmatrix}, \tag{3}$$

relative to canonically conjugate coordinates $(\mathbf{p}, \mathbf{q})$. Therefore, only the canonical form of the second Hamiltonian operator needs to be explicitly indicated.

Given a Hamiltonian pair J_1, J_2, any associated biHamiltonian systems is a solution to the linear system of partial differential equations

$$\nabla H_1 = M \nabla H_0, \qquad M = J_1^{-1} \cdot J_2, \tag{4}$$

where M is the transpose of the recursion operator, [5]. We remark here that the simple system of differential equations (4), which arises in a surprising number of different contexts, is not well understood, except when the matrix M is constant, in which case the general solution can be found in [2]. In the present case, the solutions all have a similar pattern. On any convex open subdomain, the two Hamiltonians H_0, H_1 are given as a sum of "basic" Hamiltonians $H_0^{(k)}, H_1^{(k)}$, which are individually solutions to (4):

$$H_0(\mathbf{x}) = H_0^{(0)}(\mathbf{x}) + H_0^{(1)}(\mathbf{x}) + \ldots + H_0^{(n)}(\mathbf{x}), \quad H_1(\mathbf{x}) = H_1^{(0)}(\mathbf{x}) + H_1^{(1)}(\mathbf{x}) + \ldots + H_1^{(n)}(\mathbf{x}).$$

Moreover, each basic pair $H_0^{(k)}, H_1^{(k)}$, can be most simply expressed in terms of the derivatives with respect to a parameter s evaluated at $s = 0$ of a single arbitrary analytic function $F(\xi_1(\mathbf{x}, s), \ldots, \xi_m(\mathbf{x}, s))$ depending on certain parameterized variables $\xi_j(\mathbf{x}, s)$. We can therefore summarize the general classification results in this convenient form.

I) Irreducible, Constant Eigenvalue Pairs,

Canonical coordinates: $\quad (\mathbf{p}, \mathbf{q}) = (p_0, p_1, \ldots, p_n, q_0, q_1, \ldots, q_n), \qquad n \geq 0.$

Second Hamiltonian operator:

$$J_2 = \begin{pmatrix} 0 & \lambda I + U \\ -\lambda I - U^T & 0 \end{pmatrix},$$

where $\lambda I + U$ denotes the irreducible $(n+1) \times (n+1)$ Jordan block

$$\lambda I + U = \begin{pmatrix} \lambda & 1 & & & \\ & \lambda & 1 & & \\ & & \lambda & 1 & \\ & & & \cdots & \cdots \end{pmatrix}.$$

Parametrized variables:

$$\pi(s) = p_0 + s\,p_1 + s^2\,p_2 + \ldots + s^n\,p_n\,, \qquad \varpi(s) = q_n + s\,q_{n-1} + s^2\,q_{n-2} + \ldots + s^n\,q_0\,.$$

Basic Hamiltonians:

$$H_0^{(k)}(\mathbf{x}) = \mu\,\frac{\partial^k}{\partial s^k}F_k(\pi(s),\varpi(s))\bigg|_{s=0} + k\,\frac{\partial^{k-1}}{\partial s^{k-1}}F_k(\pi(s),\varpi(s))\bigg|_{s=0}\,,$$

$$H_1^{(k)}(\mathbf{x}) = \frac{\partial^k}{\partial s^k}F_k(\pi(s),\varpi(s))\bigg|_{s=0}\,, \qquad 0 \le k \le n\,.$$

Note that these Hamiltonians are polynomials in the "minor variables" $p_1,\ldots,p_n, q_0,\ldots,q_{n-1}$, whose coefficients are certain derivatives of the arbitrary smooth functions $F_k(p_0, q_n)$ of the remaining two "major variables" p_0, q_n. This implies, cf. [6], that any biHamiltonian system corresponding to an irreducible, constant eigenvalue Hamiltonian pair is completely integrable, since it can be reduced to a single two-dimensional (planar) autonomous Hamiltonian system for the major variables, with Hamiltonian $n!\,F_n(p_0, q_n)$. (Curiously, the major variables are *not* canonically conjugate for the standard symplectic structure given by J_1, nor are they conjugate for J_2.) The time evolution of the minor variables is then determined by successively solving a sequence of orced linear planar Hamiltonian systems in the variables p_k, q_{n-k}.

II. Elementary, Constant Eigenvalue Pairs.

Canonical coordinates: $\quad (\mathbf{p}, \mathbf{q}) = (\mathbf{p}^1, \ldots, \mathbf{p}^m, \mathbf{q}^1, \ldots, \mathbf{q}^m),$

$$\mathbf{p}^i = (p_0^i, \ldots, p_{n_i}^i), \quad \mathbf{q}^i = (q_0^i, \ldots, q_{n_i}^i), \qquad \text{where} \qquad n_1 \ge n_2 \ge \ldots \ge n_k \ge 0\,.$$

Second Hamiltonian operator:

$$J_2 = \begin{pmatrix} 0 & & 0 & \lambda I + U_1 & & 0 \\ & \cdots & & & \cdots & \\ 0 & & 0 & 0 & & \lambda I + U_m \\ -\lambda I - U_1^T & & 0 & 0 & & 0 \\ & \cdots & & & \cdots & \\ 0 & & -\lambda I - U_m^T & 0 & & 0 \end{pmatrix},$$

where $\lambda I + U_i$ denotes an irreducible $(n_i + 1) \times (n_i + 1)$ Jordan block as above.

Parametrized variables:

$$\pi^i(s) = p_0^i + s\, p_1^i + s^2\, p_2^i + \dots + s^{n_i}\, p_{n_i}^i, \qquad \varpi^i(s) = q_{n_i}^i + s\, q_{n_i-1}^i + s^2\, q_{n_i-2}^i + \dots + s^{n_i}\, q_0^i .$$

Basic Hamiltonians:

$$H_0^{(k)}(\mathbf{x}) = \mu \frac{\partial^k}{\partial s^k} F_k(\pi^1(s), \varpi^1(s), \dots, \pi^{m_k}(s), \varpi^{m_k}(s)) \Big|_{s=0} +$$

$$+ \; k \frac{\partial^{k-1}}{\partial s^{k-1}} F_k(\pi^1(s), \varpi^1(s), \dots, \pi^{m_k}(s), \varpi^{m_k}(s)) \Big|_{s=0} ,$$

$$H_1^{(k)}(\mathbf{x}) = \frac{\partial^k}{\partial s^k} F_k(\pi^1(s), \varpi^1(s), \dots, \pi^{m_k}(s), \varpi^{m_k}(s)) \Big|_{s=0} . \qquad 0 \le k \le n_1 .$$

Here m_k denotes the number of n_i with $n_i \ge k$, i.e. the number of irreducible sub-pairs of dimension greater than $2k+1$; in particular $m_0 = m$.

As in the irreducible case, the Hamiltonians are polynomials in the minor variables p_j^i, $q_{n_i-j}^i$, $j \ge 1$, whose coefficients are certain derivatives of arbitrary functions of the major variables p_0^i, $q_{n_i}^i$. Thus, such a biHamiltonian system reduces to an autonomous $2m$ – dimensional Hamiltonian system in the major variables, followed by linear non-autonomous Hamiltonian systems in the appropriate minor variables p_k^i, $q_{n_i-k}^i$, $n_i \ge k \ge 1$.

III. Irreducible, Non-constant Eigenvalue Pairs.

Canonical coordinates: $(\mathbf{p}, \mathbf{q}) = (p_0, p_1, \dots, p_n, q_0, q_1, \dots, q_n)$, $\qquad n \ge 0$.

Second Hamiltonian operator:

$$J_2 = \begin{pmatrix} 0 & P(\mathbf{p}) \\ -P(\mathbf{p})^T & 0 \end{pmatrix},$$

where $P(\mathbf{p})$ denotes the $(n+1) \times (n+1)$ banded upper triangular matrix

$$P_n(\mathbf{p}) \;=\; P(\mathbf{p}) \;=\; \begin{pmatrix} p_0 & p_1 & p_2 & p_3 & \cdots & p_n \\ & p_0 & p_1 & p_2 & \cdots & p_{n-1} \\ & & p_0 & p_1 & \cdots & \cdots \\ & & & p_0 & \cdots & \cdots \\ & & & & \cdots & \cdots \\ & & & & & p_0 \end{pmatrix}. \tag{5}$$

(Interestingly, both $P(\mathbf{p})$ and its inverse determine isomorphic Hamiltonian operators!)

Parametrized variables:

$$\pi(s) \;=\; p_0 + s\,p_1 + s^2\,p_2 + \ldots + s^n\,p_n\,, \qquad \varpi(s) \;=\; q_n + s\,q_{n-1} + s^2\,q_{n-2} + \ldots + s^n\,q_0\,.$$

Basic Hamiltonians:

$$H_0^{(-1)}(\mathbf{x}) \;=\; \tilde{h}(p_0)\,, \qquad H_1^{(-1)}(\mathbf{x}) \;=\; h(p_0)\,, \qquad \text{where} \qquad \tilde{h}'(\xi) \;=\; \xi\,h'(\xi)\,,$$

$$H_0^{(k)}(\mathbf{x}) \;=\; \frac{\partial^k}{\partial s^k}\left\{\,\pi(s)\,\pi'(s)\,F_k(\pi(s),\varpi(s))\right\}\Bigg|_{s=0},$$

$$H_1^{(k)}(\mathbf{x}) \;=\; \frac{\partial^k}{\partial s^k}\left\{\,\pi'(s)\,F_k(\pi(s),\varpi(s))\right\}\Bigg|_{s=0}, \qquad 0 \le k \le n-1\,.$$

Here $\pi'(s)$ is the derivative of π with respect to s.

In this case, the eigenvalue is a constant, hence p_0 is a first integral. Once its value is fixed, the other minor variable q_n is determined by solving a single autonomous ordinary differential equation. The remaining minor variables $p_1,\ldots,p_n, q_0,\ldots,q_{n-1}$ satisfy a sequence of forced, linear planar Hamiltonian systems.

IV. Elementary, Non-constant Eigenvalue Pairs.

Canonical coordinates: $(\mathbf{p}, \mathbf{q}) = (p_0, \mathbf{p}^1, \ldots, \mathbf{p}^m, q_0, \mathbf{q}^1, \ldots, \mathbf{q}^m)\,, \qquad m \ge 2\,,$

$$\mathbf{p}^i = (p_1^i, \ldots, p_{n_i}^i), \quad \mathbf{q}^i = (q_1^i, \ldots, q_{n_i}^i), \qquad \text{where} \qquad n_1 \ge n_2 \ge \ldots \ge n_k \ge 1\,.$$

Second Hamiltonian operator:

$$J_2 = \begin{pmatrix} 0 & \mathbf{P}^*(\mathbf{p}) \\ -\mathbf{P}^*(\mathbf{p})^T & 0 \end{pmatrix},$$

where

$$\mathbf{P}^*(\mathbf{p}) = \begin{pmatrix} p_0 & \mathbf{p}^1 & \mathbf{p}^2 & \cdots & \mathbf{p}^m \\ & P_{n_1-1}(\hat{\mathbf{p}}^1) & 0 & \cdots & 0 \\ & & P_{n_2-1}(\hat{\mathbf{p}}^2) & & \\ & & & \cdots & \\ & & & & P_{n_m-1}(\hat{\mathbf{p}}^m) \end{pmatrix}.$$

Here $\hat{\mathbf{p}}^i = (p_0, p^i_1, \ldots, p^i_{n_i-1})$, and the P_{n_i-1}'s are as given in (5). Note that this particular pair is algebraically reducible, but cannot be decoupled using canonical transformations.

Parametrized variables:

$$\pi^i(s) = p_0 + s\,p^i_1 + s^2\,p^i_2 + \ldots + s^{n_i}\,p^i_{n_i}, \qquad \omega^i(s) = q^i_{n_i} + s\,q^i_{n_i-1} + \ldots + s^{n_i-1}\,q^i_1, \quad i \geq 1,$$

$$\mu^j(s) = \frac{z^j(s)}{s}, \qquad \sigma^j(s) = \varpi(z^j(s)), \qquad \text{where } z^j(s) \text{ solves } \pi^j(z) = \pi^1(s), \qquad j \geq 2.$$

Using the Lagrange inversion formula, [4], the latter two parametrized variables have the alternative expansions

$$\mu^j(s) = \sum_{n=0}^{n_j-1} \frac{s^n\,(\zeta^1(s))^{n+1}}{(n+1)!} \left.\frac{d^n}{dt^n}\,\frac{1}{(\zeta^j(t))^{n+1}}\right|_{t=0},$$

$$\sigma^j(s) = q^j_{n_j} + \sum_{n=0}^{n_j-1} \frac{s^n\,(\zeta^1(s))^n}{n!} \left.\frac{d^n}{dt^n}\,\frac{1}{(\zeta^j(t))^n}\,\frac{d\omega^j(t)}{dt}\right|_{t=0},$$

where $\zeta^i(s) = (\pi^i(s) - p_0)\,/\,s$. These expansions can be expressed in terms of the remarkable nonlinear series differential operator

$$\mathcal{D} = D^{-1} : e^{s\,D\,u} : D = 1 + \sum_{n=1}^{\infty} \frac{s^n}{n!}\,D^{n-1}\,u^n\,D, \qquad D = \frac{d}{dt}, \quad u = u(t),$$

where the colons denote *normal ordering* of the non-commuting operators D and u, which is analogous to the so-called "Wick ordering" in quantum mechanics. This operator has the surprising property that it commutes with *any* analytic function $\Phi(u)$, i.e. $\mathcal{D}\,\Phi(u) = \Phi(\mathcal{D}\,u)$! See [7] for details and applications of this operator in combinatorics, orthogonal polynomials and new higher order derivative identities.

Basic Hamiltonians:

$$H_0^{(-1)}(\mathbf{x}) = \tilde{h}(p_0)\,, \qquad H_1^{(-1)}(\mathbf{x}) = h(p_0)\,, \qquad \text{where} \quad \tilde{h}'(\xi) = \xi\, h'(\xi)\,,$$

$$H_0^{(k)}(\mathbf{x}) = \frac{\partial^k}{\partial s^k}\left\{ s\,\zeta^1(s)\,\frac{d\pi^1}{ds}\,F_k(\pi^1(s),\mu^2(s),\ldots,\mu^{m_k}(s),\omega^1(s),\sigma^2(s),\ldots,\sigma^{m_k}(s))\right\}\Bigg|_{s=0},$$

$$H_1^{(k)}(\mathbf{x}) = \frac{\partial^k}{\partial s^k}\left\{ \frac{d\pi^1}{ds}\,F_k(\pi^1(s),\,\mu^2(s),\,\ldots,\,\mu^{m_k}(s),\,\omega^1(s),\,\sigma^2(s),\,\ldots,\,\sigma^{m_k}(s))\right\}\Bigg|_{s=0},$$

for $0 \le k \le n_1 - 1$, where $m_k = \#\,\{\, n_i \ge k \,\}$.

In general, such biHamiltonian systems reduce to the integration of a $2\,m-2$ dimensional autonomous Hamiltonian system for the coordinates $p_1^i,\ q_{n_i}^i$, $i = 1, \ldots, m$, followed by a sequence of forced linear Hamiltonian systems. The final coordinate q_0 is determined by quadrature. Actually, the initial Hamiltonian system can be reduced in order to $2\,m-3$ since it only involves the homogeneous ratios of momenta $r^i = p_1^i \,/\, p_1^1$, $i \ge 2$, as can be seen from the second formula for μ^j.

Further work: The key outstanding problem in this area is to determine similar canonical forms in degenerate compatible biHamiltonian systems. Unfortunately, Turiel's approach, which is fundamentally tied to the covariant differential form framework for symplectic structures, does not appear to readily generalize, since degenerate Poisson structures can only be readily expressed in the contravariant language of bi-vector fields, [5].

Acknowledgements: I would like to thank the Institute for Mathematics and its Applications (I.M.A.) for providing additional support, and Darryl Holm, Niky Kamran, Yvette Kosmann-Schwarzbach, Franco Magri, and Francisco-Javier Turiel for encouragement and vital comments.

References

[1] Gantmacher, F.R., *The Theory of Matrices*, vol. 2, Chelsea Publ. Co., New York, 1959.

[2] Jodeit, M. and Olver, P.J., On the equation grad f = M grad g, *Proc. Roy. Soc. Edinburgh*, to appear.

[3] Magri, F., A simple model of the integrable Hamiltonian equation, *J. Math. Phys.* **19** (1978) 1156-1162.

[4] Melzak, Z.A., *Companion to Concrete Mathematics*, Wiley-Interscience, New York, 1973.

[5] Olver, P.J., *Applications of Lie Groups to Differential Equations*, Graduate Texts in Mathematics, vol. 107, Springer-Verlag, New York, 1986.

[6] Olver, P.J., Canonical forms and integrability of biHamiltonian systems, *Phys. Lett.* **148A** (1990), 177-187.

[7] Olver, P.J., A nonlinear series differential operator which commutes with any function, preprint, 1990.

[8] Turiel, F.–J., Classification locale d'un couple de formes symplectiques Poisson-compatibles, *Comptes Rendus Acad. Sci. Paris* **308** (1989), 575-578.

Part V

Maps and Cascades

Transitions from Chaotic to Brownian Motion Behaviour

Ch. Beck

Institute for Theoretical Physics,
Technical University of Aachen,
5100 Aachen, Germany.

Several transition scenarios from *ordered* to *chaotic* behaviour are known, the most popular being the period doubling scenario. Here we deal with a class of maps that exhibits a transition from *chaotic* to *Brownian motion* behaviour if a control parameter is changed. That is to say, the endpoint of the usual period doubling scenario (the fully developed chaotic state) is the starting point of this scenario leading to a Gaussian Stochastic process. We investigate the transition by means of various appropriate observables, such as the Renyi dimensions, the KS entropy, and relaxation times. Some of these quantities exhibit a phase transition like behaviour when the Gaussian limit case is approached. We prognose a hydrodynamical experiment where transition scenarios of this type might be observed.

1. Introduction

Classical theories of stochastic processes such as Einstein's theory of Brownian motion or the Ornstein-Uhlenbeck theory [1,2] deal with Gaussian processes. The theories are based on certain assumptions of statistical independence (independent increments of the Wiener process or, equivalently, the assumption of δ-correlated white noise). These assumptions, together with the central limit theorem for independent events, lead to smooth Gaussian probability distributions of the corresponding processes. In contrast to this the invariant densities of chaotic dynamical systems typically possess a complicated non-Gaussian structure: often the measure is fractal and there are singularities or even spectra of singularities. This complicated structure has to do with the fact that there are correlations between successive iterates of the deterministic dynamical system. This means that the central limit theorem in its usual form cannot be applied.

There are, however, some dynamical systems for which sums of iterates, when appropriately rescaled, possess a Gaussian invariant density ([3] and references therein). At first sight this may come as a surprise, because the iterates of a dynamical system are not independent random variables. However, statistical independence is only a sufficient, not a necessary condition for Gaussian behaviour. It is remarkable that there are even classes of nonlinear deterministic systems that exhibit a long time behaviour equivalent to that of the Langevin equation [3,4,5]. These dynamical systems can generate, for example, the Ornstein-Uhlenbeck process. This means that not only the 1-point distribution but also all higher order distributions of the corresponding process converge to a Gaussian. The maps for which this limit behaviour has been proved are of the following form:

$$\begin{aligned} x_{n+1} &= T x_n \\ y_{n+1} &= \lambda y_n + \tau^{1/2} f(x_n) \end{aligned} \tag{1}$$

Here T is a map with strong mixing properties, f a is smooth function, $\lambda = e^{-\gamma\tau}$ is a parameter, and γ and τ are positive constants. An example is the Kaplan-Yorke map [10,11] where $Tx = 2x \bmod 1$, $f(x) = \cos \pi x$ and $\tau = 1$. As was pointed out in [4], y_n can be interpreted as the stroboscobic velocity of a particle moving in a medium of viscosity γ under deterministic impulses. The kick force acting on the particle is given by the expression

$$L_\tau(t) = \tau^{1/2} \sum_{n=1}^{\infty} f(x_{n-1}) \delta(t - n\tau) \tag{2}$$

In addition, there is a damping force proportional to the velocity $Y(t)$ of the particle: the equation of motion is

$$\dot{Y} = -\gamma Y + L_\tau(t) \tag{3}$$

Integration of (3) leads to

$$Y(t) = e^{-\gamma(t-n\tau)} y_n \qquad n = \lfloor t/\tau \rfloor \tag{4}$$

($\lfloor\ \rfloor$: integer part) where $y_n = Y(n\tau + 0)$ obeys the recurrence relation (1). It has been proved that in the limit $\tau \to 0$ the process $Y(t)$ converges to the Ornstein-Uhlenbeck process provided T has the so-called φ-mixing property, f is of bounded variation, and x_0 is distributed according to some smooth probability distribution [4]. Thus under these assumptions the y-variable of the deterministic system (1) generates a "classical" Langevin process, and consequently the invariant density of y_n becomes Gaussian for $\lambda \to 1$. The proof for this asymptotic behaviour is based on a generalization of the functional central limit theorem for weakly dependent events [4,12] . Examples of φ-mixing dynamical systems are the maps $Tx = 1 - 2x^2$, $Tx = 2x \bmod 1$ and $Tx = (1/x) \bmod 1$. It is also possible to generalize the discussion for nonlinear equations of the form $\dot{Y} = A(Y) + L_\tau(t)$ (see [4]).

As a standard example let us consider the map

$$\begin{aligned} x_{n+1} &= 1 - 2x_n^2 \\ y_{n+1} &= \lambda y_n + \tau^{1/2} x_n \end{aligned} \tag{5}$$

i.e. the map (1) with $Tx = 1 - 2x^2$ and $f(x) = x$. For $\lambda = 0$ this dynamical system is effectively 1-dimensional and equivalent to the map $x_{n+1} = 1 - 2x_n^2$ exhibiting fully developed chaos [13]. The (natural) invariant density is given by

$$\rho_0(y) = \frac{1}{\pi\sqrt{1-y^2}} \qquad y \in [-1, 1] \tag{6}$$

For $\tau \to 0$, $\lambda = e^{-\gamma\tau} \to 1$ the map (5) generates the Ornstein-Uhlenbeck process in the y-variable, and the marginal invariant density of y is given by [4]

$$\rho_1(y) = \left(\frac{\pi}{2\gamma}\right)^{-1/2} \exp\left\{-2\gamma y^2\right\} \tag{7}$$

These two limit cases are fully understood. What, however, happens for the intermediate values of λ? In this case the marginal invariant density $\rho_\lambda(y)$ of the y-variable (the velocity distribution of the particle) has a complicated structure, and there is no simple analytic expression for it, although a lot of the statistical properties of the attractor can be calculated analytically [9]. We are interested in the transition scenario from a fully developed chaotic state to a Gaussian state that we obtain if we increase λ from 0 to 1. In physical terms the increase of λ can be regarded as a decrease of the time difference τ between successive impulses on the particle. This means that we observe the particle on a larger and larger time scale.

2. Transition scenarios

Fig. 1 shows how the attractor of the map (5) and the velocity distribution $\rho_\lambda(y)$ of the particle changes with λ ($\gamma = 1$). For small λ the attractor is fractal and the density has a complicated, selfsimilar structure. There are several peaks that represent singularities. These peaks disappear if λ is increased and the Gaussian limit dynamics is approached. The corresponding attractor becomes 2-dimensional.

Another example is the system (1) where the impulses are given by the iterates of the continued-fraction map $Tx = (1/x) \bmod 1$ and $f(x) = x$. T is known to be φ-mixing [12]. For $\lambda = 0$ the invariant density is given by [14,15]

$$\rho_0(y) = \frac{1}{\log 2} \frac{1}{1+y} \qquad y \in [0,1] \tag{8}$$

Fig. 2 shows the transition scenario in this case.

A further simple example is the map (1) with $Tx = 2x \bmod 1$ and $f(x) = x$, for which

$$\rho_0(y) = 1 \qquad y \in [0,1] \tag{9}$$

This model is appropriate for analytical calculations. A detailed analytic treatment has been presented in [3].

So far we have dealt with examples where the Gaussian limit behaviour for $\lambda \to 1$ has been proved. However, we empirically observed a Gaussian limit distribution for almost every choice of a chaotic map T [8]. Thus the class of dynamical systems that exhibit a transition to Gaussian random behaviour seems to be much larger than the class of systems for which the φ-mixing property can be proved. Obviously the Gaussian limit distribution can be regarded as a "universal" limit distribution for rescaled sums of iterates of a large class of chaotic maps. This is a highly non-trivial statement, because the iterates of a chaotic system are not independent random variables.

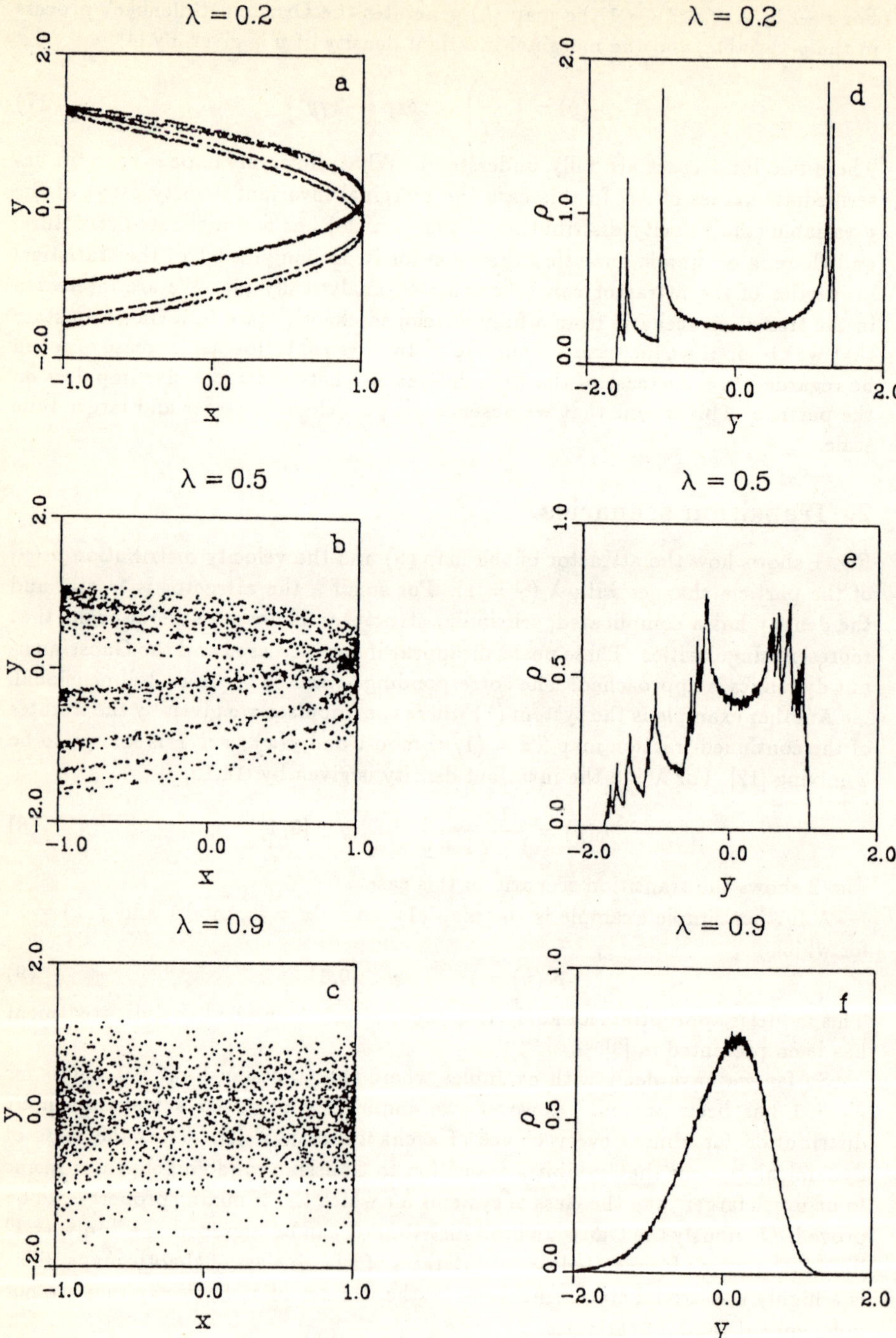

Figure 1. Transition scenario for the map (1) with $Tx = 1 - 2x^2$, $f(x) = x$. a-c: the attractor depending on λ. d-f: the corresponding velocity distribution

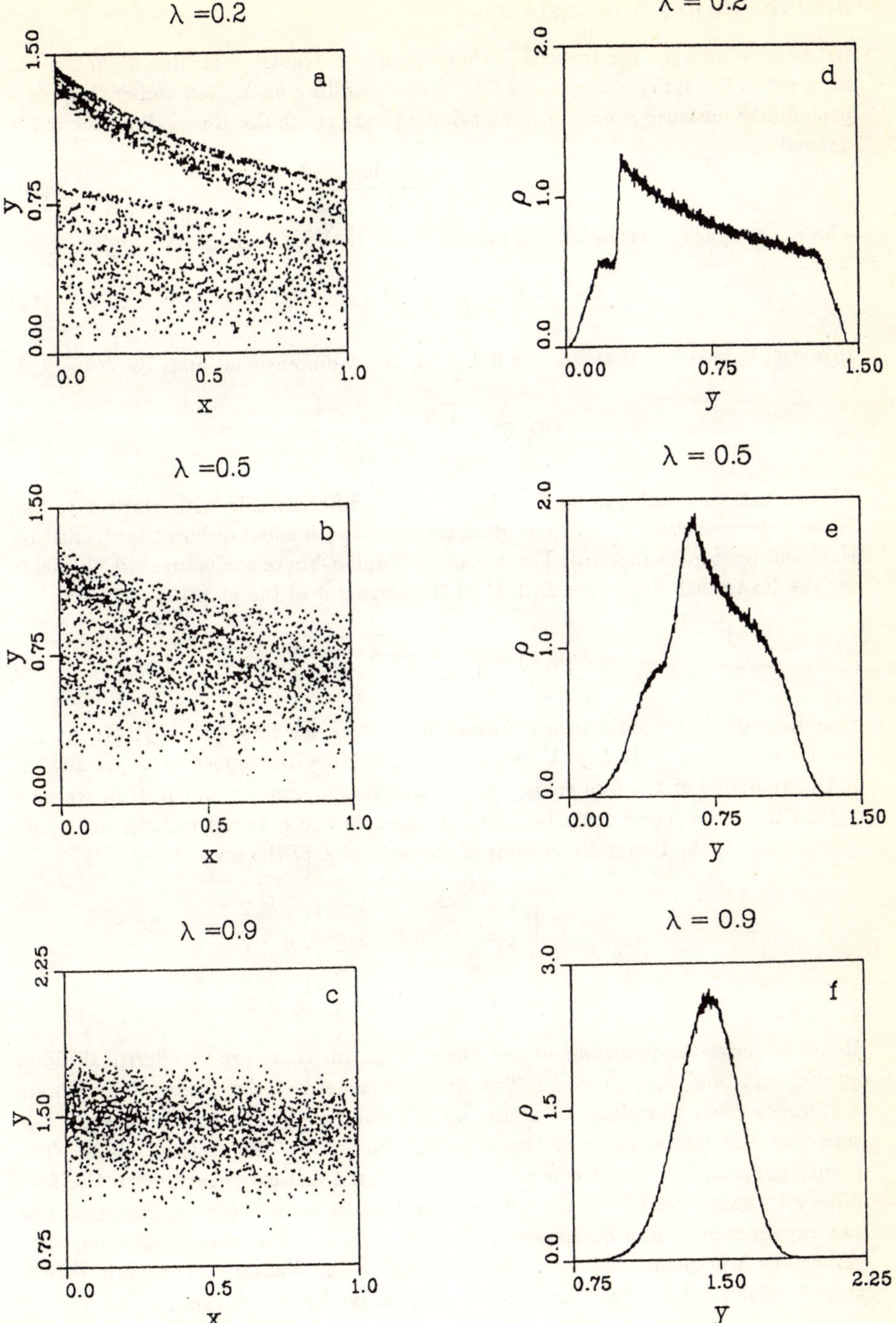

Figure 2. The same as Fig. 1, but for the continued-fraction map $Tx = (1/x) \bmod 1$

3. The Renyi dimensions

Let us now analyze the transition from chaotic to Gaussian stochastic behaviour in terms of the Renyi dimensions $D(q)$ [16] depending on λ. Remember that for a probability measure μ on a m-dimensional phase space the Renyi dimensions are defined as

$$D(q) = \lim_{\Delta \to 0} \frac{1}{q-1} \frac{\log \sum p_i^q}{\log \Delta} \tag{10}$$

The phase space is assumed to be divided into little cells of size Δ^m and

$$p_i = \int_{i-th\ cell} d\mu(x) \tag{11}$$

It is well known [17] that for $\lambda = 0$, i.e. for the 1-dimensional map $Tx = 1 - 2x^2$

$$D(q,0) = \begin{cases} 1 & q \leq 2 \\ \frac{1}{2}\frac{q}{q-1} & q \geq 2 \end{cases} \tag{12}$$

At the critical point $q_{crit} = 2$, $D(q,0)$ is not differentiable with respect to the "inverse temperature" q, which is often compared with a first order phase transition ([18] and references therein). For $\lambda > 0$ the Kaplan-Yorke conjecture [10,11] yields for the Hausdorff dimension $D(0,\lambda)$ of the attractor of the map (5)

$$D(0,\lambda) = \begin{cases} 1 + \frac{\log 2}{|\log \lambda|} & \lambda \leq \frac{1}{2} \\ 2 & \lambda \geq \frac{1}{2} \end{cases} \tag{13}$$

(the Kaplan-Yorke conjecture is known to be true for this system [11]). At the critical point $\lambda_{crit} = \frac{1}{2}$, $D(0,\lambda)$ is not differentiable with respect to λ. As $D(q,\lambda)$ is the analogue of the free energy in the thermodynamic formalism [19], we may regard this as a phase transition with respect to the external parameter λ [20]. More general, the Renyi dimensions of the attractor of the map (5) are [21]

$$D(q,\lambda) = \begin{cases} 1 + \frac{\log 2}{|\log \lambda|} & \lambda \leq \frac{1}{2},\ q \leq 2 \\ \frac{1}{2}\frac{q}{q-1} + \frac{\log 2}{|\log \lambda|} & \lambda \leq \frac{1}{2},\ q \geq 2 \\ 2 & \lambda \geq \frac{1}{2},\ q \leq 2 \\ 1 + \frac{1}{2}\frac{q}{q-1} & \lambda \geq \frac{1}{2}, q \geq 2 \end{cases} \tag{14}$$

Hence we have the following simple phase diagram: there are two "critical" lines $q_{crit}(\lambda) = 2$ and $\lambda_{crit}(q) = \frac{1}{2}$. The first line seperates the phase of maximum uniformity from the phase of minimum uniformity [6], whereas the second line seperates the fractal state of the attractor from the 2-dimensional state. The point $(q_{crit}, \lambda_{crit}) = (2, \frac{1}{2})$ can be regarded as a 4-critical point where the four different phases coexist. It should be clear that in general (for other maps) one can expect much more complicated phase diagrams. The phase transition with respect to λ is generic for all systems for which the Kaplan-Yorke conjecture is true. In general $\lambda_{crit} = e^{-\nu}$, where ν is the positive Ljapunov exponent of T.

4. The Kolmogorov-Sinai entropy

It is generally accepted that an ordered process is characterized by a negative KS entropy, a chaotic process by a finite positive KS entropy, and a "random process" by an infinite KS entropy [22]. In our case, changing λ, we have a transition from a chaotic process to "random process", namely (dynamical) Brownian motion. Hence we expect the KS entropy of our system to diverge for $\lambda \to 1$. At first sight this looks like a contradiction: How can a simple deterministic map such as eq. (5) produce an infinite KS entropy for $\lambda \to 1$? Indeed, Pesin's theorem [23] states that the KS entropy is smaller or equal to the sum of positive Ljapunov exponents (in many cases this is an equality). The map (5) has just one positive Ljapunov exponent, namely $\log 2$. Hence the KS entropy is bounded from above for arbitrary λ!

The solution to this "problem" is very simple. Indeed, the KS entropy of our system is $\log 2$ if we regard n as the time variable. However, one has to keep in mind that we *rescale* the time variable n approaching the limit case of Brownian motion: The "physical" time variable is $t = n\tau$, the Ornstein-Uhlenbeck process $U(t)$ is approached in the limit $\tau \to 0$ keeping $t = n\tau$ constant. This means that each iteration step corresponds to a physical time step of length τ. Now, the "physical" KS entropy h_{KS}^{phys} (the loss of information) must of course be related to the physical time scale. We obtain $h_{KS}^{phys} = \log 2/\tau$, and this indeed diverges for $\tau \to 0$.

5. The relaxation time

In typical cases the correlation function of a mixing dynamical system decays exponentially. The corresponding rate of decay is called relaxation time η. For the map $Tx = 1 - 2x^2$ both the inverse relaxation time and the KS entropy h_{KS} have the value $\log 2$. It is, however, a wrong prejustice to believe that $\eta^{-1} = h_{KS}$ is a general equality. A nice counter example is provided by the continued-fraction map $Tx = (1/x) \bmod 1$ where $\eta^{-1} = 1.1918$ and $h_{KS} = \pi^2/(6\log 2) = 2.3731$ [15]. Hence relaxation times are important independent quantities.

In [5] it was proved that under quite general assumptions the relaxation time of maps of type (1) is given by

$$\eta = \max((\gamma\tau)^{-1}, \xi) \tag{15}$$

Here ξ is the relaxation time of the map T, and $\gamma\tau = -\log\lambda$. We again observe an interesting phase transition like phenomenon when approaching the Brownian motion limit case by a decrease of τ: There is a critical time scale $\tau_c = 1/(\gamma\xi)$, repectively a critical parameter value $\lambda_c = e^{-1/\xi}$, where the relaxation time η depends on the control parameter λ in a nonanalytic way. For the map $Tx = 1-2x^2$ this critical point coincides with the critical point where the Hausdorff dimension exhibits its phase transition behaviour. In general, however, it does not, as we have illustrated by the example of the continued-fraction map.

6. A hydrodynamical experiment

The nice property of dynamical systems of the form (1) is that they have a direct physical interpretation: As has been pointed out in [4] the variable y_n can be regarded as the stroboscopic velocity of a kicked damped particle. Therefore, it is straightforward to conjecture that transition scenarios similar to those described in section 2 can be observed in a real physical experiment. In a liquid the dynamics of impulses acting on a test particle is determined by a very complicated dynamical system T on a high dimensional phase space representing all the molecules in the liquid. Nevertheless, by projection onto the phase space of a single molecule a simple dynamics of Langevin type may occur, from which the hydrodynamical equations can be deduced [7]. It should be clear that the dynamical system (1) is an idealization in the sense that the impulses in a real liquid are not equidistant. However, it is reasonable to assume that there is an average time $\bar{\tau}$ between successive impulses on the particle and that the average effect of all the molecules in the fluid can be described by a damping constant γ. For a classical Brownian particle $\gamma \approx 10^8 s^{-1}$ and $\bar{\tau} << \gamma^{-1}$ [1,2]. Thus $\lambda = e^{-\gamma\bar{\tau}} \approx 1$ and we observe, of course, a Gaussian velocity distribution, the well known Maxwell Boltzmann distribution. Suppose, however, that for a system far from thermodynamic equilibrium other time scales are relevant such that $\lambda = e^{-\gamma\bar{\tau}} < 1$. In this case one may expect non-Gaussian velocity distributions provided the evolution of the kick strengths is determined by a deterministic dynamical system exhibiting chaotic behaviour. This non-Gaussian behaviour is more likely to be expected on a macroscopic scale rather than on a microscopic scale, because for a very small particle noise effects will always lead to a Gaussian distribution.

As an example let us consider a Raleigh-Benard system with convection rolls. These rolls are known to show on a macroscopic scale periodic, quasiperiodic or chaotic oscillations depending on the external parameters [24,25,26]. We assume that there is a macroscopic test particle in a fixed small subregion of the liquid. The oscillating rolls are acting on the test particle similar to a kick force, where the time scale $\bar{\tau}$ between "kicks" is given by the inverse of a typical frequency associated with the rolls. If the test particle has a spherical shape, then Stokes law [2] yields for the damping constant

$$\gamma = \frac{6\pi\nu\rho a}{m} \tag{16}$$

Here a is the radius of the test particle, m its mass, ν is the kinematic viscosity of the liquid and ρ the density of the liquid. Assume that $\bar{\tau}$ and γ have comparable size such that $\lambda \approx e^{-1}$. If the dynamics is chaotic and not significantly perturbed by noise effects, we expect a non-Gaussian velocity distribution of the test particle. It is easy to change λ by choosing test particles of different size. Thus, varying the radius of the test particle, one might observe a transition from a complicated to a Gaussian velocity distribution. For water at room temperature and test particles with a density of 1 g/cm^3 we obtain $\gamma \approx 5 \cdot 10^{-6} m^2/s \cdot 1/a^2$. Suppose $\bar{\tau} \approx 1s$, then $\lambda = e^{-1}$ for $a \approx 2 \cdot 10^{-3}m$, whereas $\lambda \approx 1$ for $a \approx 10^{-2}m$. We would like to encourage experimentalists to look for such transitions from complicated to Gaussian stochastic behaviour.

References

[1] N.G. van Kampen, *Stochastic Processes in Physics and Chemistry* (North-Holland, Amsterdam, 1981)

[2] E. Nelson, *Dynamical Theories of Brownian Motion* (Princeton University Press, Princeton, 1967)

[3] C. Beck, Physica **169A**, 324 (1990)

[4] C. Beck, G. Roepstorff, Physica **145A**, 1 (1987)

[5] C. Beck, Commun. Math. Phys. **130**, 51 (1990)

[6] C. Beck, Physica **41D**, 67 (1990)

[7] C. Beck, G. Roepstorff, Physica **165A**, 270 (1990)

[8] C. Beck, A. Kleczkowski, unpublished

[9] C. Beck, *Higher Correlation Functions of Chaotic Dynamical Systems — A Graph Theoretical Approach*, Preprint (RWTH Aachen, 1990)

[10] J.L. Kaplan, J.A. Yorke, Lecture Notes in Mathematics **730**, 204 (Springer, Berlin, 1979)

[11] J.L. Kaplan, J. Mallet-Paret, J.A. Yorke, Ergod. Th. Dynam. Syst. **4**, 261 (1984)

[12] P. Billingsley, *Convergence of Probability Measures* (Wiley, New York, 1968)

[13] G. Györgyi, P. Szépfalusy, J. Stat. Phys. **34** , 451 (1984)

[14] R.O. Kuz'min, Dokl. Akad. Nauk, Ser. A. 375 (1928)

[15] D. Mayer, G. Roepstorff, J. Stat. Phys. **47**, 149 (1987), **50**, 331 (1988)

[16] A. Rényi, *Probability Theory* (North-Holland, Amsterdam, 1970)

[17] E. Ott, W. Withers, J.A. Yorke, J. Stat. Phys. **36**, 697 (1984)

[18] A. Csordás, P. Szépfalusy, Phys. Rev. **39A**, 4767 (1989)

[19] T. Tél, Z. Naturforsch. **43A**, 1154 (1988)

[20] R. Badii, G. Broggi, B. Derighetti, M. Ravani, S. Ciliberto, A. Politi, M.A. Rubio, Phys. Rev. Lett. **60**, 979 (1988)

[21] A. Chennaoui, J. Liebler, H.G. Schuster, J. Stat. Phys. **59**, 1311 (1990)

[22] H.G. Schuster, *Deterministic Chaos* (Physik-Verlag, Weinheim, 1984)

[23] D. Ruelle, Bol. Soc. Bras. Mat. **9**, 83 (1978)

[24] M. Giglio, S. Musazzi, U. Perini, Phys. Rev. Lett. **47**, 243 (1981)

[25] A. Libchaber, C. Laroche, S. Fauve, J. Physique Lett. **43**, L-211 (1982)

[26] P. Bergé, M. Dubois, P. Manneville, Y. Pomeau, J. Physique Lett. **41**, L-341 (1980)

Kinetic Theory for the Standard Map

H.H. Hasegawa and W.C. Saphir.

Center for Studies in Statistical Mechanics and Complex Systems
The University of Texas at Austin, Austin, TX 78712 USA.

Abstract In this paper we apply the methods of kinetic theory (Subdynamics) to the Chirikov-Taylor Standard Map. We reinterpret the mechanism for the Lyapounov instability of a map and obtain a criterion for chaos within the framework of kinetic description.

1. Introduction

One of the reasons for studying chaos is to understand the basis of non-equilibrium statistical mechanics. For this purpose we are interested in the time evolution of globally chaotic dynamical systems in which kinetic description is natural[1].

The Lyapounov exponent is one of the most important concepts in chaos theory and is usually described in terms of trajectories. But the concept is also important in globally chaotic systems in which a description in terms of trajectories is physically meaningless. Therefore we want to understand it in terms of the distribution function.

In section 2 we apply the methods of kinetic theory to the Chirikov-Taylor Standard Map

$$\begin{aligned} p_{n+1} &= p_n + K \sin x_n \\ x_{n+1} &= x_n + p_{n+1} , \end{aligned} \qquad (1-1a, 1b)$$

using a resolvent-like operator for the time evolution operator for the map. We estimate the position of poles for this operator to obtain the diffusion equation. We also introduce the projection operator Π which decomposes the distribution function into the part which obeys the diffusion equation and the remainder. In section 3 we show that this decomposition naturally explains how the exponential growth of p and x appears for a point-like initial distribution. Through this argument we obtain a criterion for chaos in kinetic theory.

2. Kinetic Theory for the Standard Map

We start with the following "time" evolution of the distribution function under the standard map

$$\rho_n = e^{-\imath \mathcal{L} n} \rho_0 \tag{2-1}$$

where $e^{-\imath\mathcal{L}} \equiv e^{-p\frac{\partial}{\partial x}} e^{-K \sin x \frac{\partial}{\partial p}}$.

By analogy to the ordinary construction of kinetic theory [2] we consider the following resolvent-like formalism,

$$\rho_n = \frac{1}{2\pi} \int_{-\pi+\imath\epsilon}^{\pi+\imath\epsilon} dz e^{-\imath z(n+1)} \frac{1}{e^{-\imath z} - e^{-\imath\mathcal{L}}} \rho_0 \tag{2-2}$$

To calculate the time evolution of ρ, we estimate the position of poles of the resolvent-like operator. Since we obtain a diffusion equation by estimating these poles, we hereafter refer them as *diffusion poles.*

Following the usual kinetic theory approach, we define the "state" $|x,p)$ to be an operator which gives the (x,p) representation of a function, ie, $(x,p|\rho) \equiv \rho(x,p)$. Then we introduce the states $|k,q)$ Fourier conjugate to $|x,p)$ and defined through the inner product $(x,p|k,q) \equiv e^{\imath kx+\imath qp}/\sqrt{2\pi\Omega}$ where $-\Omega/2 < p < \Omega/2$ ($\Omega/2\pi$:integer) and $-\pi < x < \pi$. We first consider Ω finite and finally take the $\Omega \to \infty$ limit to obtain a diffusion process.

For these states the matrix elements of the time evolution operator are

$$(k,q|e^{-\imath\mathcal{L}}|k',q') = \sum_{l=-\infty}^{\infty} \delta_{k,k'-l\mathrm{sgn}q'} \delta_{q,q'-k'+l\mathrm{sgn}q'} \mathcal{J}_l(|q'|K) \tag{2-3}$$

where $\mathcal{J}_l$ is the l-th Bessel function. When we decompose q (the Fourier conjugate of p) into integer and fractional parts, $q = j + \lambda$ ($|\lambda| < 1/2$), there is no transition between q and q' which have different λ and λ' in Eq.(2-3). Since the time evolution for each λ evolves independently, we define the following projection operators:

$$\begin{aligned} P_\lambda &\equiv |0,\lambda)(0,\lambda| \\ I_\lambda &\equiv \sum_{k,j} |k,j+\lambda)(k,j+\lambda| \\ Q_\lambda &\equiv I_\lambda - P_\lambda \,. \end{aligned} \tag{2-4a, 4b, 4c}$$

In order to estimate the diffusion poles, we introduce the following decompositon for the resolvent-like operator[2]

$$\frac{1}{e^{-iz}-e^{-i\mathcal{L}}} = \sum_{\lambda}[P_\lambda + \mathcal{C}_\lambda(z)]\frac{1}{e^{-iz}-e^{-i\psi_\lambda(z)}}[P_\lambda + \mathcal{D}_\lambda(z)] + \mathcal{P}_\lambda(z) \qquad (2-5)$$

where

$$e^{-i\psi_\lambda(z)} \equiv P_\lambda e^{-i\mathcal{L}}P_\lambda + P_\lambda e^{-i\mathcal{L}}Q_\lambda \frac{1}{e^{-iz}-Q_\lambda e^{-i\mathcal{L}}Q_\lambda}Q_\lambda e^{-i\mathcal{L}}P_\lambda \qquad (2-6a)$$

$$\begin{aligned} \mathcal{C}_\lambda(z) &\equiv \frac{1}{e^{-iz}-Q_\lambda e^{-i\mathcal{L}}Q_\lambda}Q_\lambda e^{-i\mathcal{L}}P_\lambda \\ \mathcal{D}_\lambda(z) &\equiv P_\lambda e^{-i\mathcal{L}}Q_\lambda \frac{1}{e^{-iz}-Q_\lambda e^{-i\mathcal{L}}Q_\lambda} \qquad (2-6b, 6c, 6d) \\ \mathcal{P}_\lambda(z) &\equiv \frac{Q_\lambda}{e^{-iz}-Q_\lambda e^{-i\mathcal{L}}Q_\lambda} . \end{aligned}$$

We denote by θ_λ the position of the pole of $1/(e^{-iz}-e^{-i\psi_\lambda(z)})$. Since it is not easy to estimate θ_λ for general λ, we estimate the explicit form of θ_λ only for small λ and expanding with respect to the following two small parameters:(1) $\lambda K \ll 1$ and (2) $1/\sqrt{K} \ll 1$. The first expansion is the ordinary pertubation expansion and the second gives non-perturbative effects which suppress the correlation between interactions for different times.

After explicit calculation we obtain the diffusion pole for small λK,

$$\begin{aligned} e^{-i\theta_\lambda} &= \mathcal{J}_0(|\lambda|K) + \mathcal{J}_0^{-2}(|\lambda|K)\sum_{l\neq 0}\mathcal{J}_l^2(|\lambda|K)\mathcal{J}_{-2|l|}(|l|K) + O(\frac{1}{K}) \\ &= 1 - \lambda^2 \mathrm{D} + O(\lambda^3, \frac{1}{K}) \end{aligned} \qquad (2-7)$$

By picking up the effect of the diffusion pole we obtain that $(0,\lambda|\rho_n)$ damps as $\exp[-\lambda^2 \mathrm{D} n]$. Since λ corresponds to $-i\partial/\partial p$ we can interpret D as a diffusion coefficient:

$$\mathrm{D} = \frac{K^2}{4}[1 - \frac{1}{2}\sqrt{\frac{2}{\pi K}}\cos(K - \frac{5\pi}{4}) + O(\frac{1}{K})] . \qquad (2-8)$$

This result agrees with that of Rechester and White[3]. In this paper we neglect effects of the accelerator modes[1].

In the remainder of this section we will introduce the projection operator which projects out the part of distribution function which obeys the diffusion equation.

In the theory of unstable particles in quantum mechanics we can interprete a complex pole of the resolvent operator $1/(z-H)$ as a complex eigenvalue of Hamiltonian H[4]. By analogy we interpret the pole of $1/(e^{-\imath z}-e^{-\imath\mathcal{L}})$ as a complex eigenvalue of the operator $\mathcal{L}$. We denote a complex eigenstate of $\mathcal{L}$ by $|\theta_\lambda>$. Since the time evolution of each $|\theta_\lambda>$ is decoupled from that of the others, it is natural to introduce the following projection operator which projects out the complex eigenstate $|\theta_\lambda>$,

$$\Pi_\lambda \equiv |\theta_\lambda><\theta_\lambda| \qquad (2-9)$$

where Π_λ is not Hermitian and we need to introduce an indefinite metric[4]. From the definition Π_λ satisfies

$$\begin{aligned}\Pi_\lambda e^{-\imath\mathcal{L}} &= e^{-\imath\mathcal{L}}\Pi_\lambda = e^{-\imath\theta_\lambda}\Pi_\lambda \\ \Pi_\lambda\Pi_{\lambda'} &= \Pi_\lambda\delta_{\lambda,\lambda'}\ .\end{aligned} \qquad (2-10a,10b)$$

Π_λ can be derived by picking up the effect from the complex pole of the resolvent-like operator:

$$\Pi_\lambda = \oint_{z=\theta_\lambda} dz\frac{-1}{2\pi}e^{-\imath\theta_\lambda}\frac{1}{e^{-\imath z}-e^{-\imath\mathcal{L}}} \qquad (2-11)$$

Since we do not have enough space to show the explicit form of Π_λ, we will give the explicit form of Π_λ in Ref.5.

Finally we introduce the projection operators Π and $\hat{\Pi}$ defined as follows

$$\Pi \equiv \sum_{\lambda\ll 1/K}\Pi_\lambda \qquad (2-12a)$$

$$\hat{\Pi} \equiv I-\Pi \qquad (2-12b)$$

Π projects out states which obey the diffusion equation. In the lowest order approximation, $\Pi_\lambda \sim P_\lambda \equiv |0,\lambda)(0,\lambda|$, so that Π projects out modes of the distribution function which are almost uniform with respect to x and p. As mentioned above, these modes are damped according to: $(0,\lambda|\rho_n) \sim \exp[-\lambda^2 \mathrm{D}n](0,\lambda|\rho_0)$. Thus the almost-uniform part of the distribution function obeys the diffusion equation.

We note that the definition in Eq.(2-12a) of Π is not unique. This ambiguity is not important for the argument in the next section, since we take the $\lambda\to 0$ limit in the calculation of $\langle(p-\langle p\rangle)^2\rangle$ and $\langle(x-\langle x\rangle)^2\rangle$.

Using the Π_λ, the time evolution of ρ_n is decomposed as

$$\rho_n = \sum_{\lambda \ll 1/K} e^{-i\theta_\lambda n} \Pi_\lambda \rho_0 + \hat{\Pi} \rho_n. \qquad (2-13)$$

This type formulation is called subdynamics and was first introduced by the Brussels group[6].

3. A Reinterpretation of the Lyapounov Instability

In this section we will show how our kinetic theory can explain the Lyapounov instability of the standard map.

First we show the results of a computer simulation of the time evolution of the variance of p for a point-like initial distribution. For short time scales $\langle (p - \langle p \rangle)^2 \rangle$ grows exponentially as illustrated in Figures 1a. When the variance in position is comparable to the size of the system, behaviour changes abruptly and $\langle (p - \langle p \rangle)^2 \rangle$ grows linearly. This can be seen in Figures 1b. Although we do not show them explicitly, the results of a computer simulation show that $\langle (x - \langle x \rangle)^2 \rangle$ grows exponentially for short time and saturates at $\pi^2/3$.

We will study how this exponential growth appears for a point-like initial

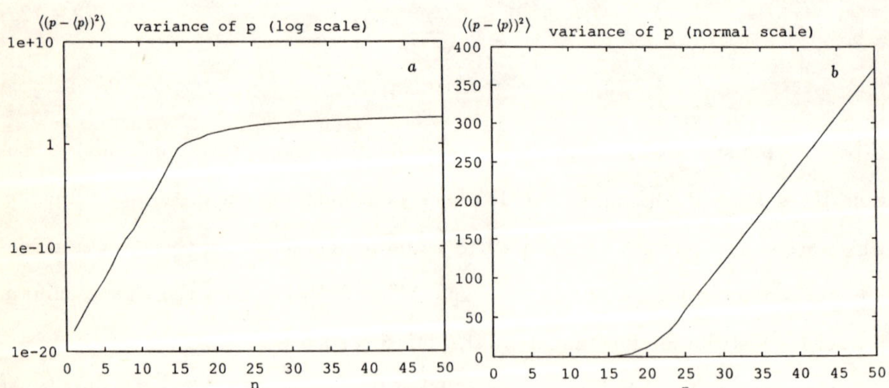

Figures 1a,b. Results of computer simulation showing $\langle (p - \langle p \rangle)^2 \rangle$ for a point-like initial condition as a function of time. 1a, which has a logarithmic scale, shows exponential growth for short times. 1b, with a linear scale, shows diffusive behaviour for long times distribution using the kinetic theory and show that the exponent which characterizes it is the same as the Lyapounov exponent.

In the computer simulation we chose the following point-like distribution,

$$\rho_0 = \frac{1}{\Delta x_0}[\theta(x - x_0 + \frac{\Delta x_0}{2}) - \theta(x - x_0 - \frac{\Delta x_0}{2})]\frac{1}{\Delta p_0}[\theta(p - p_0 + \frac{\Delta p_0}{2}) - \theta(p - p_0 - \frac{\Delta p_0}{2})]. \tag{3-1}$$

Since $\langle (p - \langle p \rangle)^2 \rangle$ and $\langle (x - \langle x \rangle)^2 \rangle$ for ρ_0 are not same as $(\Delta p_0)^2$ and $(\Delta x_0)^2$ in Eq.(3-1), we define $(\Delta p_n)^2 \equiv 12(\langle p^2 \rangle_n - \langle p \rangle_n^2)$ and $(\Delta x_n)^2 \equiv 12(\langle x^2 \rangle_n - \langle x \rangle_n^2)$ where $\langle f \rangle_n \equiv \int dpdx f \rho_n$ and the facter 12 was determined for consistency with both Δp_0 and Δx_0.

First we consider the extreme case in which the initial distribution is a single point. Since a point remains a point under time evolution, both Δp_n and Δx_n are always zero. In order to understand this behavior in the context of our kinetic theory, we decompose the distribution into

$$\rho_n = \Pi \rho_n + \hat{\Pi} \rho_n. \tag{3-2}$$

Since $\Pi \rho_n$ obeys the diffusion equation, the contribution from $\Pi \rho_0$ to Δp_1 grows according to the diffusion equation. This growth should be canceled out by the contribution from $\hat{\Pi} \rho_0$ to Δp_1 so that the anti-diffusion terms compensate the diffusion term for all times.

Now we consider the case of the point-like initial distribution defined in Eq.(3-1). As was mentioned in section 2, the part of the distribution function which is almost uniform with respect to p and x obeys the diffusion equation. Since Π projects out the almost-uniform part of ρ_0, the diffusion effect from $\Pi \rho_0$ does not depend in detail on ρ_0. On the other hand $\hat{\Pi} = I - \Pi$, $\hat{\Pi}$ projects out non-uniform parts of ρ_0. Therefore the finiteness of Δp_0 and Δx_0 slightly suppresses the anti-diffusion effects of $\hat{\Pi} \rho_0$ on Δp_1. Therefore the diffusion effect of $\Pi \rho_0$ on Δp_1 is not completely cancelled. This makes Δp_1 slightly larger than Δp_0. The slightly larger Δp_1 makes Δx_1 larger than Δx_0. Therefore at $n = 1$ the distribution function ρ_1 has larger mean deviation in both p and x than does the initial distribution ρ_0. These larger Δp_1 and Δx_1 more strongly suppress the anti-diffusion effects of $\hat{\Pi}\rho_1$ on Δp_2 so that Δx_2 becomes much larger than Δx_1. By repeating this auto-catalytic process, Δp_n and Δx_n grow exponentially. When Δp_n and Δx_n become large enough to suppress almost completely the anti-diffusion effects from $\hat{\Pi}\rho_n$, Δp_n starts obeying the diffusion equation and Δx_n saturates at 2π.

In order to show this auto-catalytic process explicitly, we first calculate Δp_1 and Δx_1 for ρ_0 in Eq.(3-1). As is shown in Figure 2, we approximate ρ at the next iteration by

$$\rho_1 = \frac{1}{\delta x_1}[\theta(x - x_1(p) + \frac{\delta x_1}{2}) - \theta(x - x_1(p) - \frac{\delta x_1}{2})]\frac{1}{\Delta p_1}[\theta(p - p_1 + \frac{\Delta p_1}{2}) - \theta(p - p_1 - \frac{\Delta p_1}{2})] \tag{3-3}$$

where in order to keep phase volume constant we choose

$$\begin{aligned} \delta x_1 &\equiv \frac{\Delta x_0 \Delta p_0}{\Delta p_1} \\ x_1(p) &\equiv \frac{\sqrt{\Delta x_1^2 - \delta x_1^2}}{\Delta p_1}(p - p_1) + x_1 \end{aligned} \tag{3-4a,4b}$$

Using this ρ_1 we calculate Δp_2 and Δx_2. Similarly since ρ_n can be determined from Δp_n and Δx_n, we can repeat the explicit calculations whenever the approximation of Eq.(3-3) is valid. The computer simulation shows that this approximation is valid until folding of the distribution function starts[5].

After simple calculations we obtain the contributions from both $\Pi\rho_n$ and $\hat{\Pi}\rho_n$ to $\Delta\ p_{n+1}$,

$$\begin{aligned} \Delta p_{n+1}^2|_\Pi &= 2D + \Delta p_n^2|_\Pi \\ \Delta p_{n+1}^2|_{\hat{\Pi}} &\simeq (\Delta p_n + K\cos x_n \Delta\ x_n\ )^2 - 2D - \Delta p_n^2|_\Pi \end{aligned} \tag{3-5a,5b}$$

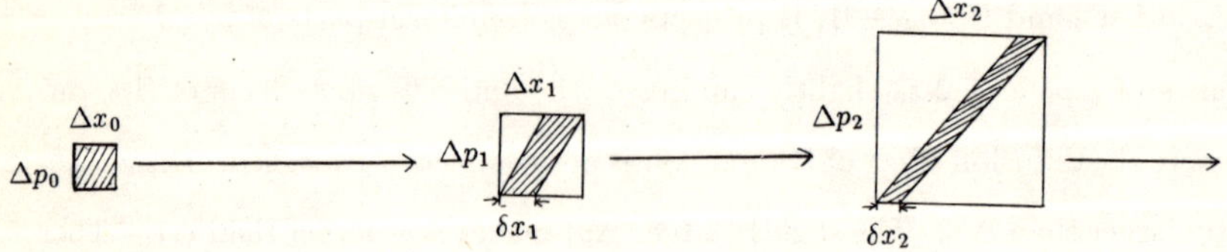

Figure 2. Schematic illustration of the map of a point-like distribution. It stretches while its area remains constant

Since $\Pi\rho_n$ obeys the diffusion equation, $\Pi\rho_n$ gives the constant increasing of $\Delta\ p_{n+1}$ which is not affected by Δp_n and Δx_n. On the other hand the contribution from $\hat{\Pi}\rho_n$ is suppressed by Δp_n and Δx_n. In Eq.(3-5b) we neglected δx_n and terms of higher order in Δp_n and Δx_n.

Similarly we obtain the contributions from both $\Pi\rho_n$ and $\hat{\Pi}\rho_n$ to $\Delta\ x_{n+1}$,

$$\begin{aligned} \Delta x_{n+1}^2|_{\Pi} &= (2\pi)^2 \\ \Delta x_{n+1}^2|_{\hat{\Pi}} &\simeq (\Delta p_n + (1+K\cos x_n)\Delta\ x_n\ )^2 - (2\pi)^2 - \Delta x_n^2. \end{aligned} \qquad (3-6a, 6b)$$

Finally we obtain the following iterative equations:

$$\begin{aligned} \Delta p_{n+1} &\simeq \Delta p_n\ +\ K\cos x_n \Delta x_n \\ \Delta x_{n+1} &\simeq \Delta p_n\ +\ (1+K\cos x_n)\Delta x_n \end{aligned} \qquad (3-7a, 7b)$$

The coefficients of Δp_n and Δx_n in Eqs.(3-7a and 7b) are the same as the components of Jacobian of the standard map. This is expected, since we keep only terms which are $O(\Delta p^2, \Delta x^2)$ in Eqs.(3-5a,5b, 6a and 6b). Therefore we obtain the exponent

$$\sigma_{\pm}\ \simeq\ \pm\log\frac{K}{2} \qquad (3-8)$$

which is same as the Lyapounov exponent obtained by Chirikov for the standard map[7].

From this argument we obtain the following qualitative criterion for chaos in our kinetic description: *if the projection operator for the almost uniform part of the distribution function is associated with a diffusion pole, the system has an instability characterized by a Lyapounov exponent.* In other words, a system which obeys a kinetic equation near equilibrium has a Lyapounov instability at the microscopic level. Ordinarily we understand that diffusion is caused by the instability of a trajectory. But the opposite point of view is possible. In this paper we showed that the diffusion poles of the resolvent-like operator cause the exponential growth of the variances of p and x. Since the properties of the resolvent-like operator in the complex plane completely determine the time evolution of the distribution function, the Lyapounov exponent can be explained in terms of kinetic theory.

Acknowledgement We thank Professor Prigogine for his constant interest, encouragement and helpful suggestions during this work. We also thank the members of both Austin and Brussels groups for fruitful discussions. We acknowledge the U.S. Department of Energy, Grant N° FG05-88ER13897, the Robert A. Welch Foundation and the European Communities Commission (contract n° PSS*0143/B) for support of this work.

References

[1] Lichtenberg and Lieberman, *Regular and Stochastic Motion* Springer-Verlag, 1983

[2] I. Prigogine, *Non-Equilibrium Statistical Mechanics* Wiley-interscience, New York, 1962.
R. Balescu, *Statistical Mechanics of Charged Particles* Wiley-interscience, New York, 1963.

[3] A.B. Rechester and R.B. White, Phys. Rev. Lett. **44** (1980) 1586.

[4] E.C.G. Sudarshan, C.B. Chiu and V. Gorini, Phys. Rev. **D18** (1978) 2914.
H.H. Hasegawa, T.Y. Petrosky, I. Prigogine and S. Tasaki, to be published in Foundation of Physics.
T.Y. Petrosky, I. Prigogine and S. Tasaki, to be published in Physica.

[5] H.H. Hasegawa and W.C. Saphir, in preparation.

[6] I. Prigogine, Cl. George, F. Henin and L. Rosenfeld, Chem. Scr. **4** (1973) 5.
T.Y. Petrosky and H. Hasegawa, Physica **160A** (1989) 351.

[7] B.V.Chirikov, Phys. Reports **52** (1979) 256

Probabilistic Description of Deterministic Chaos: A Local Equilibrium Approach

D. Mac Kernan and G. Nicolis

Faculté des Sciences,
Université Libre de Bruxelles, Campus Plaine,
C.P. 231, 1050 Bruxelles, Belgium.

Extreme sensitivity to small changes in initial conditions together with inherent limited experimental precision imply that after a time of the order of the Liupanov time, a point like deterministic description of chaotic systems loses much of its meaning. One typically then resorts to speaking about the long time averaged properties of the system in the spirit of ergodic theory, forsaking a description of the dynamics of the evolving state during a large time interval. Now a realistic initial condition is not a point in phase space but, rather, a highly peaked probability distribution. It is, therefore, tempting to describe a chaotic system before the onset of equilibrium but after the Liupanov time by defining the initial state as a probabilistic distribution(sharply peaked if desired) and then describing the subsequent evolution of the distribution. A number of results have already been obtained for conservative systems in the framework of such a description, but little is known about dissipative systems which are intrinsically more complicated because of their semi-group character(i.e. their evolution operator is not invertible). For example it is well known that if a concervative system posseses a topological Markov partition of its phase space then the system also defines an associated Markov process. Unfortunately this is not necessarily the case for dissipative systems(consider for instance, the cusp map[11] $f(x) = 2(|x|)^{1/2}$).

The object of our work so far has been to investigate the possibility of setting up a workable probabilistic description of one dimensional chaotic iterative systems. In our description the phase space of the system(which is assumed to possess a non fractal asymptotic invariant density) is partitioned into cells which define the local regions. In each of them one considers coarse grained initial probability distributions given by a local constant factor multiplying the invariant("equilibrium") distribution, see figure. On our coarse grained scale the probability distribution can be represented as a finite dimensional vector. By then demanding that the probability of any cellular trajectory(series of cell to cell transitions) be generated by a time independent matrix

$$\overrightarrow{P(n)} = M^n \overrightarrow{P(0)}$$

one obtains a Chapman-Kolmogorov condition/equation.If this condition is satisfied(which imposes strong constraints upon the choice of partition) then the initial distribution evolves according to a master equation obeying a strictly monotonic H-theorem. Furthermore, as it evolves it retains the "local equilibrium" form of the initial distribution.

THE PHASE SPACE IS PARTITIONED INTO CELLS AND THE INITIAL DENSITY IS IN LOCAL EQUILIBRIUM

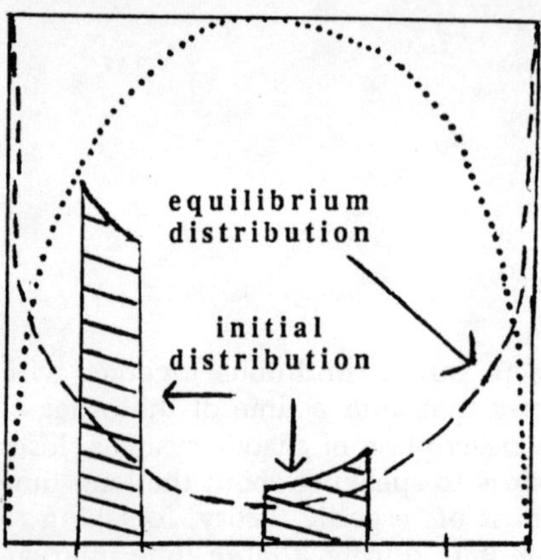

The figure illustrates what is meant by "an initial density in local equilibrium" for the logistic map $f(x) = 4x(1-x)$.

Four approaches towards achieving such a description have been developed in our recent work:

(a), a sufficient and necessary condition for one dimensional chaotic iterative systems to be described probabilistically as n^{th} order Markov processes(the future state/probability distribution of the system is fully determined if one knows the present and n-1 preceding states of the system) has been derived. If the nth order Chapman-Kolmogorov condition is satisfied then for each self-refinement of the partition (which is a specific way of constructing a finer partition using inverse images of the cells of the original) the Markov order of the process drops. Thus the system can also be described using the $(n-1)^{th}$ self-refined partition as a first order Markov process. Indeed if the process is already 1st order Markov for a given partition then it will remain first order as the partition is self-refined, this enables us to ,approximate very well many initial probability distributions as being in local equilibrium. Intuitively it seems reasonable that many chaotic systems should be described probabilistically as finite order Markov processes because one does not expect them to have an infinite memory of their past.

(b), a very large class of chaotic systems having piecewise linear iterative maps, whose "angular points" are mapped in a finite number of iterations on to an unstable orbit /fixed point(implying that there exists a Markov partition on the attractor), has been studied and found to be mapped into first order Markov processes for appropriate initial probability distributions.

(c), preliminary numerical and analytical results indicate that piecewise linear iterative maps, whose "angular points " are not mapped in a finite number of iterations onto an unstable periodic orbit/fixed point do indeed define Markov processes for appropriate partitions and associated initial densities, but at an approximative rather than exact level. In this analysis we have developed a approximation scheme which for the sytems looked at so far converges rapidly.

(d), topological conjugacy(two functions/vector fields are topologically conjugate if their exists an invertible transformation h such that $f = h^{-1} \circ g \circ h$) and its relation to metric isomorphism has been investigated because it defines, when differentiable, a vast equivalence class of systems having identical probabilistic properties. For example the preceeding logistic map and the tent map(with slope 2) are differentiably conjugate, thus they have identical probabilistic properties(on the coarse grained scale) for equivalent(conjugate) partitions if the initial densities are in local equilibrium. An important open question then is how smooth does a conjugacy have to be for two conjugate systems to be approximatively metrically isomorphic.

Research in this area is very much at its infancy and many problems remain open. For instance, generalization of the probabilistic properties found for piecewise linear maps to smooth endomorphisms may be facilitated by a better understanding of the relation between metric isomorphism and topological conjugacy. An extension of the formalism to higher dimensional systems and asymptotic invariant distributions over fractal sets, and a further development of an approximation scheme for systems which are known not to satisfy the Chapman Kolmogorov equation exactly would be very interesting. Finally we would like to attempt to apply these ideas to non-equilibrium conservative statistical mechanics.

ACKNOWLEDGEMENTS

This research is supported in part by the Commission of the European Communities under contract ST2J-299-B.

REFERENCES

1.D.Mac Kernan and G.Nicolis(1990), submitted to J.Stat.Phys.
2.G. Nicolis and C.Nicolis(1988),Phys. Rev. **A38**,427.
3.G.Nicolis, S.Martinez and E. Tirapegui(1989),J.Stat.Phys.
4M.Courbage and G. Nicolis(1989), Europhysics letters.
5.G. Nicolis,C. Nicolis and J.S. Nicolis(1989),J.Stat.Phys.**54**,915.
6.Stochastic methods in physics and chemistry,N.G. Van Kampen (1981),North Holland.
7.Iterated maps on the interval as dynamical systems,P.Collet and J.P. Eckmann(1980) ,Birkhauser.
8.Nonlinear oscillations, dynamical systems, and bifurcations of vector fields, J. Guckenhiemer and P. Holmes(1986),Springer-Verlag.
9J.P.Eckmann & D. Ruelle(1985),Rev. Mod.Phys,**57**,617.
10.Introduction to ergodic Theory, Y.G. Sinai(1977), Princeton University Press.
11.P.C. Hemmer(1984),J.Phys:Math.Gen.**17**,L247.

State Prediction for Chaotic 1-D-Maps

B. Pompe

Department of Physics,
E.-M.-Arndt-Universität Greifswald,
F.-L.-Jahn-Str. 16.
0-2200 Greifswald, FRG.

1. Motivation

In order to predict future states of a dynamical system which is assumed to be well described by a deterministic law of motion we must first measure the initial state and then look to the development in time of this initial state under the action of the law of motion which is usually done by using an analytical mathematical expression of the solution of the equations of motions or, if such expression is not known, by simulating the dynamics on a computer. Beside the fact that in general any equation of motion can approximate only within a certain finite precision the motion of a real system there remains the problem that initial states cannot be measured exactly. Nowadays it is widely known that alone this missing knowledge of the exact initial state can make the above procedure of state prediction questionable or even without support. This is the case if we have a chaotic system which is characterized by an exponential grow of small errors of initial states in the time mean.

There are several quantities chracterizing the degree of chaos resp. unpredictability of a dynamical system (f, μ), where f is the phase flow and μ the natural f-invariant measure. Some of them are the Lyapunov exponents $(\lambda_1, \lambda_2, \ldots, \lambda_d)$ /1/ measuring the exponential separation $(\lambda_i > 0)$ or convergence $(\lambda_i < 0)$ of nearby orbits in several directions of the d-dimensional phase space and the Kolmogorov entropy $h(f,\mu)$ /2/ measuring the *most possible* uncertainty on a future state after one time step provided that the whole history of the development of states of the dynamical system is known. In this interpretation the word state is understood as the index of a box of a finite partition of the state space of the dynamical system. Henceforth such a partition is assumed to be induced by the measuring instrument. (It should be noted that if the sequence of measured states represents a Markov chain of order n then only the last n states must be known in the above interpretation but not the whole history.)

The positive Lyapunov exponents and the Kolmogorov entropy

are related to each other /3/,

$$h(f,\mu) = \sum_{i:\lambda_i>0} D_i \lambda_i \leq \sum_{i:\lambda_i>0} \lambda_i, \tag{1}$$

where D_i are "partial information dimensions" of μ corresponding to the unstable directions. Both, the entropy and the exponents fulfil some invariance properties. Thus they are "nice" quantities from a mathematical point of view. Moreover, they can be used for instance to estimate the period within which certain predictions on future states are possible /4/: Assume that the motion of the dynamical system takes place on a subset of state space (e. g. chaotic attractor) which is described by the ergodic measure μ. Then a measurement with precision δ provides for small enough values of δ the information $D(\mu)\mathrm{ld}(c/\delta)$, where $D(\mu)$ is the information dimension of μ and c is a certain constant ($\mathrm{ld} \equiv \log_2$). Any forecastings on future states, on the basis of the knowledge on the initial state, should become impossible if the information on the initial state is completely replaced by the uncertainty which is produced by the chaotic dynamics. The rate of uncertainty production of the phase flow f is given by the Kolmogorov entropy $h(f,\mu)$ and hence predictions on future states should become impossible if the prediction period exceeds the time

$$\tau \approx \frac{D(\mu)}{h(f,\mu)} \mathrm{ld}(c/\delta). \tag{2}$$

A doubling of the prediction period τ requires essentially a refinement of the measuring precision from c/δ to $(c/\delta)^2$ which reflects the exponential divergence of nearby trajectories of chaotic systems. The factor of proportionality $D(\mu)/h(f,\mu)$ between the prediction period and the logarithm of the measuring precision is a characteristic invariant (at least under diffeomorphic transformations) of the dynamcical system.

Unfortunately the nice formula (2) gives in general only rough estimates of the prediction period in real experiments /5/, even if the system is mixing. This could be for instance due to the fact that in real measurements the measuring precision is not high enough especially in such regions of phase space where the local instability of the phase flow is high (which is expressed by relatively great values of some local Lyapunov exponents /6/). But there may be also some other reasons. It is the purpose of this paper to explain in somewhat more detail what happens with the information flow in chaotic systems on the basis of several partitioning concepts. Our results are valid for 1D maps. However, some of our ideas could be applied to higher-dimensionsional systems as well.

2. State Prediction Using Boxes of Equal Size

Consider a 1D map $f: x \longrightarrow y$ acting on the unit interval $[0, 1]$. f should have a continuous second derivate in this interval, except possibly at a finite number of points. Let μ be an invariant probability measure of f which is assumed to be absolutely continuous with respect to Lebesgue measure. Let ϱ denote the corresponding probability density. Moreover, let β be a finite partition covering the interval $[0, 1]$ such that each box B_i of β has equal size:

$$\beta = \{B_i\}_{i=1}^{n}, \quad B_i \cap B_j = \emptyset \text{ for } i \neq j, \quad \bigcup_i B_i \supseteq [0, 1],$$
$$\int_{B_i} dx = \text{const.} = \delta \qquad \forall B_i \in \beta. \tag{3}$$

The boxes of β are considered now as states of the system which could be identified at any time t by a measuring instrument. Hence a measurement would provide the (Shannon-) information

$$I(\mu,\beta) = -\sum_i p_i \operatorname{ld} p_i \qquad \text{with } p_i = \mu(B_i). \tag{4}$$

This information contains a certain amount of information on a state $B_{i(t)}$ which is attained t time steps in the future (called mutual information)

$$I(t,\mu,\beta) = I(\mu,\beta) - H(t,\mu,\beta). \tag{5}$$

Here $H(t,\mu,\beta)$ is the uncertainty on the future state provided that the initial state is known,

$$H(t,\mu,\beta) = \sum_i p_i H_i(t,\mu,\beta), \tag{6}$$

with the local uncertainty production of $f^t \equiv f \circ f^{t-1}$ ($t = 1, 2, 3, \ldots; f^0 \equiv 1$)

$$H_i(t,\mu,\beta) = -\sum_j p_{j/i}(t) \operatorname{ld} p_{j/i}(t) \tag{7}$$

and the transition probabilities

$$p_{j/i}(t) = \begin{cases} \dfrac{\mu(f^{-t} B_j \cap B_i)}{\mu(B_i)} & \text{if } \mu(B_i) > 0 \\ 0 & \text{otherwise.} \end{cases} \tag{8}$$

The uncertainty $H(t,\mu,\beta)$ on a future state attains positive values if there is at least one box, say B_i, such that there are at least two disjoint subsets of B_i each of them having positive measure μ and which are mapped by f^t to different boxes of β. Under the assumptions made above this can happen if

(i) there is an expanding action of f^t in certain subsets of the interval,
(ii) box boundaries are not mapped to box boundaries, which is henceforth called overlappings, or
(iii) there are points of discontinuity of f^t that are not box boundaries.

We try now to determine the contributions of each of the above points to the uncertainty $H(t,\mu,\beta)$ in the limit when the measuring precision goes to infinity (box length $\delta \longrightarrow 0$):

(i) Under the conditions made above the local uncertainty production (7) can be approximated by the logarithm of the absolute value of the slope of f^t at a point $x_i \in B_i$, provided that there is a locally expanding action of f^t. Otherwise there would be locally no uncertainty on a future state. Hence we are motivated to approximate the mean uncertainty production (6) in the limit $\delta \longrightarrow 0$ by the quantity

$$h_{exp}(t) = \int_{x:\ |df^t(x)/dx| > 1} \varrho(x)\ \mathrm{ld}\ |df^t/dx|\ dx. \qquad (9)$$

If f^t is expanding for μ-a.e. point x then $h_{exp}(t) = t\lambda$ where λ is the Lyapunov exponent of f (which equals the Kolmogorov entropy because μ is no fractal under the assumptions made above). However, in general we have

$$h_{exp}(t) \geq t\lambda . \qquad (10)$$

This means that for high resolution measurements the actually observed uncertainty is in general greater than that given by the Lyapunov exponent multiplied by the period t within which forecastings want to be made. This is due to the fact that uncertainty that is produced locally by an expanding action of f^t is not diminished by an locally contracting action of f^t in other regions of phase space.

(ii) The uncertainty due to overlappings represents a somewhat arctificial reason for uncertainty on a future state because it is first of all due to the choice of the parti-

tion β, but it does not reflect intrinsic properties of the dynamical system (f,μ). However, in general they occur in real measurements, especially if we do not know enough about the dynamical system in order to avoid overlappings by making choice of special β. This has motivated us to investigate the influence of overlappings on the uncertainty production in some more detail for our situation where we have a certain class of partitons as defined by (3). For finite resolution ($\delta > 0$) the influence of overlappings is rather complex and we can only find some useful upper and lower bounds of the contibutions of overlappings to the uncertainty production (for more details see /7/).
However, assuming that the slope of f^t varies continously within supp(μ) ($d^2 f^t/dx^2 \neq 0$), except possibly at a finite number of points, then we obtain in the limit as δ goes to zero a simple expression for the uncertainty production due to overlappings /8/:

$$h_{ov}(t) = \frac{1}{2 \ln 2} \left[\int_{x:\ |df^t(x)/dx| > 1} \varrho(x) \frac{1}{|df^t/dx|} dx + \int_{x:\ |df^t(x)/dx| < 1} \varrho(x)\ |df^t/dx|\ dx \right]. \tag{11}$$

(iii) The influence of a finite number of points of discontinuity of f^t on the uncertainty production vanishes in the limit $\delta \longrightarrow 0$ if μ is absolutely continuous with respect to the Lebesgue measure.

Summarizing we thus obtain for the uncertainty on a future state $B_{i(t)}$ provided that an initial state $B_{i(0)}$ is known:

$$H(t,\mu,\beta(\delta)) \xrightarrow[\delta \longrightarrow 0]{} h_{exp}(t) + h_{ov}(t). \tag{12}$$

It should be noted that in (12) β is a partition of boxes of equal size δ. Otherwise this would, in general, not hold. Moreover, note that if f is chaotic then the iterate f^t becomes expanding for μ-a.e. point x as t goes to infinity. On the other hand $h_{ov}(t)$ is bounded above by $1/(2 \ln 2) = 0.721...$bit, which is independent of t. Thus we obtain

$$\lim_{t \longrightarrow \infty} t^{-1}[h_{exp}(t) + h_{ov}(t)] = \lambda. \tag{13}$$

We also see immediately that $h_{ov}(t) \longrightarrow 0$ as $t \longrightarrow \infty$.

For an illustration of the validity of (12) we consider now the logistic equation $f(x) = 4x(1-x)$ with the invariant density $\rho(x) = 1/(\pi\sqrt{x(1-x)})$. Thus we can calculate $H(1,\mu,\beta(\delta))$ via equations (4, 6, 7, 8) for decreasing values of δ. The results are illustrated in Figure 1 (λ= 1bit).

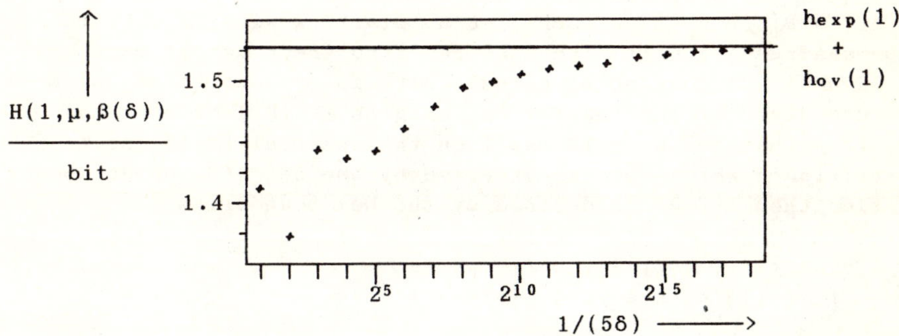

Figure 1:
Uncertainty on a future state after one iteration of the logistic equation, provided that the initial state is known, as a function of the measuring precision $1/\delta$ using the partitions $\beta(\delta) = \{[(i-1)\delta,\ i\delta[\}_{i=1,2,...,1/\delta}$;

$$h_{exp}(1) = 2 - \frac{2}{\pi \ln 2}\int_{1/4}^{1} x^{-1} \arcsin x \, dx = 1.230423\ldots \text{ bit};$$

$$h_{ov}(1) = 4 - \sqrt{15} + \frac{1}{8\pi} \,\mathrm{ld}\, \frac{4+\sqrt{15}}{4-\sqrt{15}} = 0.295224\ldots \quad \text{bit}$$

3. Information Flow Without Overlappings

We assume now that the flow f is known at the box boundaries with high accuracy, and we ask for future states knowing initial states, each of them within a finite precision. f could be approximated e. g. by the solution of the equations of motion modelling the real dynamics or by any measurements. On the basis of this knowledge on the dynamics we would not use, in general, a measuring instrument which is characterized by the same precision everywhere in phase space as assumed above. On the basis of this idea we develop now a certain partitioning concept which allows us to measure the Kolmogorov entropy for nonivertible 1D maps.

To be more precise, the future states should be represented by a partition $\beta = \{B_j\}_{j=1,2,\ldots,m}$ and the initial states by a partition $\alpha = \{A_i\}_{i=1,2,\ldots,n}$ which are allowed to differ from each other. Each of the partitions need not to contain only boxes of equal size any longer.

Due to the assumed knowledge of f it is easy to choose α such that from the kowledge of an initial state A_i we can predict exactly a future state B_j. Obviously this happens when $f(A_i) = B_{j(i)} \in \beta$ for each $A_i \in \alpha$. Starting with an arbitrary (μ-measureable) partition β we can find the roughest partition $\alpha(f,\beta)$ guaranteeing that there is no uncertainty on a future state by looking for the origins $f^{-1}(B_j)$ of each box of β. If we have e. g. a 1D map f on the interval [0,1] and a partition β which is characterized by the set $D(\beta)$ of box boundaries then $\alpha(f,\beta)$ is defined by the box boundaries

$$D(\alpha) = \{\{f^{-1}(b_j)\}\}_j \cup X_d, \text{ where}$$
$$\{f^{-1}(b_j)\} = \{x \in [0,1] \mid f(x) = b_j\}$$

are the origins of the box boundary $b_j \in D(\beta)$, and X_d denotes the points x_d of discontinuity of f that fulfil the constraint

$$\left[\lim_{\delta\downarrow 0} f(x_d \overset{(-)}{+} \delta), \lim_{\delta\downarrow 0} f(x_d \overset{(+)}{-} \delta)\right] \cap D(\beta) \neq \emptyset.$$

Possibly this construction does not provide a unique, roughest partition $\alpha(f,\beta)$ which ensures the exact prediction of future states because there may be e. g. boxes $A_i \in \alpha$ such that $\mu(A_i) = 0$, and boxes of measure zero have no influence on the information flow. Moreover, there may be a box boundary a_i of α such that there is an open ball K around a_i with $\mu(K) = 0$, and then a shift of a_i within K also has no influence on the information flow decribed in the following. To cover such situations we consider all partitions to be equivalent that differ only in boxes of measure zero.

Henceforth we call this partitioning concept Backward-Partitioning, denoted by $\alpha = \alpha(f,\beta)$. Obviously it is characterized on the one hand by avoiding overlappings as described in section 2, and on the other hand by avoiding uncertainty on a future state presuming the initial state is known.

If f is not invertible, then, on the basis of Backward-Partitioning, there remains, generally, a certain uncertainty on an initial state A_i, provided the final state B_j is known (looking backward in time). This uncertainty is given by

$$H_I(\mu,\beta,\alpha(f,\beta)) = -\sum_j q_j \sum_i q_{i/j} \operatorname{ld} q_{i/j} \qquad (14)$$

with the probabilities $q_j = \mu(B_j)$ and

$$q_{i/j} = \begin{cases} \dfrac{\mu(f^{-1}B_j \cap A_i)}{\mu(B_j)} & \text{if } \mu(B_j) > 0 \\ 0 & \text{otherwise.} \end{cases}$$

Due to the Backward-Partititoning we have

$$\mu(A_i) = \mu(f^{-1}B_j \cap A_i), \tag{15}$$

and hence (14) can be rewritten as (take notice of (4))

$$H_1(\mu,\beta,\alpha(f,\beta)) = I(\mu,\alpha(f,\beta)) - I(\mu,\beta). \tag{16}$$

An initial and a final state measurement provides the Information $I(\mu,\alpha(f,\beta))$ and $I(\mu,\beta)$, respectively. Due to the Backward-Partitioning $I(\mu,\alpha(f,\beta))$ contains all information on the final state, i. e. $I(\mu,\beta)$ equals the mutual information between initial and final state measurement. (For a more general discussion see /9/.) Formula (16) is valid also for high-dimensional maps with a suitable Backward-Partitioning. However, to be not trivial f must be noninvertible, otherwise the uncertainty (14) would vanish. We continue now with the assumption that f is a 1D map which is noninvertible and differentiable, exept possibly at a finite number of points, with an ergodic measure μ which is characterized by the information dimension $D(\mu)$, and we are interested in the uncertainty (14) when the diameter of β,

$$\emptyset(\beta) = \max_{B_i \in \beta} \{\sup\{|x-y|: x, y \in B_i\}\},$$

goes to zero. Due to the Backward-Partitioning the diameter of α also vanishes if $\emptyset(\beta)$ goes to zero and the conditional probabilities in (14) can be expressed by (take notice of (15))

$$\frac{\mu(A_i)}{\mu(B_j)} \approx \frac{\delta_i^{D(x_i)}}{(\delta_i \; |df(x_i)/dx|)^{D(x_j)}}$$

where x_i and x_j are points in $A_i \in \alpha(\beta)$ and $B_j \in \beta$, respectively, with $f(x_i) = x_j$. δ_i is the length (Lebesgue measure) of A_i, and $D(x_i)$ and $D(x_j)$ are the pointwise dimensions of μ at x_i and x_j, respectively. In the presumed ergodic case for μ-a.e. point x the pointwise dimension equals the information (Hausdorff) dimension $D(\mu)$ /10/. Hence we can approximate (14) by

$$H_1(\mu,\beta,\alpha(f,\beta)) \approx - \sum_i \mu(A_i) \text{ ld } |df(x_i)/dx|^{D(\mu)},$$

and finally we obtain

$$H_1(\mu,\beta,\alpha(f,\beta)) \xrightarrow[\phi(\beta)\,\longrightarrow\,0]{} \int \mathrm{ld}\,|df(x)/dx|^{D(\mu)}\,d\mu \qquad (17)$$

$$= D(\mu)\lambda\ ,$$

or equivalently (via (1) and (16))

$$I(\mu,\alpha(f,\beta)) - I(\mu,\beta) \xrightarrow[\phi(\beta)\,\longrightarrow\,0]{} h(f,\mu). \qquad (18)$$

Thus we have found a partitioning concept to calculate the Kolmogorov entropy as the difference of two simple (first order) entropies which might sometimes facilitate its computation. Moreover, from our construction follows that the Kolmogorov entropy of an noninvertible 1D map can be considered as the *minimum amount of information, obtained in an optimum initial state measurement, that has no predictive power.* In this interpretation the word state is understood as a box of sufficiently fine partitions β and $\alpha(f,\beta)$ that are related to each other by Backward-Partitioning. Moreover, the word optimum means that on the one hand from an initial state we can predict exactly future states, and on the other hand this prediction is realized with a minimum of experimental expense because any further refinement of $\alpha(f,\beta)$ cannot improve our predictions (future states are already determined from the knowledge of an initial state) whereas any essential coarse-graining of $\alpha(f,\beta)$ would prevent exact state prediction.

References

1. V. I. Oseledec: Tr. Mosk. Mat. Ob-va 19, 179 (1968)
2. A. N. Kolmogorov: Dokl. Akad. Nauk SSSR 119, 861 (1958)
   Dokl. Akad. Nauk SSSR 124, 861 (1959)
3. Ya. B. Pesin: Russian Math. Surveys 32, 55 (1977);
   D. Ruelle: Bol. Soc. Bras. Mat. 9, 83 (1978);
   F. Ledrappier and L.-S. Young: Ann. Math. 122, 509, 540 (1985)
4. R. Shaw: Z. Naturforsch. 36a, 80 (1981)
5. B. Pompe and R. W. Leven: Physica Scripta 34, 8 (1986)
6. J. S. Nicolis, G. Mayer-Kress, and G. Haubs: 38a, 1157 (1983).
7. B. Pompe: Dissertation, Greifswald 1986.
8. B. Pompe, J. Kruscha, and R. W. Leven: Z. Naturforsch. 41a, 801 (1986)
9. K. J. G. Kruscha and B. Pompe: Z. Naturforsch. 43a, 93 (1988)
10. L.-S. Young: Ergod. Th. & Dynam. Sys. 2, 109 (1982)

Exact and Approximate Reconstruction of Multifractal Coding Measures

G. Mantica[1] and Daniel Bessis

Service de Physique Théorique, CEN-SACLAY,
F-91191 Gif-sur-Yvette, Cedex, France.

Abstract. We show that multifractal measures arising from the symbolic dynamics of chaotic systems can be reproduced using iterated functions involving Möbius maps. This technique is exact for a family of hyperbolic billiards and can be succesfully applied to any chaotic system. The application to the anisotropic Kepler problem is developed as an example.
PACS numbers: 02.20.+b, 02.60.+y, 03.20.+i, 05.45.+b.

Multifractal measures for the coding of chaotic dynamical systems where originally studied by Martin Gutzwiller [1,2]. A simple example of coding is obtained for the geodesic motion in a singular triangle $\mathcal{T}$ in the hyperbolic upper half plane $(ds^2 = (dx^2 + dy^2)/y^2)$ [3]. Let us first describe this motion, and then introduce the coding.

$\mathcal{T}$ is bounded by the circle $\mathcal{R}$ of radius 1, centered at $x = \frac{1}{2}$, and the lines $x = 0$ (to be denoted by $\mathcal{O}$), and $x = 1$ ($\mathcal{L}$). Geodesics in this space are given by the upper half of circles centered on the real axis. They intersect this latter at ξ (the infinite past), and η (the infinite future) (see Fig.1). Periodic boundary conditions are imposed by mapping $\mathcal{R}$ on $\mathcal{O}$ via $z \rightarrow z/(1-z)$ and $\mathcal{L}$ on $\mathcal{O}$ via $z \rightarrow z - 1$.

To construct a "coding" of a trajectory of such *hyperbolic billiard* we take a pencil, look at the motion, and when this hits the side $\mathcal{R}$ we record an r; when on the contrary the trajectory encounters the side $\mathcal{L}$, we write down an l. The symbolic code (s.c) of a trajectory is the doubly-infinite (past and future) sequence of symbols $\{s_i\}$, $s_i = r, l$, $i = -\infty, \infty$. There is a one-to-one relation (modulo sets of zero measure) between s.c. and geodesics [3].

A multifractal function can be associated to the coding. We take $\eta \leq 1$, and we consider only the motion following an arbitrary initial time: $(s_0, s_1, \ldots)$, with $s_0 = r$. This sequence depends *only* upon the infinite future point η, in an arithmetical

[1]Notes of a talk delivered by the first author. BITNET: MANTICA@FRSAC11

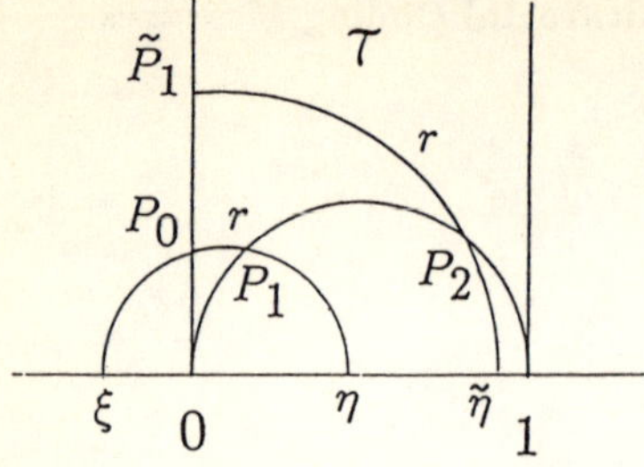

Figure 1: The hyperbolic triangular billiard $\mathcal{T}$

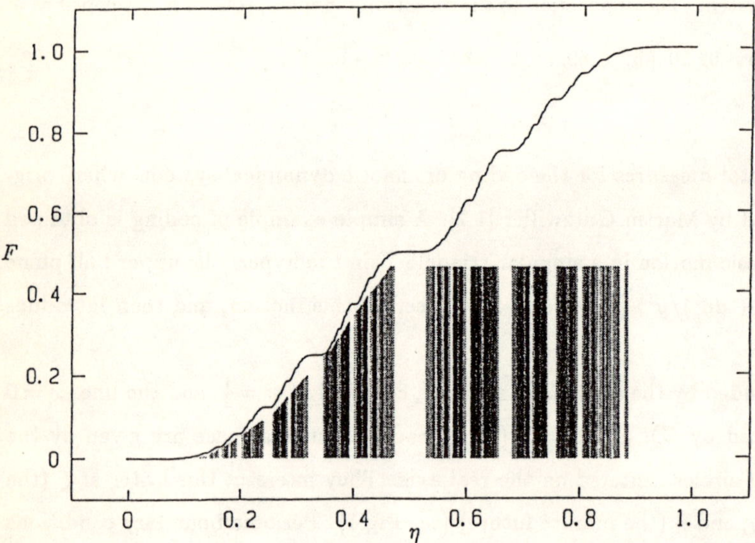

Figure 2: Coding function for the hyperbolic billiard $\mathcal{T}$. A 256-pixels discretization of μ is shown superimposed

fashion [1]. Let us now define a *coding function* F so that $F = F(\text{code}) = F(\eta))$ and that it be possible to uniquely recover the s.c. from the value of the function:

$$F(\eta) = \sum_{i=1}^{\infty} \varepsilon(s_i)\, 2^{-i}, \tag{1}$$

where $\varepsilon(r) = 0$, $\varepsilon(l) = 1$. The function defined in this way (Fig. 2) has interesting multifractal properties. We now show that they follow from a semigroup of exact symmetries, which offer an equivalent definition of F. To do this, let us consider the Möbius mappings of the unit interval into itself given by

$$\begin{aligned} M_0(x) &= 1-x, & P_0(y) &= 1-y, \\ M_1(x) &= \tfrac{x}{1+x}, & P_1(y) &= \tfrac{y}{2}, \\ M_2(x) &= \tfrac{1}{2-x}, & P_2(y) &= \tfrac{y+1}{2}. \end{aligned} \tag{2}$$

We can prove (see Ref. [4]) the crucial functional relations

$$F(M_i(\eta)) = P_i(F(\eta)), \;\; i = 0,1,2, \tag{3}$$

which can be also verified by visual inspection of Fig. 2. Eqs. (3) can be extended to the full semigroups generated by the M_i, and P_i.

These similarity properties can be used to provide an alternative definition of F, via the formalism of Iterated Functions Systems [5]. First of all, we note that F is best seen as the *distribution function* of a measure, μ : $F(\eta) = \int_0^\eta d\mu(x)$. Consider the random process over $[0,1]$ given by

$$x_n \longrightarrow x_{n+1} = M_{\sigma n}(x_n), \;\; \sigma = 1,2, \tag{4}$$

where σ_n are random variables which take the values 1, and 2, with equal probability. This is an I.F.S. generated by the two maps M_1, M_2. Its equilibrium measure can be shown to be μ; as a consequence, one can compute the density of F by taking an arbitrary x_0 and sampling a realization of the random process (4): Fig. 2.

Using I.F.S. we can also reproduce the function F itself. To do this, we use the mappings w_i from $[0,1] \times [0,1]$ into itself given by $w_i(x,y) = (M_i(x), P_i(y))$, for $i = 1,2$. These mappings possess a *unique* invariant set, the attractor G, which verifies $G = \cup_i \; w_i(G)$. Because of eqs. (3) G coincides with the *graph* of the function F, i.e. the set of points $(x, F(x))$. A random process analogous to (4) can now be introduced: $(x_{n+1}, y_{n+1}) \rightarrow w_{\sigma n}(x_n, y_n)$. The points (x_n, y_n) draw the graph of F. Fig. 1 has been obtained in this way.

Periodic trajectories can be easily computed: let p be the coding number computed via eq. (1) from the symbolic sequence of a given trajectory. The η-value leading to this trajectory is easily obtained from the equation $F(\eta) = p$.

An interesting consequence of the above formalism regards the Hölder exponent $\alpha(\eta)$ of $F(\eta)$: $F(\eta + h) - F(\eta) \sim h^{\alpha(\eta)}$, as $h > 0$ tends to zero [6]. Due to eqs. (3), and the fact that M_i are smooth maps, one finds the crucial relation

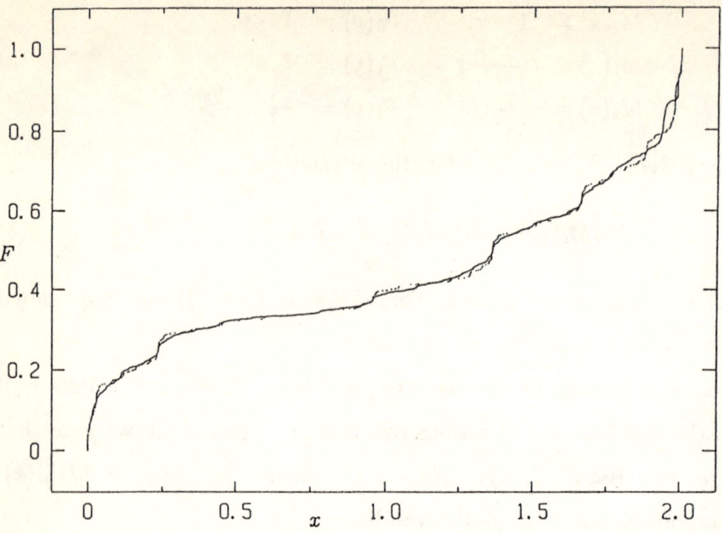

Figure 3. Exact (continuous line) and approximate (dots, I.F.S. with 4 maps) coding functions for the A.K.P., $\frac{\mu}{\nu} = 3$

$\alpha(\eta) = \alpha(M_i(\eta))$: $\alpha(\eta)$ is therefore invariant under the semigroup of Möbius maps generated by M_i, $i = 0, 1, 2$.

Not only Möbius maps I.F.S. provide the exact symmetries of hyperbolic billiards with angles at infinity, but they can also approximate the coding measures of arbitrary chaotic systems. Consider for instance the well known anisotropic Kepler problem [2], whose Hamiltonian reads

$$H = \frac{p_x^2}{2\mu} + \frac{p_y^2}{2\nu} - \frac{1}{\sqrt{x^2 + y^2}}. \tag{5}$$

In the A.K.P., the effective masses μ, and ν, are different. This system is effectively chaotic when the mass ratio $\frac{\mu}{\nu}$ is sufficiently high (≥ 5). A symbolic coding can be obtained following Gutzwiller [7]: we take a set of trajectories starting on the x positive axis, with zero initial p_x momentum. We fix a constant energy surface, e.g. $H = -1/2$. Then x labels an unique trajectory, for any $0 \leq x \leq 2$. Following the time evolution, one records a bit sequence b_j: $b_i = 0$ if the i-th intersection with the x-axis occurs for $x \leq 0$, and $b_i = 1$ otherwise. A coding function F is then defined via $F(x) = \sum b_i 2^{-i}$. This function is non-decreasing, and shows multifractal features.

We now find suitable coefficients of new Möbius maps M_i and linear maps P_i so that eqs. (3) (and the associated I.F.S.) define a good approximation of F [8,9,10]. The results are appreciably good already with a small number of maps: Fig. 3. One can indeed prove that increasing the number of maps the accuracy can be refined arbitrarily well.

The multifractal functions F are transformations from the *trivial* coding given by initial coordinates (η,x) to the *dynamica* coding induced by appropriate partitions of phase space. Hence, they resume the relevant dynamical information. I.F.S. are a powerful tool to code and decode this information. Potential applications of this technique concern image compression, forecasting and modelling of chaotic phenomena, and semiclassical quantization.

References

[1] M.C. Gutzwiller, and B.B. Mandelbrot, *Phys. Rev. Lett.* **60** (1988) 673-676.

[2] M.C. Gutzwiller, *Physica* **38 D** (1989) 160-171.

[3] C. Series, *Ergodic Theory of Dynamical Systems* **6**, (1986), 601, *J. London Math. Soc.* (2), **31**, (1985) 69, and references therein.

[4] D. Bessis and G. Mantica, "Construction of Multifractal Measures in Dynamical Systems from their Invariance Properties", SPhT 90-146, submitted to *Phys. Rev. Lett.*

[5] M.F. Barnsley and S.G. Demko, *Proc. R. Soc. London* **A 399** (1985) 243-275.

[6] B.B. Mandelbrot, *The Fractal Geometry of Nature*, (Freeman, New York, 1982).

[7] M.C. Gutzwiller, *J. Math. Phys.* **18**, (1977), 806.

[8] M.F. Barnsley, *Fractals Everywhere*, (Academic Press, New York, 1988)

[9] C.R. Handy, and G. Mantica, *Physica* **D 43**, 17-36 (1990).

[10] G. Mantica and Allan Sloan, *Complex Systems* **3**, (1989) 37-62.

Conservative Versus Reversible Dynamical Systems

J.A.G. Roberts[1], T. Post[1], H.W. Capel[1] and G.R.W. Quispel[2]

[1]Institute for Theoretical Physics, University of Amsterdam,
Valckenierstraat 65, 1018 XE Amsterdam, The Netherlands.
[2]Department of Mathematics, La Trobe University,
Bundoora, Victoria 3083, Australia.

Abstract. The study of mappings of the plane is one of the simplest, yet most usefull, ways of becoming acquainted with the behaviour of large classes of dynamical systems. The purpose of this paper is to study the relation between the two classes of planar mappings: *measure preserving* mappings and *reversible* mappings.

1. Introduction

Time-reversal symmetry has played, and still plays, an important role in physics and many of the differential equations of physics are time-reversible (for a readable account, see [1]). The classical concept of time-reversal symmetry refers to the invariance of equations under the transformation $t \to -t$. A dynamical system that is invariant under time-reversal will be called *reversible.*

A well-known example of a reversible dynamical system is a Hamiltonian system with Hamiltonian $H(\underset{\sim}{x}, \underset{\sim}{p})$, $\underset{\sim}{x}, \underset{\sim}{p} \in \mathbf{R}^n$, which is even in the momenta. In such a system, the reversibility manifests itself as invariance of the 2n equations of motion under $t \to -t$ together with a change of sign in the momenta i.e. $(\underset{\sim}{x}, \underset{\sim}{p}) \to (\underset{\sim}{x}, -\underset{\sim}{p})$. Devaney [2] generalised this invariance by defining a dynamical system as reversible if there is an involution G in phase space which reverses the direction of time. Recall that an involution is its own inverse i.e. $G \circ G = Id$, the identity mapping. In the Hamiltonian example, we have $G : (\underset{\sim}{x}, \underset{\sim}{p}) \to (\underset{\sim}{x}, -\underset{\sim}{p})$.

Significantly, the generalised time-reversal invariance is not restricted to Hamiltonian, or *conservative*, dynamical systems. Therefore there can exist reversible conservative systems, reversible non conservative systems as well as conservative non reversible systems [2-6]. In the present contribution, we illustrate this at the level of mappings of the plane, which are dynamical systems with a 2-dimensional (2D) phase space and with discrete time. The study of mappings of the plane has numerical and conceptual advantages over the study of differential equations with many of the dynamical features of the mappings being analogues of features of the continuous time flows [7-9].

The 2D mapping analogue of a Hamiltonian flow is a *measure preserving mapping*; the analogue of a reversible flow is a *reversible mapping*.

A mapping of the plane $L : \mathbf{R}^2 \to \mathbf{R}^2$ given by

$$L: \quad x' = f(x, y) \quad (1.1a)$$

$$\quad\;\; y' = g(x, y) \quad (1.1b)$$

is called measure preserving if there is a function m(x, y) such that

$$\int_D m(x, y)\, dx\, dy \;=\; \int_{L(D)} m(x', y')\, dx'\, dy' \quad (1.2)$$

for any region of the plane D. That is, the double integral of m(x, y) has the same value when extended over any region as it has when extended over the image of the region. The function m(x, y) is called a measure for the mapping L. The Jacobian determinant $J = \text{Det}\, dL(x,y)$ of a measure preserving mapping can be written $J = m(x,y) / m(x', y')$, so that at a fixed point of the mapping $J = 1$. The most frequently-studied measure preserving mappings are *area preserving mappings* for which $m(x, y) \equiv 1$ so that $J = 1$ everywhere [7]. Measure preserving, or conservative, mappings cannot possess attractors.

A mapping of the plane (1.1) is called reversible [5] if there exists an involution $G : \mathbf{R}^2 \to \mathbf{R}^2$

$$G: \quad \begin{aligned} x' &= u(x, y) \\ y' &= v(x, y) \end{aligned} \quad (1.3)$$

satisfying

$$L \circ G \circ L = G. \quad (1.4)$$

G is called a (reversing) *symmetry* of L. Usually G is also assumed to be orientation reversing i.e. $\text{Det}\, dG < 0$.

This definition implies that a reversible mapping can be written as the product (composition) of two involutions $H := L \circ G$ and G. Conversely, any mapping that decomposes as a product of involutions U and R i.e. $L = U \circ R$, is reversible with e.g. $G = U$ or $G = R$, so that this is an equivalent way of defining a reversible mapping. The reversibility property ensures that a mapping is invertible. In fact, from (1.4) with G an involution, the inverse mapping L^{-1} can be expressed as

$$L^{-1} = G \circ L \circ G, \quad (1.5)$$

that is to say, the mapping and its inverse are conjugate. This conjugacy has the important consequence that the reflection by G of the forward (backward) orbit of a point $\underset{\sim}{x}_0$ gives the backward (forward) orbit of the point $G\underset{\sim}{x}_0$. This (discrete) time-reversal [1, 6] is illustrated in

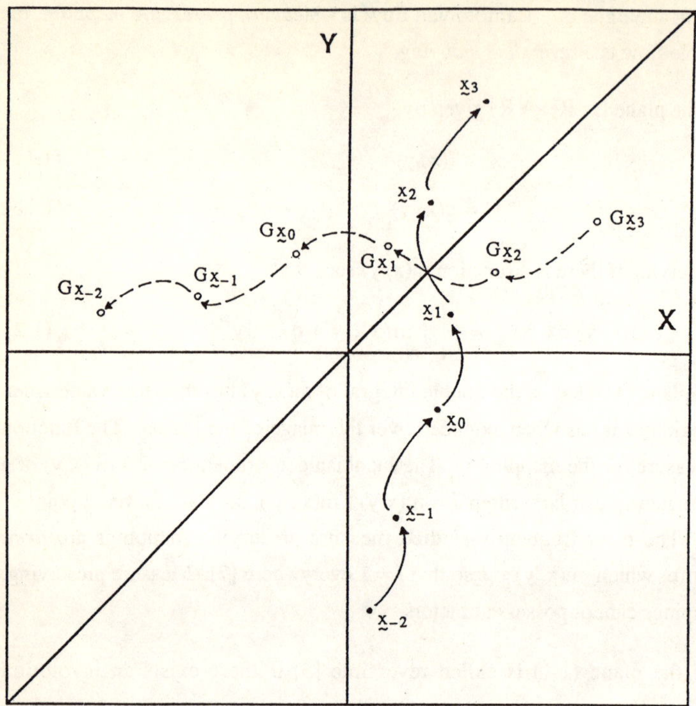

Figure 1. Illustration of the nature of the motion in a reversible mapping L of the plane with symmetry G: x' = y, y' = x. Shown is part of the trajectory of a point $\underset{\sim}{x}_0$ with unbroken arrows indicating the action of L on each point. Reversibility implies that the trajectory of $G\underset{\sim}{x}_0$ is found by reflecting the trajectory of $\underset{\sim}{x}_0$ by G. The forward and backward trajectories are interchanged on reflection as the broken arrows indicate

Figure 1. Thus if the motion in one part of the plane is known, the motion in another part of the plane (i.e. in the region reflected by G) can be deduced.

The vast majority of reversible mappings studied to date have been conservative, and in particular area preserving. Conversely, almost without exception, all explicit area preserving mappings studied numerically in the literature have been reversible. For example, the mapping

$$M: \begin{array}{ll} x' = x + f(y) & \quad (1.6a) \\ y' = y + g(x') & \quad (1.6b) \end{array}$$

is readily verified to be area preserving. If f(y) or g(x') is an odd function, the mapping is also reversible. In the former case an involution is

$$G: \begin{array}{l} x' = x + f(y), \quad f \text{ odd} \\ y' = -y. \end{array} \tag{1.7}$$

The choice $f(y) = K \sin y$, $g(x') = x'$ in (1.6) gives the well known Chirikov-Taylor or standard mapping. As far as we know, up to now no-one has been able to analytically show that an explicitly given conservative mapping of the plane is not reversible.

A possible relation between the two sets of conservative and reversible mappings is depicted in the Venn diagram in Figure 2. Region II comprising measure preserving reversible mappings has been thoroughly investigated. Here we show that such a Venn diagram is valid since there exist reversible non measure preserving mappings (Region I) and measure preserving non reversible mappings (Region III) cf. also [6].

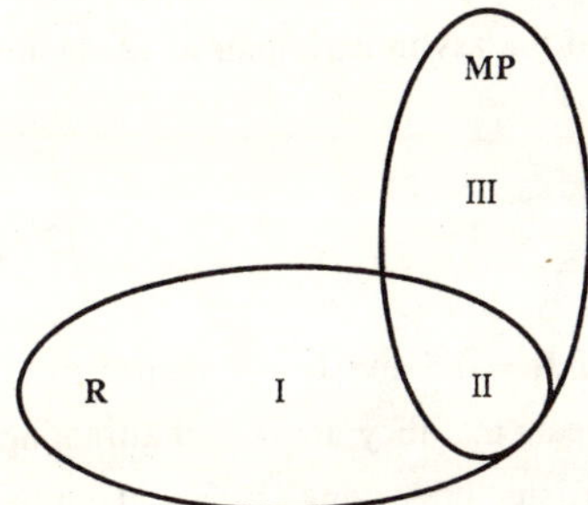

Figure 2. Venn diagram illustrating the relation between measure preserving mappings (denoted by MP) and reversible mappings (denoted by R)

2. Reversible Mappings need not be Measure Preserving

We have been able to construct reversible non conservative mappings and study their properties. We do this by creating a non area preserving involution and composing it with an area preserving involution. We illustrate the method with a simple example. We create the non area preserving involution H_1

$$H_1: \quad \begin{aligned} x' &= y\,(1+ (y'-1)^2) \\ y' &= \frac{x}{1+ (y-1)^2}, \end{aligned} \tag{2.1}$$

which we compose with the simple one parameter area preserving involution

$$G_1: \quad \begin{aligned} x' &= x \\ y' &= C - y. \end{aligned} \tag{2.2}$$

The result $H_1 \circ G_1$ is the reversible mapping :

$$L_1: \quad \begin{aligned} x' &= (C-y)(1+ (y'-1)^2) \\ y' &= \frac{x}{1+ (C-y-1)^2}. \end{aligned} \tag{2.3}$$

Note that H_1 and G_1 are orientation reversing and that their fixed points form lines in the plane e.g. $y = C/2$ for G_1. These lines are called symmetry lines (actually, once given one decomposition of L into symmetries H and G, one can find an infinite family of other symmetries with their own symmetry lines). A fixed point (periodic orbit) of a reversible mapping is called symmetric if it is also invariant under the symmetries. Otherwise, it is asymmetric and its reflection under the symmetries is another fixed point (periodic orbit) of the mapping.

One finds that for a substantial range of the parameter C the mapping L_1 above has a symmetric fixed point and a pair of asymmetric fixed points, e.g., at $C = 3$ these are readily calculated: the symmetric fixed point is at (15/8, 3/2) and the asymmetric pair at (2, 1) and (2, 2). The Jacobian determinant of (2.3) is:

$$J_1(x, y) = \frac{1+(y'-1)^2}{1+(C-y-1)^2}. \tag{2.4}$$

So at the asymmetric fixed points when $C = 3$, we find $J_1 = 0.5$ and $J_1 = 2$ respectively. Further calculation of the linearisation at these points reveals that they are in fact attracting, respectively repelling, fixed points. Therefore L_1 is not measure preserving. In fact, the time-reversal property of a reversible mapping implies that the presence of an attractor implies the presence of a repellor. The symmetric fixed point of L_1 has $J = 1$ and this approximate area preservation associated with symmetric periodic orbits is also a general property of reversible mappings.

The fact that reversible mappings of the plane can have both conservative and dissipative behaviour is illustrated pictorially in the phase portrait of another of our mappings in Figure 3. The origin in the picture is an elliptic symmetric fixed point and the curves and island chains around it are reminiscent of an area preserving mapping (in fact these are KAM curves because a KAM theorem exists for reversible mappings cf. refs. [3-5]). In the bottom right hand corner, there is a spirally attracting fixed point at A surrounded by an attracting 5 cycle. The reflection of these features in the symmetry line $y = x$ gives a repelling fixed point R and a repelling 5 cycle.

Numerical investigations of our mappings (cf. [6], [10-11]) have revealed that :

* Symmetric fixed points period double symmetrically with the scalings previously found in area preserving mappings [7] e.g. $\delta_{PD,SYM} = 8.721..$.
* Asymmetric fixed points period double with the scalings previously found in dissipative ($|J| < 1$) mappings [8] e.g. $\delta_{PD,ASYM} = 4.669..$.
* As a parameter is varied, the KAM curves with 'noble' winding number surrounding symmetric fixed points break up with the scalings previously found in area preserving mappings [7].

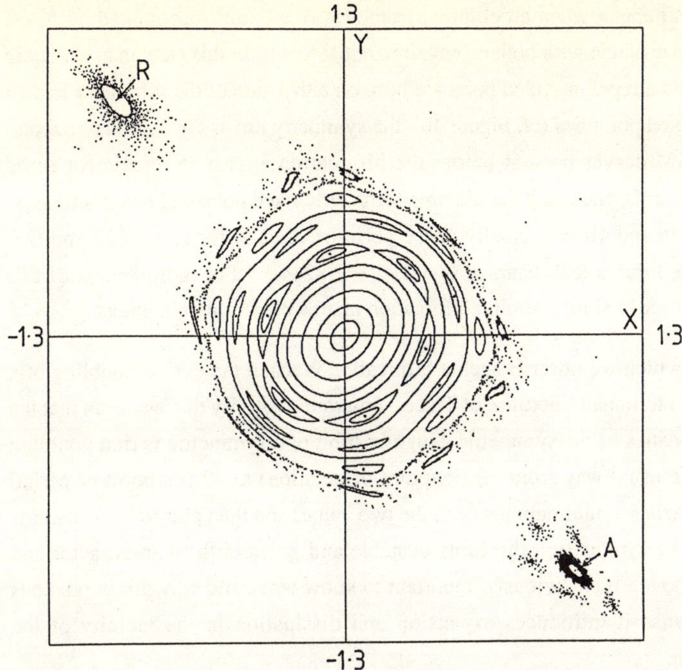

Figure 3 : This picture illustrates some of the important properties of a non measure preserving reversible mapping L. Here the mapping has a symmetry G : x' = y, y' = x. The origin is an elliptic symmetric fixed point and is surrounded by invariant curves. These curves intersect the symmetry line y = x and their symmetry with respect to reflection about this line is obvious. Also shown are elliptic symmetric 6, 7 and 8-cycles at the centres of the concentric 'rings' of 6, 7 and 8 islands. External to these islands is a hyperbolic 10-cycle which nearby points move around before escaping to infinity (creating the 10 fuzzy islands). Away from the origin, the fixed point A at (1, –1) is attracting and its reflection GA = R at (–1, 1) is repelling. A trajectory that spirals in towards A is shown; its relection by G spirals outwards from R. An attracting 5-cycle is found near A and a trajectory that spirals towards it is also shown. The trajectories of many of the points not enclosed by curves around the origin or near A escape to infinity

These results extend the universality classes previously associated with these exponents and show that a reversible mapping can have two period doubling cascades in different parts of the plane occurring with different rates, one conservative and one dissipative. Other observations include the appearance of strange attractors (and therefore, because of reversibility, strange repellors) beyond the point of accumulation of the asymmetric period doubling cf. [6, 11].

3. Attractors and Repellors born from Conservative Features

In Figure 3, we see that the attractor and repellor in our reversible mapping lie on the outside of the KAM curves. It is also possible to have the attractor and repellor encircled by these curves. This can happen when an elliptic symmetric fixed point surrounded by KAM curves (Figure 4a) turns unstable with both eigenvalues equal to +1. In this case an asymmetric attracting fixed point and a repelling fixed point are born on either side of the symmetry line on which the symmetric fixed point lies (cf. Figure 4b - the symmetry line is the dashed horizontal line shown). The KAM curves present before the bifurcation appear to remain for some parameter interval afterwards, enclosing the unstable symmetric fixed point and two asymmetric fixed points. Figures 4a and 4b are actually phase portraits of (2.3) for C = 2.827 and C = 2.83, respectively. The bifurcation occurs at $C = 2\sqrt{2} = 2.8284..$. For more details of this bifurcation cf. [12], and see [13] for a similar bifurcation in area preserving mappings.

One situation in which we observe such a bifurcation is when the period doubling of a symmetric periodic orbit terminates because of Trace Turnaround [6]. By this we mean that the Trace of the Jacobian matrix of the symmetric daughter orbit of a symmetric period doubling does not proceed in the usual way from +2 (its point of creation) to –2 (its point of period doubling), but rather reaches a minimum between the two values and then climbs back towards +2. On reaching +2, the symmetric orbit turns unstable and gives birth to an attractor and repellor as described above. It is obviously important to know when and why this anomalous behaviour occurs because it introduces expansion and dissipation in the vicinity of the symmetric periodic orbit.

4. Measure Preserving Mappings need not be Reversible

We have been able to construct measure preserving and area preserving mappings which we can show are not reversible.

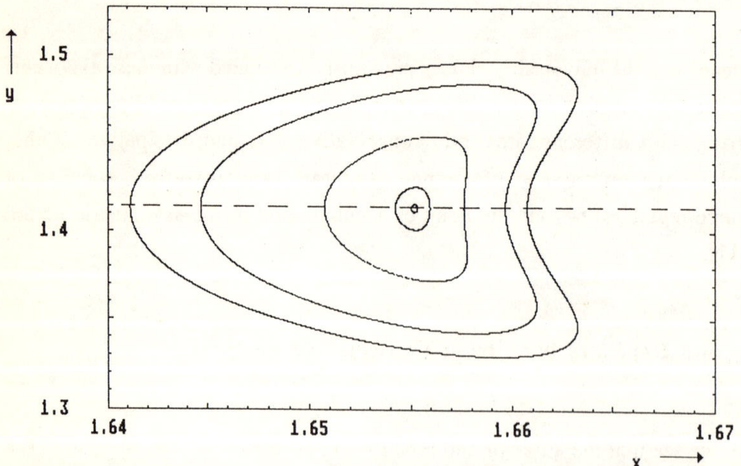

Figure 4a. Before the bifurcation of an attractor and a repellor from a symmetric fixed point

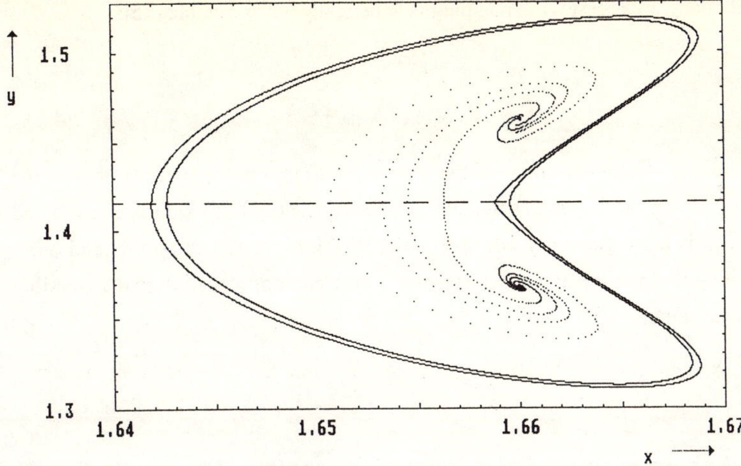

Figure 4b. After the bifurcation of an attractor and a repellor from a symmetric fixed point

To do this we use the idea of *local reversibility* (cf. [14]). The idea is that if a mapping is reversible, then the relation (1.4) holds throughout the plane. In particular, if we have a symmetric fixed point, which by definition is fixed under G *and* L, then (1.4) holds in a neighbourhood of the point. For a given measure preserving mapping which is not known *a priori* to be reversible, we can identify a fixed point as necessarily symmetric if the mapping is reversible by comparing its eigenvalues with those of the other fixed points of the mapping. This uses the fact that asymmetric fixed points of a reversible mapping necessarily have inverse eigenvalues. Consequently if the fixed point has no 'partner' with inverse eigenvalues it is symmetric if the mapping is reversible. By now taking the expansions of L and a possible G around the fixed point and substituting in (1.4), we can determine at the nth order of the expansion necessary conditions on the coefficients of L in order that there exist some mapping satisfying (1.4) which is an involution to order n. If these conditions are not satisfied then L is not reversible because not locally reversible around the fixed point implies not globally reversible.

Application of this method to third order around hyperbolic and elliptic fixed points of measure preserving mappings and fixed points with linearisation (1, b, 0, 1) with $b \neq 0$ revealed that local measure preservation always implies local reversibility [14].

We have applied the method to expansions of measure preserving mappings around fixed points which have linear part given by the identity matrix (1, 0, 0, 1) and no second order terms [15]. At fourth order we find an additional necessary condition for reversibility that is not implied by measure preservation.

As a result, the following area preserving mapping is generically not reversible [15]:

$$P: \quad \begin{aligned} x' &= x + ay^3 + by^4 && (4.1a) \\ y' &= y + cx'^3 + dx'^4 && (4.1b) \end{aligned}$$

when $ac \neq 0$ and $bd \neq 0$. Note that this is of the form of (1.6) except that here neither of the functions f(y) nor g(x') are odd. It raises the question as to whether (1.6) with both even and odd terms in f and g is more generally not reversible. We hope to report on this and other questions concerning the similarities and differences between conservative and reversible mappings in the near future.

References

[1] R G Sachs, *The Physics of Time Reversal* (Univ. of Chicago Press, Chicago, 1987).

[2] R L Devaney, Trans. Am. Math. Soc. 218 (1976) 89.

[3] V I Arnol'd, in: *Nonlinear and Turbulent Processes in Physics* Vol 3, ed. R Z Sagdeev (Harwood, Chur, 1984) p. 1161.

[4] V I Arnol'd and M B Sevryuk, in: *Nonlinear Phenomena in Plasma Physics and Hydrodynamics*, ed. R Z Sagdeev (Mir, Moscow, 1986) p. 31.

[5] M B Sevryuk, *Reversible Systems*, Lecture Notes in Mathematics Vol 1211 (Springer, Berlin, 1986).

[6] J A G Roberts and G R W Quispel, *Chaos and Time-Reversal Symmetry*, submitted to Physics Reports (1991).

[7] R S MacKay and J D Meiss (eds.), *Hamiltonian Dynamical Systems : A Reprint Selection* (Adam Hilger, Bristol, 1987).

[8] P Cvitanovic (ed.), *Universality in Chaos* (Adam Hilger, Bristol, 1984).

[9] A J Lichtenberg and M A Lieberman *Regular and Stochastic Motion* (Springer, New York, 1983).

[10] G R W Quispel and J A G Roberts, Phys. Lett. A 132 (1988) 161.

[11] G R W Quispel and J A G Roberts, Phys. Lett. A 135 (1989) 337.

[12] T Post, H W Capel, G R W Quispel and J P van der Weele, Physica A 164 (1990) 625.

[13] R J Rimmer, Mem. Am. Math. Soc. 41 (1983) 1.

[14] G R W Quispel and H W Capel, Phys. Lett. A 142 (1989) 112.

[15] J A G Roberts and H W Capel, *Area Preserving Mappings that are not Reversible,* submitted to Phys. Lett. A (1991).

A Simple Method to Generate Integrable Symplectic Maps

O. Ragnisco

Dipartimento di Fisica dell'Università degli Studi di Roma "La Sapienza",
P.le Aldo Moro, 2, Roma, Italy,
Istituto Nazionale di Fisica Nucleare, Sezione di Roma,Italy.

Abstract. We show that by taking stationary flows of integrable evolution equations on lattices one obtains integrable symplectic maps. We also tersely discuss an alternative method based on the so-called nonlinearization of a scattering problem, and elucidate its intimate connections with the previous one. A few examples of possibly interesting integrable maps are presented.

In the last decade, most of the research on integrable systems has been devoted to investigate integrability in more than one space dimension. Only recently, the opposite path has been also pursued, and the classical topic of integrable systems with finitely many degrees of freedom has been somehow rediscovered by people working on soliton theory. In this context, the pioneering work by Bogoyavlenski and Novikov on stationary KdV flows [1] has been attracting larger and larger attention, and other techniques have been proposed, for instance by chinese scientists [2], to generate integrable hamiltonian systems with finitely many degrees of freedom out of integrable partial differential equations.

To our knowledge, Quispel [3] was the first who pointed out that by taking stationary versions of discrete integrable evolution equations one should have been able to produce integrable «discrete-time» hamiltonian systems, i.e. integrable maps. Integrable maps were later on constructed by several people: here, we quote for instance Moser and Veselov [4] and Capel-Nijhoff-Papageorgiou [5].

Using the general theory developed by Maeda [6], Quispel's claim has been rigorously proven by the author in a joint work with Bruschi, Santini and Tu [7].

In the following, the main results derived in [7] are briefly summarized.

As a starting point, one considers integrable discrete evolution equations arising as compatibility conditions between the linear problems:

$$\Psi_{n+1} = U_n \Psi_n \qquad (n \in \mathbb{Z}, t \in \mathbb{R}) \tag{I.1a}$$

$$\Psi_{n,t} = V_n \Psi_n \tag{I.1b}$$

namely:

$$U_{n,t} = U_n V_{n+1} - V_n U_n \tag{I.2}$$

where U_n, V_n are NxN matrices depending on the fields $u_\alpha(n,t)$ ($\alpha=1, \ldots M \leq N^2$) and (rationally) on the spectral parameter $\lambda \in \mathbb{C}$. For U_n an invertible matrix, the stationary version of (I.2) reads:

$$V_{n+1} = U_n V_n U_n^{-1} \tag{I.3}$$

and thus is nothing but a similarity transformation for V: in other words, with respect to the discrete time "n", the matrix V evolves in such a way that $I_k := \mathrm{Tr}\, V^k$ are preserved. Note that (I.3) can be viewed as the «evolution» equation arising from the Lax pair:

$$V_n \Psi_n = \mu \Psi_n \tag{I.4a}$$

$$\Psi_{n+1} = U_n \Psi_n \tag{I.4b}$$

Therefore, it is reasonable to expect that stationary flows of integrable discrete evolution equations are integrable discrete-time systems (i.e. integrable maps). In ref.[7], under rather "mild" technical assumptions, one has proven that this is indeed the case. The key-points are the following ones:

(i) The evolution equations (I.2) are a hierarchy of commuting flows, possessing a common (infinite) sequence of integrals of motions in involution.

(ii) Property (i) implies that each stationary flow is a Lagrangian map and thus (under some regularity conditions) a symplectic map, possessing infinitely many conserved quantities.

(iii) These conserved quantities are in involution with respect to the canonical Poisson bracket and enough of them are functionally independent. Hence, each stationary flow (I.3) can be linearized by introducing appropriate action-angle variables.

We present a single nontrivial example of a completely integrable map of "standard type", that comes out as a reduction of the stationary flow of the discrete Nonlinear Schroedinger equation. It reads:

$$u_{n+1} + u_{n-1} - \frac{2u_n}{1-u_n^2} = 0 \tag{I.5a}$$

By introducing the canonical coordinates $q=u_n$, $p=u_{n-1}$, Eq.(I.5a) takes the form:

$$p' = q \tag{I.5b}$$

$$q' = -p + \frac{2q}{1-q^2} \tag{I.5c}$$

The map (I.5) has the invariant: $I = (q-p)^2 - q^2 p^2$ and can be explicitly linearized in terms of elliptic functions. Indeed we have:

$$q(k) = u_k = \frac{a}{\sqrt{a^2-1}} \text{sn}\,(k\,\text{sn}^{-1}\,a + \varphi_o) \tag{I.5d}$$

where $I: = \dfrac{a^2}{\sqrt{a^2-1}}$

I would like to add a few words about a different approach, widely exploited in the continuous case to get integrable systems of nonlinear ODEs. It is based on the so-called «nonlinearization» of a given eigenvalue problem, obtained, at least in its simplest version, by assuming the «potentials» to be restricted to the «finite gap» sector. Actually, this approach can be reduced to the «stationary flows» approach. In fact, given a linear (differential or difference) operator L[u], the eigenvalue equations:

$$L[u]\,\psi^{(j)} = \lambda_j\,\psi^{(j)} \qquad j = 1, \ldots, N \tag{I.6}$$

entails, as a (differential or difference) consequence, the following eigenvalue equations for (the adjoint of) the associated recursion operator N*:

$$N^* \frac{\delta\lambda_j}{\delta u} = \lambda_j \frac{\delta\lambda_j}{\delta u}\,, \tag{I.7}$$

and thus:

$$(N^*)^k \frac{\delta\lambda_j}{\delta u} = \lambda_j^k \frac{\delta\lambda_j}{\delta u}\,, \tag{I.8}$$

As is well known, N* generates an infinite sequence of gradients of commuting Hamiltonians out of the starting one, through the recursive relation:

$$y^{(k)} = N^*\,y^{(k-1)} \qquad k=1,2,\ldots \quad \left(y^{(k)} = \frac{\delta \mathcal{H}^{(k)}}{\delta u};\ \{\mathcal{H}^{(k)}, \mathcal{H}^{(\ell)}\} = 0\right)$$

Thus, by constraining the field u on the finite-gap manifold:

$$y^{(0)}\,[u] = \sum_{j=1}^{N} \frac{\delta\lambda_j}{\delta u}\,, \tag{I.9}$$

Eq.(I.8) implies:

$$y^{(k)}[u] = \sum_{j=1}^{N} \lambda_j^{\kappa} \frac{\delta\lambda_j}{\delta u} \qquad k=0, \ldots, N-1 \tag{I.10}$$

and eq.(I.7) can be identically rewritten in the «stationary flow» form:

$$(N^*-\lambda_1)(N^*-\lambda_2) \ldots (N^*-\lambda_N)\, y^{(0)} := \sum_{k=0}^{N} c_k\, y^{(k)} = 0 \tag{I.11}$$

To be more concrete, I shall derive an integrable map that is obtained by «non-linearizing» the Toda-lattice eigenvalue problem, and provides a discrete-time Garnier system. For the Toda lattice, (I.6) reads:

$$a_{n-1} \Psi^{(j)}_{n-1} + b_n \Psi^{(j)}_n + a_n \Psi^{(j)}_{n+1} = \lambda_j \Psi^{(j)}_n \tag{I.12}$$

and we have: $\dfrac{\delta\lambda_j}{\delta a_n} = 2\, \Psi^{(j)}_n\, \Psi^{(j)}_{n+1}$, $\dfrac{\delta\lambda_j}{\delta b_n} = (\Psi^{(j)}_n)^2$

Choosing $y^{(0)}$ as $\binom{2a_n}{b_n}$, the constraint (I.9) yields, from (I.12), the nonlinear map:

$$(\Psi_{n-1}, \Psi_n)\, \Psi^{(j)}_n + (\Psi_n, \Psi_n)\, \Psi^{(j)}_n + (\Psi_n, \Psi_{n+1})\, \Psi^{(j)}_{n+1} = \lambda_j \Psi^{(j)}_n \tag{I.13}$$

when $(\Psi_n, \Psi_m) := \sum_{j=1}^{N} \Psi^{(j)}_n \Psi^{(j)}_m$

The map (I.6) comes from the Lagrangian:

$$\mathcal{L}(x,y) = \frac{1}{2}(x,y)^2 + \frac{1}{4}(x,x)^2 - \frac{1}{2}(x, \Lambda x) \tag{I.14}$$

$$\Lambda = \mathrm{diag}\,(\lambda_1, \ldots, \lambda_N)\ ;\ x_j = \Psi^{(j)}_n,\ y_j = \Psi^{(j)}_{n+1}$$

By introducing the canonical coordinates:

$$q_j = \psi^{(j)}_n\ ; \quad p_j = \frac{\partial \mathcal{L}}{\partial y_j} \Big|\, x_j = \psi^{(j)}_{n-1},\ y_j = \psi^{(j)}_n$$

Eq.(I.13) takes the elegant canonical form:

$$q'_j = g^{-1} (\Lambda_j q_j - p_j - (q,q)\, q_j) \qquad (I.15a)$$

$$p'_j = g\ p_j \qquad (I.15b)$$

$$g = \pm \sqrt{(q, \Lambda q) - (p,q) - (q,q)^2}$$

Being equivalent to a stationary flow of type (I.11), the map (I.13) (or (I.15)) is integrable. Indeed, we have proven that it has N functionally independent invariants in involution, given by:

$$K_j = \sum_{k\neq j} \frac{\ell_{jk}^2}{\lambda_j - \lambda_k} - p_j^2 - (p,q)\, q_j^2 + \lambda_j p_j q_j \quad (j=1, \ldots, N) \qquad (I.16)$$

where $\ell_{jk} = p_j q_k - p_k q_j$.

References

[1] O.I.Bogoyavlenski and S.P.Novikov, Funct. Anal. Appl. 10 (1976), 9-13.

[2] See for instance: Cao Ce Wen, Chinese Science A7 (1989), 701-707.

[3] G.R.W.Quispel, J.A.G.Roberts and C.J.Thompson, Physica 34D (1989), 183.

[4] J.Moser and A.P.,Veselov, "Discrete Version of Some Classical Integrable Systems and Factorization of Matrix Polynomials", preprint Forschung Institut fuer Mathematik, ETH Zurich 1989.

[5] V.G.Papagerogiau, F.W.Nijhoff and H.W.Capel, "Integrable Mapping and Nonlinear Integrable Lattice Equations", Preprint INS 143, Clarkson University 1990.

[6] S.Maeda, Math. Jap. 25 (1980), 405-420.

[7] M.Bruschi, O.Ragnisco, P.M.Santini and Gui Zhang Tu, "Integrable Symplectic Maps", Physica D (to appear).

Integrable Mappings and Soliton Lattices

H.W. Capel[1], F.W. Nijhoff[2], V.G. Papageorgiou[2] and G.R.W. Quispel[3]

[1]Instituut voor Theoretische Fysica, Universiteit van Amsterdam,
Valckeniersstraat 65, 1018 XE Amsterdam, The Netherlands.
[2]Deparment of Mathematics and Computer Science and
Institute for NonlinearStudies, Clarkson University,
Potsdam, NY 13699-5815,USA.
[3]Department of Mathematics, La Trobe University,
Bundoora, Victoria 3083, Australia.

1 Introduction

Solitons are found as solutions of integrable partial differential equations (PDE's) which have been extensively studied during the last three decades. More recently there has been increasing interest in partial difference equations with fields defined at the sites of 2- and 3-dimensional (2D and 3D) lattices. They are the discrete-time analogues of PDE's as well as of partial difference equations depending on time as well as on 1 or 2 discrete variables. During the last years a variety of integrable 2D and 3D lattices has been found, by the direct linearization method (DLM) which is based on a linear integral equation with arbitrary measure and contour [1]-[4], as well as by the bilinearization method [5] and the τ-function approach [6]. By taking continuum limits the lattice equations yield hierarchies of integrable partial difference equations (or PDE's) with one or more continuous väriables together with a Poisson structure and an infinite number of conserved quantities in involution [7,8], see also ref. [9].

On the other hand there has been a widespread interest in dynamical mappings with $1, 2, 3, \cdots$ degrees of freedom (dimensions) as time-discrete analogues of sets of coupled ordinary differential equations, mainly in connection with chaotic phenomena. A very special type of mappings are the *integrable* mappings [10]-[15]. These mappings are *symplectic*, i.e. they have a Poisson-bracket structure that is invariant under the mapping, and $2N$ dimensional mappings have N integrals that are in involution with respect to the Poisson bracket. In spite of the progress on the theory of integrable mappings there are still many open problems. An 18-parameter family of 2 dimensional mappings with 1 integral generalizing the McMillan mapping [10] has been given in ref. [16], but the symplectic structure of this family has not been investigated so far. In this report we show how integrable mappings with $2P$ degrees of freedom, $P = 1, 2, \cdots$, are obtained by systematic reduction of 2D lattices [17]-[19] taking the lattice version of the Korteweg-de Vries (KdV) equation as an example.

In section 2 we consider a rather general class of difference equations with solutions that are found from periodic initial conditions on a staircase consisting of subsequent horizontal and vertical steps and which can be described in terms of $2P$ dimensional mappings. In section 3 we introduce the $2P$ dimensional mappings associated with the lattice KdV and the integrals of the mapping are evaluated in section 4 from the Lax representation of the lattice KdV. In section 5 we give an action principle for the lattice KdV which leads in a natural way to the Poisson-bracket structure for the KdV-type of mappings. The involution property of the integrals is treated in section 6 and some concluding remarks are given in section 7.

2 Periodic solutions of difference equations

We consider a difference equation on the 2D lattice of the type

$$\Phi(u, Hu, Vu, HVu) = 0 \tag{2.1}$$

where Φ is a function of the variable u at the 4 sites of an elementary square on the 2D (square) lattice. At some site of the lattice we have the field u, the field at the site that is obtained by a horizontal (vertical) shift is denoted as Hu (Vu). We assume that from the function Φ in (2.1) any of the 4 fields in the argument can be solved as function of the other 3 ones. Then a complete initial condition for eq. (2.1) is provided by specifying the fields $\cdots, a_0, a_1, a_2, \cdots$ at the sites of a staircase consisting of successive horizontal and vertical steps as indicated in Figure 1.

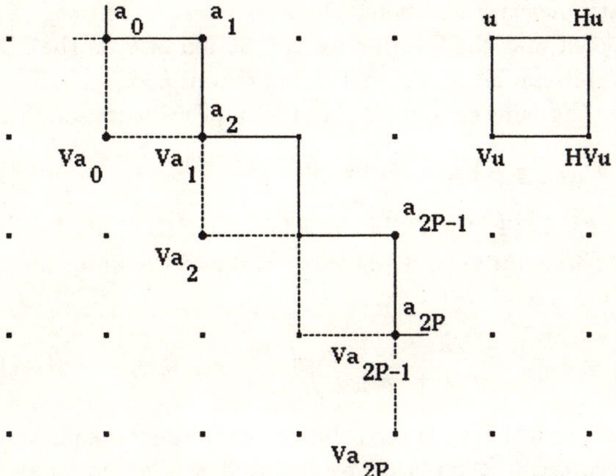

Figure 1. Difference equation, staircase (——) and shifted staircase (- - -) on the 2D lattice

By completing squares and using (2.1) one can solve the difference equation iteratively at all sites at the left and below the staircase.

We now consider the case that the initial data on the staircase satisfy the periodicity property $(HV)^P u = u$, $P = 1, 2, \cdots$. Then by completing squares it is clear that the same property holds for the complete solution of (2.1). In the periodic case it is sufficient to specify the fields $a_0, a_1, \cdots, a_{2P-1}$ on a finite part of the staircase consisting of P horizontal and P vertical steps ($a_j = a_{j+2P}$). To solve the difference equation we can consider the $2P$ dimensional mapping corresponding to the vertical shift V. The mapping is given by

$$Va_{2j+1} = a_{2j+2} \quad , \quad \Phi(a_{2j}, a_{2j+1}, Va_{2j}, a_{2j+2}) = 0 \quad , \tag{2.2}$$

$j = 0, 1, \cdots, P-1$, as is clear from (2.1) and Fig.1 considering the elementary square with $u = a_{2j}$, $Hu = a_{2j+1}$, $HVu = a_{2j+2}$.

3 Lattice KdV

The solution of the lattice KdV can be found from the linear integral equation

$$u_k + \rho_k \int_C d\lambda(l) \frac{u_l}{k+l} = \rho_k \tag{3.1}$$

in which ρ_k is a free-wave basis function satisfying [1]

$$H\rho_k = \frac{p+k}{p-k}\rho_k \quad , \quad V\rho_k = \frac{q+k}{q-k}\rho_k \tag{3.2}$$

(The integral equation with $\rho_k = \exp(kx + k^3t)$ was introduced in ref. [20] to study the solutions of the KdV.) C is an arbitrary contour in the space of the complex spectral parameter k and $d\lambda(k)$ an arbitrary measure satisfying the *uniqueness condition*, i.e. for the given ρ_k C and $d\lambda(k)$ are such that the solution u_k as function of k on the 2D lattice is unique. Starting from (3.1) it can be shown that the potential $u = \int_C d\lambda(k) u_k$ obtained by an integration of u_k over the same contour C with the same measure satisfies

$$(p + q + u - HVu)(p - q + Vu - Hu) = p^2 - q^2 \quad , \tag{3.3}$$

which is the 2D lattice version of the KdV. In the DLM the solutions of (3.3) are found solving the linear integral equation (3.1). From eq. (3.1) with (3.2) and the uniqueness condition one finds the linear relation

$$(p-k)H \begin{pmatrix} u_k \\ v_k \end{pmatrix} = \begin{pmatrix} p - Hu & 1 \\ k^2 - p^2 + * & p + u \end{pmatrix} \begin{pmatrix} u_k \\ v_k \end{pmatrix} \tag{3.4}$$

corresponding to the horizontal shift, in which v_k is the solution of an integral equation similar to (3.1) but with the source term ρ_k replaced by $k\rho_k$ and $*$ is a short-hand notation for the product of the diagonal elements $(p - Hu)(p + u)$. For the vertical shift we have a similar relation that can be found from (3.4) by the replacements $H \to V$, $p \to q$. The compatibility of both relations (3.4) yields the lattice KdV (3.3).

Assuming the periodicity property $(HV)^P u = u$ one obtains from (3.3) and (2.2) an explicit expression for the $2P$ dimensional KdV mapping in terms of the fields a_j with $j = 0, 1, \cdots, 2P-1$ on the staircase. This mapping can be reduced to a $2P-2$ dimensional mapping in terms of the fields $v_j = \varepsilon + a_{j-1} - a_{j+1}$ with $\varepsilon = p+q$, $\delta = p-q$. We have [17]

$$\begin{aligned} Vv_{2j} &= v_{2j+1} \\ Vv_{2j+1} &= v_{2j+2} - \frac{\varepsilon\delta}{v_{2j+3}} + \frac{\varepsilon\delta}{v_{2j+1}} \quad , \quad j = 0, 1, \cdots, P-2 \end{aligned} \tag{3.5}$$

with $v_{2P-1} = P\varepsilon - (v_1 + v_3 + \cdots + v_{2P-3})$, $v_{2P-2} = P\varepsilon - (v_0 + v_2 + \cdots + v_{2P-4})$. For simplicity we consider here only periodic conditions ($a_j = a_{j+2P}$). More general situations with $a_{2j+2P} = a_{2j} + C_0$, $a_{2j+1+2P} = a_{2j+1} + C_1$ can be treated as well using a slightly different Lax representation [18].

4 Integrals

To evaluate the integrals of (3.5) we compare the basis functions u_k, v_k of the *Lax representation* (3.4) at the points $(0,0)$ and (P,P) at the beginning and at the end of the staircase. We have

$$(p-k)^P(q-k)^P \begin{pmatrix} u_k(P,P) \\ v_k(P,P) \end{pmatrix} = \mathcal{T}_k \begin{pmatrix} u_k(0,0) \\ v_k(0,0) \end{pmatrix} \tag{4.1}$$

in which the *monodromy matrix* is the product of all Lax matrices along the staircase, i.e.

$$\mathcal{T}_k(a_0, a_1, \cdots, a_{2P-1}) = \overleftarrow{\prod_{j=0}^{2P-1}} L_k(a_j, a_{j+1}, p_j) \tag{4.2}$$

$$L_k(a_j, a_{j+1}, p_j) = \begin{pmatrix} p_j - a_{j+1} & 1 \\ k^2 - p_j^2 + * & p_j + a_j \end{pmatrix} \tag{4.3}$$

In eq. (4.3) $p_j = p$ for $j =$ even, $p_j = q$ for $j =$ odd and $*$ denotes the product of the diagonal elements, $\longleftarrow$ in eq. (4.2) means that the matrices in the product are ordered from the right to the left.

But there is another way to connect the basis functions at $(0,0)$ and (P,P) as shown in Fig.1. From a_0 we go down to Va_0, next we go through the shifted staircase until we reach Va_{2P} and finally we go to a_{2P}. Thus $\mathcal{T}_k$ is also given by

$$\mathcal{T}_k = L_k^{-1}(a_{2P}, Va_{2P}, q) \left(\overleftarrow{\prod_{j=0}^{2P-1}} L_k(Va_j, Va_{j+1}, p_j) \right) L_k(a_0, Va_0, q) \tag{4.4}$$

Because of the periodicity the Lax matrices involving a_0, Va_0 and a_{2P}, Va_{2P} are the same. Hence, the trace of the product of Lax matrices along the staircase is invariant under the mapping [17]

$$\operatorname{tr} \mathcal{T}_k = \operatorname{tr} \overleftarrow{\prod_{j=0}^{2P-1}} L_k(a_j, a_{j+1}, p_j) = \sum_{j=0}^{P} k^{2j} I_j = \text{invariant} \tag{4.5}$$

Eq. (4.5) holds for arbitrary values of the spectral parameter k and all coefficients $I_j, j = 0, 1, \cdots, P$, of the expansion in powers of k^2 must be invariant. The coefficients I_P, I_{P-1} associated with k^{2P}, k^{2P-2} turn out to be trivial, but the coefficients $I_0, I_1, \cdots, I_{P-2}$ give $P-1$ nontrivial integrals for the $2P-2$ dimensional mapping in terms of the v_j (3.5).

5 Poisson bracket structure

For the lattice KdV we have the action

$$S = \sum_{n,m \in \mathbf{Z}} V^n H^m \mathcal{L} \tag{5.1}$$

with the Lagrangian $\mathcal{L}$ given by

$$\mathcal{L} = (Vu)(HVu - u) + \varepsilon\delta \log(\varepsilon + u - HVu) \tag{5.2}$$

for fields u at the sites of the $2D$ lattice. Assuming S to be invariant under infinitesimal variations of the fields u at the different sites of the $2D$ lattice, we have the Euler-Lagrange equations

$$\frac{\partial \mathcal{L}}{\partial u} + V^{-1}\frac{\partial \mathcal{L}}{\partial Vu} + (HV)^{-1}\frac{\partial \mathcal{L}}{\partial HVu} = 0 \tag{5.3}$$

Eq. (5.3) is automatically satisfied for any solution u of the lattice KdV (3.3).

For the $2P$ dimensional KdV-type of mapping we introduce the action [18]

$$S = \sum_{n\in\mathbf{Z}} V^n L(\{a_{2j}\},\{Va_{2j}\}) \tag{5.4}$$

in which the Lagrangian L is a function of the even a's $a_0, a_2, \cdots, a_{2P-2}$ on the staircase and the shifted values $Va_0, Va_2, \cdots, Va_{2P-2}$ given by

$$L(\{a_{2j}\},\{Va_{2j}\}) = \sum_{j=0}^{P-1}\left[(Va_{2j})(a_{2j+2} - a_{2j}) + \varepsilon\delta\log(\varepsilon + a_{2j} - a_{2j+2}) + \frac{1}{2}(Va_{2j})^2 - \frac{1}{2}a_{2j}^2\right] \ , \tag{5.5}$$

where we have added 2 terms which do not affect the action but which are convenient for the introduction of canonical momenta. The Euler-Lagrange equations

$$\frac{\partial L}{\partial a_{2j}} + V^{-1}\frac{\partial L}{\partial Va_{2j}} = 0 \quad , \quad j = 0, \cdots, P-1 \tag{5.6}$$

with (5.5) yield the mapping (3.5).

Having established the Lagrangian property of the mapping we can follow refs. [14,15] to introduce canonical momenta and a discrete-time hamiltonian. We have the relations

$$Vp_{2j} \equiv \frac{\partial L}{\partial Va_{2j}} \quad , \quad p_{2j} = a_{2j} + a_{2j+1} - a_{2j-1} \tag{5.7}$$

$$\mathcal{H}(\{Vp_{2j}\}, a_{2j}) = \sum_{j=0}^{P-1}(Vp_{2j})(Va_{2j} - a_{2j}) - L$$

$$= \sum_{j=0}^{P-1}\left[\frac{1}{2}(Vp_{2j} - a_{2j+2})^2 + \frac{1}{2}(a_{2j+2} - a_{2j})^2 - \varepsilon\delta\log(\varepsilon + a_{2j} - a_{2j+2})\right] \tag{5.8}$$

The Hamiltonian acts as the generating functional of the mapping, i.e. one has the discrete hamiltonian equations

$$Va_{2j} - a_{2j} = \frac{\partial \mathcal{H}}{\partial Vp_{2j}} \quad , \quad Vp_{2j} - p_{2j} = -\frac{\partial \mathcal{H}}{\partial a_j} \tag{5.9}$$

The hamiltonian is not invariant under the mapping, but is the generating function of the canonical transformation associated with the mapping. In fact, the standard Poisson brackets

$$\{p_{2j}, a_{2j'}\} = \delta_{jj'} \quad , \quad \{a_{2j}, a_{2j'}\} = \{p_{2j}, p_{2j'}\} = 0 \tag{5.10}$$

are invariant under the mapping.

6 Involution

For the trace of the monodromy matrix given by (4.3) and (4.5) one can derive the explicit expression

$$\operatorname{tr} \mathcal{T}_k = \left(\prod_{j=0}^{2P-1} v_j\right) \left\{ 1 + \sum_{\substack{J_{\nu+1} - J_\nu \geq 2 \\ J_f \leq 2P-1 \\ J_1 - J_f + 2P \geq 2}} \prod_{\nu=1}^{f} \frac{k^2 - p_{J_\nu}^2}{v_{J_\nu} v_{J_\nu+1}} \right\} \tag{6.1}$$

with v_j and p_j as in (3.5) and (5.7). From (5.10) we have the Poisson brackets

$$\{v_j, v_{j'}\} = \delta_{j',j-1} - \delta_{j',j+1} \tag{6.2}$$

To prove the involution property use can be made of the determinant formula

$$\operatorname{tr} \mathcal{T}_k = \det Y_k + 1 + \left(p^2 - k^2\right)^P \left(q^2 - k^2\right)^P \tag{6.3}$$

with the $2P \times 2P$ matrix Y_k given by

$$\left(Y_k\right)_{jj'} = \left(p_j^2 - k^2\right) \delta_{j',j+1}(\mathrm{mod}2P) + \delta_{j,j-1}(\mathrm{mod}2P) + v_j \delta_{j'j} \tag{6.4}$$

$j, j' = 0, 1, \cdots, 2P-1$, where $\delta_{j,k}(\mathrm{mod}2P) = 1$, if $j - k$ is a multiple of $2P$, and 0 otherwise. From (6.2) and (6.4) we find

$$\{\operatorname{tr} \mathcal{T}_k, \operatorname{tr} \mathcal{T}_{k'}\} = \sum_{J=0}^{2P-1} \det \left(Y_k^{(J)} \cdot Y_{k'}^{(J-1)}\right) - (k \leftrightarrow k') \tag{6.5}$$

with

$$\left(Y_k^{(J)}\right)_{jj'} = \left(p_{j+J}^2 - k^2\right) \delta_{j',j+1} + \delta_{j',j-1} + v_{J+j} \delta_{j'j} \tag{6.6}$$

$j, j' = 1, 2, \cdots, 2P-1, p_{j+2P} = p_j \cdot (k \leftrightarrow k')$ in (6.5) denotes the previous term with k and k' interchanged. The matrix product $Y_k^{(J)} \cdot Y_{k'}^{J-1}$ is symmetric in k and k', apart from the $(1,1)$ and $(2P-1, 2P-1)$ elements, but it can be shown that the sum of the contributions from these elements to the first term in (6.5) is symmetric in k and k' as well. Therefore,

$$\{\operatorname{tr} \mathcal{T}_k, \operatorname{tr} \mathcal{T}_{k'}\} = 0 \tag{6.7}$$

implying that the integrals of the mapping are in involution.

7 Concluding remarks

i) We have established complete integrability in the Liouville-Arnol'd sense for a family of $2P$ dimensional mappings associated with a vertical shift V for periodic solutions of the lattice KdV (3.3). One can also derive the complete integrability for the mappings associated with a diagonal shift $D = H^{-1}V$.

ii) The involution property has been proved directly on the basis of a determinant formula [18], but a more fundamental justification is obtained via an r-matrix structure of a rather unusual non-ultralocal structure [19]. In ref. [21] the complete integrability for a class of discrete-time Toda lattices has been obtained using the usual r-matrix formalism.

iii) The lattice KdV has been treated as an example, but completely integrable mappings can also be found starting from other integrable 2D lattice equations, cf. [17,18] for some results concerning a (mixed) lattice version of the modified Korteweg-de Vries (MKdV) and Toda equation. More complicated mappings not included in the considerations of section 2 arise from lattice versions of the Gel'fand-Dikii hierarchy [19].

iv) On the 2D lattice with sites (l,m) one can investigate similarity solutions satisfying $u_{l,m} = u_n$ with $n = z_1 l - z_2 m$, z_1 and z_2 being relatively prime. The initial data can be chosen on a so called standard staircase [22] with z_2 horizontal and z_1 vertical steps. The similarity reduction amounts to a $z_1 + z_2$ dimensional mapping. A sufficient number of integrals has been found for the mappings associated with the lattice versions of the KdV, the MKdV and the sine-Gordon (SG) equations [22]. We expect that an invariant Poisson structure for these mappings can be found.

v) After a continuum limit the 2D lattice equations yield hierarchies of partial difference equations with time-dependent fields at the sites of a 1D chain, together with an infinite number of conserved quantities in involution [7]. Taking stationary solutions (or a slightly different simple time-dependence) one can obtain a variety of mappings. The simple two-dimensional examples belong to the 18-parameter family of ref. [16]. For some of these mappings complete integrability has been established on the basis of a Poisson structure with a sufficient number of integrals in involution [14].

vi) Starting from a 3D lattice version of the Kadomtsev-Petviashvili (KP) equation one can derive a variety of 2D difference equations as well as mappings more general than the ones arising from the 2D lattice equations. In simple cases the mappings have been identified with known integrable cases. Although one may anticipate to obtain a larger class of integrable mappings, none of the underlying ideas (like the staircase of initial data, the evaluation of integrals and the Poisson structure) has been established with a satisfactory amount of generality.

References

[1] F.W. Nijhoff, G.R.W. Quispel and H.W. Capel, Phys.Lett. **97A**(1983)125.

[2] G.R.W. Quispel, F.W. Nijhoff, H.W. Capel and J. van der Linden, Physica **125A**(1984)344.

[3] F.W. Nijhoff, H.W. Capel, G.L. Wiersma and G.R.W. Quispel, Phys.Lett. **103A**(1984)293; **105A**(1984)267.

[4] H.W. Capel, G.L. Wiersma and F.W. Nijhoff, Physica **138A**(1986)76.

[5] R.Hirota, J.Phys.Soc. Japan **43**(1977)1424,2074,2079; **50**(1981)3785.

[6] E. Date, M. Jimbo and T. Miwa, J.Phys.Soc. Japan **51**(1982)4125; **52**(1983)388,766.

[7] G.L. Wiersma and H.W. Capel, Physica **142A**(1987)199.

[8] G.L. Wiersma and H.W. Capel, Physica **147A**(1988)49,75.

[9] F. Kako and M. Mugibayashi, Progr.Theor.Phys. **60**(1978)975; **61**(1979)778.

[10] E.M. McMillan, in: *Topics in Physics*, eds. W.E. Britten and H. Odabasi, (Colorado Associated Univ. Press, Boulder, 1971) p.219.

[11] A.P. Veselov, Funct.Anal.Appl. **22**(1988)83; Theor.Math.Phys. **71**(1987)446.

[12] J. Moser and A.P. Veselov, Preprint ETH (Zürich), 1989.

[13] P.A. Deift and L.C. Li, Commun.Pure Appl.Math. **42**(1989)963.

[14] S. Maeda, Proc.Japan Acad. **63A**(1987)198; Math. Japonica **25**(1988)405.

[15] M. Bruschi, O. Ragnisco, P.M. Santini and G.-Z. Tu, *Integrable Symplectic Maps*, Preprint Università di Roma I, June 1990.

[16] G.R.W. Quispel, J.A.G. Roberts, and C.J. Thompson, Phys.Lett. **A128** (1988)419; Physica **D34**(1989)183.

[17] V.G. Papageorgiou, F.W. Nijhoff and H.W. Capel, Phys.Lett. **A147** (1990)106.

[18] H.W. Capel, F.W. Nijhoff and V.G. Papageorgiou, *Complete Integrability and Lattices of the KdV type*, Preprint INS # 165/90.

[19] F.W. Nijhoff, V.G. Papageorgiou and H.W. Capel, *Integrable time-discrete systems: Lattices and Mappings*, Preprint INS # 166/90. Proceedings Second International Workshop on Quantum Groups, Leningrad, November 1990.

[20] A.S. Fokas and M.J. Ablowitz, Phys.Rev.Lett. **47**(1981)1096.

[21] Yu.B. Suris, Phys.Lett. **A145**(1990)113.

[22] G.R.W. Quispel, H.W. Capel, V.G. Papageorgiou and F.W. Nijhoff, *Integrable Mappings derived from Soliton Equations*, Preprint La Trobe University, October 1990, to be published in Physica A.

Part VI

Direct Methods Applicable to Soliton Systems

Integrable Higher Nonlinear Schrödinger Equations

B. Grammaticos[1] and A. Ramani[2]

[1]LPN, Université Paris VII, Tour 24-14,
5 étage, 75251 Paris, France.
[2]CPT, Ecole Polytechnique,
91128 Palaiseau, France.

Abstract. We investigate a form of a higher nonlinear Schrödinger equation using the tools of singularity analysis and show that only the equations obtained by Hirota and by Satsuma and Sasa pass the integrability test.

The nonlinear Schrödinger equation (NLS):

$$iq_t+q_{xx}+q^*q^2=0 \qquad (1)$$

has met with particular success in the description of the propagation of solitons in optical fibers [1]. Numerical simulations [2,3] and experimental [4] studies have amply demonstrated the feasibility of this novel mode for information transmission. However recent detailed experiments have revealed the existence of phenomena that cannot be explained by the simple NLS equation. This lead Kodama and Hasegawa [5] into proposing an extension of the NLS equation (HNLS) containing higher dispersive terms:

$$iq_t+q_{xx}+\lambda q^*q^2+i(q_{xxx}-Aq^2q^*_x-Bqq^*q_x)=0 \qquad (2)$$

(See the contribution of Kaup to the present conference for a comparison between experiment and HNLS-based theoretical predictions). This higher NLS is not, in general, integrable. Thus the question of its possible integrable forms arises naturally.

Two cases must be distinguished at the outset: either the term q_{xxx} is present or not. In the latter case eq. (2) is just the well-known derivative NLS equation (λ=1), of which two integrable cases exist:

a) A=B/2 (See Ref. [6])

b) A=0 (See Ref. [7])

Here we will limit ourselves to the study of the HNLS which includes the higher dispersive q_{xxx} term and investigate its integrability in the framework of singularity analysis. We proceed, as usual, by writting two distinct equations for q and $r \equiv q^*$, and look for the dominant behaviour. It is straightforward to convince oneself that the dominant part of the equations can only be the following, where the prime (') denotes the derivative with respect to x:

$$q''' = Aq^2r' + Bqrq'$$
$$r''' = Ar^2q' + Bqrr' \qquad (3)$$

Two singular behaviours emerge: (I) either both q and r diverge as Φ^{-1}, around a singular manifold $\Phi = x + \phi(t)$, or (ii) one of them is subdominant (but note that qr must behave as Φ^{-2}). In the first case we find: $q=\alpha\Phi^{-1}$, $r=\beta\Phi^{-1}$ with $\alpha\beta=6/(A+B)$. For the study of the resonances we expand the dependent variables: $q=\alpha\Phi^{-1}(1+\gamma\Phi^n)$ and $r=\beta\Phi^{-1}(1+\delta\Phi^n)$. Substituting back into (3) and expanding to first order in γ, δ we find readily the resonance equation:

$$n(n+1)(n-3)(n-4)[n^2-6n+11+6(A-B)/(A+B)]=0 \qquad (4)$$

Thus the condition for the Painlevé property to hold is that the trinomial within brackets possesses integer roots. This introduces a first condition on the A, B. We turn next to singular behaviour of type (II): $q=a\Phi^{p-1}$, $r=b\Phi^{-p-1}$ with integer p. Substituting back into (3) we find a relation between A, B and a, b of the form $Aab= 5(p^2-1)/2$ and $Bab= (7p^2+17)/2$. We can use these values of A/B into equation (4) that now becomes:

$$n(n+1)(n-3)(n-4)[n^2-6n+10p^2/(p^2+1)]=0 \qquad (5)$$

We find readily that the only possibilities for the Painlevé property to hold are $p^2=0$, 1, 4, 9. The case p=0 can be ruled out immediately. In fact for p=0 we have n=0 as double resonance and since the product $\alpha\beta$ is fixed only one free parameter enters, a priori, at this order. Thus the expansion has a logarithm at n=0 and this case does not satisfy the Painlevé criterion. On the other hand we remark that for $p^2=9$ the resonance n=3 is triple! This cannot be realized without the introduction of a logarithm in the singular expansion. Thus only two cases remain:

a) $p^2=1$ with A=0

b) $p^2=4$ with A=B/3

These cases are precisely the only known integrable forms of NLS: the first corresponds to the equation obtained by Hirota [8] while the second one is due to Satsuma and Sasa [9].

One remaining point concerns the precise value of the coefficients A and B. It is clear that they can be fixed only by comparing to the lower order terms i.e. at the resonance compatibility conditions. We will illustrate this point for Hirota's HNLS for which the resonances are -1, 0, 1, 3, 4, 5 for singularities of type (I). We rewrite (3) introducing subdominant terms:

$$\begin{aligned} q''' &= Bqrq' + i(q''+\lambda q^2 r) \\ r''' &= Bqrr' - i(r''+\lambda qr^2) \end{aligned} \qquad (6)$$

and expand q, r to just first order (in order to investigate the compatibility at n=1): $q=\alpha(\Phi^{-1}+\gamma)$ and $r=\beta(\Phi^{-1}+\delta)$ with $\alpha\beta=6/B$. Substituting back into (6) leads to the compatibility condition $\lambda=-B/3$. This is precisely the correct value for the parameter λ for Hirota's equation. (Similar conclusions can be reached for Satsuma's equation, where $\lambda=-2B/9$). It is straightforward to verify that, once the coefficients have been fixed at this order, the remaining compatibility conditions are satisfied and thus the full equation passes the Painlevé test. Thus we can conclude, following the ARS conjecture, that among the various HNLS equations only the two obtained by Hirota and Satsuma-Sasa are integrable.

References.

1. V. E. Zakharov and A. B. Shabat, Sov. Phys. JEPT **34** (1972) 62.
2. A. Hasegawa and F. D. Tappert, Appl. Phys. Lett. **23** (1973) 142.
3. B. Dorizzi et al., contribution at the RCP 759, Dijon 1984.
4. L. F. Mollenauer et al., Phys. Rev. Lett. **45** (1980) 1095.
5. Y. Kodama and A. Hasegawa, IEEE Jour. Quant. Elec. **23** (1987) 510.
6. D. Anderson and M. Lisak, Phys. Rev. A **27** (1983) 1393.
7. H. H. Chen et al., Phys. Scripta **20** (1979) 490.
8. R. Hirota, J. Math Phys. **14** (1973) 805.
9. N. Sasa and J. Satsuma, preprint.

The authors are grateful to professor J. Satsuma for communicating them his results prior to publication.

Nonclassical Symmetry Reductions of a Generalized Nonlinear Schrödinger Equation

P. A. Clarkson

Department of Physics, University of Exeter,
Exeter, EX4 4QE, England.

Abstract. In this paper new symmetry reductions and exact solutions for a generalized nonlinear Schrödinger are presented. These are obtained using the direct method, originally developed by Clarkson and Kruskal [*J. Math. Phys.*, **30** (1989) 2201] to study similarity reductions of the Boussinesq equation, which involves no group theoretical techniques.

1. Introduction

The classical method for finding similarity reductions of PDEs is the Lie method of infinitesimal transformations [**1**], for which symbolic manipulation programs have been developed, e.g. in MACSYMA [**2**], REDUCE [**3**] and MUMATH [**4**]. Bluman and Cole [**5**] proposed a generalization of Lie's method called the "*nonclassical method of group-invariant solutions*," in their study of similarity reductions of the linear heat equation (see also [**6**]). This has been extended by Olver and Rosenau [**7**] who concluded that "the unifying theme behind finding special solutions of PDEs is not, as is commonly supposed, group theory, but rather the more analytic subject of overdetermined systems of PDEs." It is known that there do exist PDEs which possess similarity reductions that are *not* obtained using the classical Lie method [**7,8**].

Recently, Clarkson and Kruskal [**10**] developed an direct algorithmic method for finding similarity reductions (hereafter referred to as the *direct method*) and using it obtained previously unknown similarity reductions of the Boussinesq equation

$$u_{tt} + uu_{xx} + u_x^2 + u_{xxxx} = 0. \tag{1.1}$$

The novel characteristic about this direct method in comparison to the others mentioned above, is that it involves *no* use of group theory; additionally, for many equations, it appears to be simpler to implement than either the classical or nonclassical methods. The basic idea is to seek a reduction of a given PDE (with two independent and one dependent variables) in the form $u(x,t) = U(x,t,w(z(x,t)))$, which is the most general form for a similarity reduction (cf. [**1**]). Substituting this into the PDE and requiring that the result be an ODE for $w(z)$ imposes conditions upon $U(x,t,w)$, $z(x,t)$ and their derivatives in the form of an overdetermined system of equations, whose solution yields the similarity reductions. (This also provides further evidence to support the aforementioned remark by Olver and Rosenau [**7**].)

Clarkson and Kruskal [**10**] show that the generic similarity reduction of the Boussinesq equation (1.1) is given by

$$u(x,t) = \theta^2(t)w(z) - \frac{1}{\theta^2(t)}\left(x\frac{\mathrm{d}\theta}{\mathrm{d}t} + \frac{\mathrm{d}\phi}{\mathrm{d}t}\right)^2, \qquad z(x,t) = x\theta(t) + \phi(t),$$

where $\theta(t)$, $\phi(t)$ are any solutions of

$$\frac{\mathrm{d}^2\theta}{\mathrm{d}t^2} = A\theta^5, \qquad \frac{\mathrm{d}^2\phi}{\mathrm{d}t^2} = (A\phi + B)\theta^4, \tag{1.2}$$

with A, B arbitrary constants, and $w(z)$ satisfies

$$w'''' + ww'' + (w')^2 + (Az+B)w' + 2Aw = 2(Az+B)^2, \tag{1.3}$$

with $' := \mathrm{d}/\mathrm{d}z$. Depending upon the choice of the constants, equation (1.3) is solvable in terms of the first, second and fourth Painlevé equations **[11]**

$$\frac{\mathrm{d}^2y}{\mathrm{d}x^2} = 6y^2 + x, \tag{PI}$$

$$\frac{\mathrm{d}^2y}{\mathrm{d}x^2} = 2y^3 + xy + \alpha, \tag{PII}$$

$$\frac{\mathrm{d}^2y}{\mathrm{d}x^2} = \frac{1}{2y}\left(\frac{\mathrm{d}y}{\mathrm{d}x}\right)^2 + \frac{3}{2}y^3 + 4xy^2 + 2(x^2+\alpha)y + \frac{\beta}{y}, \tag{PIV}$$

with α, β arbitrary constants, respectively. Solving (1.2) yields six classes of similarity reductions:

(i) $u(x,t) = w_1(z)$, $\qquad z = x + \lambda_1 t$;

(ii) $u(x,t) = t^2 w_2(z) - x^2/t^2$, $\qquad z = xt$;

(iii) $u(x,t) = w_3(z) - 4\lambda_3^2 t^2$, $\qquad z = x + \lambda_3 t^2$;

(iv) $u(x,t) = t^2 w_4(z) - (x + 6\lambda_4 t^5)^2/t^2$, $\qquad z = xt + \lambda_4 t^6$;

(v) $u(x,t) = t^{-1} w_5(z) - (x - 3\lambda_5 t^2)^2/(4t^2)$, $\qquad z = xt^{-1/2} + \lambda_5 t^{3/2}$;

(vi) $$u(x,t) = \wp^{-1}(t;0,g_3)\left\{w_6(z) - \left(\tfrac{1}{2}z\frac{\mathrm{d}\wp}{\mathrm{d}t}(t;0,g_3) + \lambda_6\wp^{3/2}(t;0,g_3)\right)^2\right\},$$

$$z = \wp^{-1/2}(t;0,g_3)\left[x + \lambda_6\zeta(t;0,g_3)\right],$$

with $\lambda_1, \lambda_3, \ldots, \lambda_6$ constants and where $\wp(t;g_2,g_3)$ and $\zeta(t;g_2,g_3)$, with g_2, g_3 constants, are the Weierstrass elliptic and zeta functions, respectively. $w_1(z)$ and $w_2(z)$ are solvable in terms of PI; $w_3(z)$ and $w_4(z)$ in terms of the PII; and $w_5(z)$ and $w_6(z)$ in terms of PIV. (Only the similarity reductions (i) and (v) in the special case $\lambda_5 = 0$, can be obtained using the classical Lie method **[8]**.) Levi and Winternitz **[12]** have given a group theoretical explanation of these results by showing that all these similarity reductions of the Boussinesq equation can also be obtained using Bluman and Cole's nonclassical method **[5]**.

By using known rational and special solutions of the Painlevé equations **[13]**, these new nonclassical similarity reductions can be used to generate new rational and special solutions of the Boussinesq equation expressible in terms of elementary functions, Weierstrass elliptic and zeta functions, Airy functions and Weber-Hermite functions (see **[14]** for details). Further applications of the direct method are given in **[14-17]**.

There is much current interest in the determination of similarity reductions of PDEs which reduce the equations to ODEs. Frequently one then checks if the resulting ODE is of Painlevé type (i.e., its solutions have no movable singularities other than poles). It appears to be the case that whenever the ODE is of Painlevé type then it can be solved explicitly (and so obtain exact solutions to the original PDE); however, if it is not of Painlevé type, then often one is unable to solve it explicitly.

2. The Generalized Nonlinear Schrödinger Equation

In this section we discuss similarity reductions of the generalized nonlinear Schrödinger (GNLS) equation

$$\mathrm{i}u_t + u_{xx} + (a_1 + \mathrm{i}a_2)(|u|^2u)_x + (b_1 + \mathrm{i}b_2)u(|u|^2)_x + c|u|^4u + d|u|^2u = 0, \tag{2.1}$$

with a_1, a_2, b_1, b_2, c, d real constants. There are several special cases of this equation which have been studied previously: (i), if $a_1 = a_2 = 0$, $b_1 = b_2 = 0$, $c = 0$, then (2.1) is the cubic nonlinear Schrödinger (NLS) equation which is completely integrable [**18**]; (ii), if $a_1 = a_2 = 0$, $b_1 = b_2 = 0$, $c \neq 0$, then (2.1) is the quintic nonlinear Schrödinger (QNLS) equation, though this appears to be non-integrable [**19**]; (iii), if $a_1 = b_1 = 0$, $d = 0$, then (2.1) is a generalized derivative nonlinear Schrödinger (GDNLS) equation, which is completely integrable if and only if additionally $c = \frac{1}{4}b_2(a_2 + b_2)$ [**20**] (this integrable equation is the DNLS equation, special cases of which include DNLSI for $b_2 = 0$, $c = 0$ [**21**], DNLSII for $b_2 = -a_2$, $c = 0$ [**22**] and DNLSIII for $b_2 = -2a_2$, $c = \frac{1}{2}a_2^2$ [**23**]); (iv), if $a_1 = 0$, $b_1 = b_2 = 0$, $c = 0$, then (2.1) is the mixed nonlinear Schrödinger equation which also is completely integrable [**24**]; (v), if $a_1 = a_2 = 0$, $b_2 = 0$, $c = \frac{1}{2}b_1^2$, $d = 0$, then (2.1) is the Eckhaus equation, which is linearizable [**25**]. The GNLS equation also arises in several physical applications including quantum field theory [**26**], weakly nonlinear dispersive water waves [**27**], and nonlinear optics [**28**].

To determine similarity reductions of the GNLS equation (2.1) using the classical Lie method we write it as the system

$$\mathrm{i}u_t + u_{xx} + (a_1 + \mathrm{i}a_2)(u^2v)_x + (b_1 + \mathrm{i}b_2)u(uv)_x + cu^3v^2 + du^2v = 0, \tag{2.2a}$$
$$-\mathrm{i}v_t + v_{xx} + (a_1 - \mathrm{i}a_2)(uv^2)_x + (b_1 - \mathrm{i}b_2)v(uv)_x + cu^2v^3 + duv^2 = 0. \tag{2.2b}$$

Following the procedure described in [**1**], consider the one-parameter (ε) Lie group of infinitesimal transformations

$$\tilde{x} = x + \varepsilon\xi(x,t,u,v) + O(\varepsilon^2), \tag{2.3a}$$
$$\tilde{t} = t + \varepsilon\tau(x,t,u,v) + O(\varepsilon^2), \tag{2.3b}$$
$$\tilde{u} = u + \varepsilon\eta(x,t,u,v) + O(\varepsilon^2), \tag{2.3c}$$
$$\tilde{v} = v + \varepsilon\zeta(x,t,u,v) + O(\varepsilon^2). \tag{2.3d}$$

Requiring that equation (2.2) is invariant under this transformation yields an overdetermined set of equations for the *infinitesimals* $\xi(x,t,u,v)$, $\tau(x,t,u,v)$, $\eta(x,t,u,v)$ and $\zeta(x,t,u,v)$ and the associated Lie algebra is realized by the vector field

$$\mathrm{X} = \xi(x,t,u,v)\partial_x + \tau(x,t,u,v)\partial_t + \eta(x,t,u,v)\partial_u + \zeta(x,t,u,v)\partial_v.$$

Florjańczyk and Gagnon [**28**] show that equation (2.2) is invariant under the following five vector fields: $\mathrm{X}_1 = \partial_x$ (corresponding to space translational invariance); $\mathrm{X}_2 = \partial_t$ (time translation); $\mathrm{X}_3 = u\partial_u - v\partial_v$ (constant change of phase); $\mathrm{X}_4 = 2x\partial_x + 4t\partial_t - u\partial_u - v\partial_v$ provided that $d = 0$ (scaling or dilation); $\mathrm{X}_5 = -2\mathrm{i}t\partial_x + xu\partial_u - xv\partial_v$ provided that $a_1 = a_2 = 0$ (Galilean boost). Therefore $\mathrm{X} = \sum_{j=1}^{5} \alpha_j\mathrm{X}_j$, with α_j, $j = 1, 2, \ldots, 5$, constants such that $d\alpha_4 = 0$ and $\alpha_5 = 0$ unless $a_1 = a_2 = 0$. Hence the infinitesimals are

$$\begin{aligned}
\xi(x,t,u,v) &= \alpha_1 + 2x\alpha_4 - 2\mathrm{i}t\alpha_5,\\
\tau(x,t,u,v) &= \alpha_2 + 4t\alpha_4,\\
\eta(x,t,u,v) &= u\alpha_3 - u\alpha_4 + xu\alpha_5,\\
\zeta(x,t,u,v) &= -v\alpha_3 - v\alpha_4 - xv\alpha_5.
\end{aligned}$$

Similarity reductions are then obtained by solving the *characteristic equations*

$$\frac{\mathrm{d}x}{\xi(x,t,u,v)} = \frac{\mathrm{d}t}{\tau(x,t,u,v)} = \frac{\mathrm{d}u}{\eta(x,t,u,v)} = \frac{\mathrm{d}v}{\zeta(x,t,u,v)},$$

or equivalently the *invariant surface conditions* $\xi u_x + \tau u_t - \eta = 0$ and $\xi v_x + \tau v_t - \zeta = 0$.

There are three types of (classical) similarity reductions (see [**28**] for details):

Case I. A travelling wave solution

$$u(x,t) = p(z)\exp\{-\mathrm{i}(\tfrac{1}{2}\lambda x + \mu t)\}, \qquad v(x,t) = q(z)\exp\{\mathrm{i}(\tfrac{1}{2}\lambda x + \mu t)\}, \tag{2.5}$$

with $z = x + \lambda t$ and where $p(z)$ and $q(z)$ satisfy the system of equations

$$\begin{aligned}
&p'' + (a_1 + \mathrm{i}a_2)\left(2pqp' + p^2q'\right) + (b_1 + \mathrm{i}b_2)\left(pqp' + p^2q'\right)\\
&\qquad + cp^3q^2 + dp^2q + (\mu - \tfrac{1}{4}\lambda^2)p = 0,\\
&q'' + (a_1 - \mathrm{i}a_2)\left(2pqq' + q^2p'\right) + (b_1 - \mathrm{i}b_2)\left(pqq' + q^2p'\right)\\
&\qquad + cp^2q^3 + dpq^2 + (\mu - \tfrac{1}{4}\lambda^2)q = 0,
\end{aligned}$$

which are solvable in terms of Jacobi and Weierstrass elliptic functions if $a_1 = 0$.

Case II. An accelerating wave solution for $a_1 = a_2 = 0$

$$u(x,t) = p(z)\exp\{-\mathrm{i}(\kappa xt + \tfrac{2}{3}\kappa^2t^3 + \mu t)\}, \tag{2.6a}$$
$$v(x,t) = q(z)\exp\{\mathrm{i}(\kappa xt + \tfrac{2}{3}\kappa^2t^3 + \mu t)\}, \tag{2.6b}$$

with $z = x + \kappa t^2$ and where $p(z)$ and $q(z)$ satisfy the system of equations

$$\begin{aligned}
&p'' + (b_1 + \mathrm{i}b_2)\left(pqp' + p^2q'\right) + cp^3q^2 + dp^2q + [\kappa z + \mu]p = 0,\\
&q'' + (b_1 - \mathrm{i}b_2)\left(pqq' + q^2p'\right) + cp^2q^3 + dpq^2 + [\kappa z + \mu]q = 0.
\end{aligned}$$

If $b_1 = 0$ and $c = \frac{1}{4}b_2^2$, then $p(z)$ and $q(z)$ are solvable in terms of PII (if $d \neq 0$) or a linear equation (if $d = 0$).

Case III. A scaling reduction for $d = 0$

$$u(x,t) = t^{-1/4}p(z)\exp\{\mathrm{i}(\tfrac{1}{8}z^2 + \mu\ln t)\}, \tag{2.7a}$$
$$v(x,t) = t^{-1/4}q(z)\exp\{-\mathrm{i}(\tfrac{1}{8}z^2 + \mu\ln t)\}, \tag{2.7b}$$

with $z = x/t^{1/2}$ and where $p(z)$ and $q(z)$ satisfy the system of equations

$$\begin{aligned}
&p'' + (a_1 + \mathrm{i}a_2)\left[\left(2pqp' + p^2q'\right) + \tfrac{1}{4}\mathrm{i}zp^2q\right]\\
&\qquad + (b_1 + \mathrm{i}b_2)\left(pqp' + p^2q'\right) + cp^3q^2 + [\tfrac{1}{16}z^2 - \mu]p = 0,\\
&q'' + (a_1 - \mathrm{i}a_2)\left[\left(2pqq' + q^2p'\right) - \tfrac{1}{4}\mathrm{i}zpq^2\right]\\
&\qquad + (b_1 - \mathrm{i}b_2)\left(pqq' + q^2p'\right) + cp^2q^3 + [\tfrac{1}{16}z^2 - \mu]q = 0.
\end{aligned}$$

If $a_1 = b_1 = 0$ and $c = \frac{1}{4}b_2(a_2 + b_2)$, then $p(z)$ and $q(z)$ are solvable in terms of PIV; if $a_1 = a_2 = 0$ and $c = \frac{1}{4}(b_1^2 + b_2^2)$, then $p(z)$ and $q(z)$ are solvable in terms of a linear ordinary differential equation.

To apply the direct method to equation (2.2), we seek a solution in the form

$$u(x,t) = R(t)p(z)\exp\{\mathrm{i}\Omega(x,t)\}, \qquad v(x,t) = R(t)q(z)\exp\{-\mathrm{i}\Omega(x,t)\},$$

with $z = x\theta(t) + \phi(t)$, hence substituting in (2.2) yields

$$\theta^2 Rp'' + (a_1 + \mathrm{i}a_2)\left[\theta R^3(p^2q)' + \mathrm{i}R^3\Omega_x p^2 q\right] + (b_1 + \mathrm{i}b_2)\theta R^3 p(pq)' + cR^5p^3q^2 + dR^3p^2q$$
$$+ \mathrm{i}R\left(x\frac{\mathrm{d}\theta}{\mathrm{d}t} + \frac{\mathrm{d}\phi}{\mathrm{d}t} + 2\theta\Omega_x\right)p' + \left[\mathrm{i}\left(R\Omega_{xx} + \frac{\mathrm{d}R}{\mathrm{d}t}\right) - (\Omega_x^2 + \Omega_t)R\right]p = 0, \qquad (2.8)$$

together with an analogous equation obtained by changing $\mathrm{i} \to -\mathrm{i}$ and $p \leftrightarrow q$. By requiring that equation (2.8) is an ODE we see necessarily that: (i), either $d = 0$, or $R = k$, with k a constant; (ii), $R(t) = \theta^{1/2}(t)$; (iii),

$$\Omega(x,t) = -\frac{1}{4\theta}\left(x^2\frac{\mathrm{d}\theta}{\mathrm{d}t} + 2x\frac{\mathrm{d}\phi}{\mathrm{d}t} + \psi(t)\right), \qquad (2.9)$$

for some function $\psi(t)$; and (iv), $(a_1 + \mathrm{i}a_2)\Omega_x = \theta F_1(z)$ and $\Omega_x^2 + \Omega_t = \theta^2 F_2(z)$ for some functions $F_1(z)$ and $F_2(z)$ (to be determined). It is easily shown that if either $a_1 \neq 0$ or $a_2 \neq 0$, then we just obtain the classical similarity reductions (2.5,2.7) given above and so henceforth we shall assume that $a_1 = a_2 = 0$. Since $\Omega_x^2 + \Omega_t$ is quadratic in x and z is linear in x, then necessarily $F_2(z) = \alpha z^2 + \beta z + \gamma$, with α, β, γ arbitrary constants. Thus $\theta(t)$, $\phi(t)$ and $\psi(t)$ satisfy the following equations

$$\theta\frac{\mathrm{d}^2\theta}{\mathrm{d}t^2} - 2\left(\frac{\mathrm{d}\theta}{\mathrm{d}t}\right)^2 + 4\alpha\theta^6 = 0, \qquad (2.10a)$$

$$\theta\frac{\mathrm{d}^2\phi}{\mathrm{d}t^2} - 2\frac{\mathrm{d}\theta}{\mathrm{d}t}\frac{\mathrm{d}\phi}{\mathrm{d}t} + (4\alpha\phi + 2\beta)\theta^5 = 0, \qquad (2.10b)$$

$$\theta\frac{\mathrm{d}\psi}{\mathrm{d}t} - \frac{\mathrm{d}\theta}{\mathrm{d}t}\psi - \left(\frac{\mathrm{d}\phi}{\mathrm{d}t}\right)^2 + 4(\alpha\phi^2 + \beta\phi + \gamma)\theta^4 = 0. \qquad (2.10c)$$

Note that if Ω is given by equation (2.9) and $R = \theta^{1/2}$, then $R\Omega_{xx} + \mathrm{d}R/\mathrm{d}t \equiv 0$.

By solving equations (2.10a,b,c) in succession we see that, as well as the classical similarity reductions (2.5–2.7), there are three additional reductions:

Case IV. For $\theta(t) = 1/t$, $\phi(t) = -\beta/t^2$, $\psi(t) = (3\kappa t^3 + 12\gamma t^2 + 8\beta^2)/(3t^4)$ with $\alpha = 0$, the similarity reduction is

$$u(x,t) = t^{-1/2}p(z)\exp\left\{\frac{\mathrm{i}(3x^2t^2 - 12\beta xt - 3\kappa t^3 - 12\gamma t^2 + 8\beta^2)}{12t^3}\right\}, \qquad (2.11a)$$

$$v(x,t) = t^{-1/2}q(z)\exp\left\{-\frac{\mathrm{i}(3x^2t^2 - 12\beta xt - 3\kappa t^3 - 12\gamma t^2 + 8\beta^2)}{12t^3}\right\}, \qquad (2.11b)$$

with $z = (xt - \beta)/t^2$ and where $p(z)$ and $q(z)$ satisfy

$$p'' + (b_1 + \mathrm{i}b_2)(pqp' + p^2q') + cp^3q^2 - (\beta z + \gamma)p = 0,$$
$$q'' + (b_1 - \mathrm{i}b_2)(pqq' + q^2p') + cp^2q^3 - (\beta z + \gamma)q = 0.$$

Setting $p(z) = \sqrt{w(z)}\exp\{i\Theta(z)\}$ and $q(z) = \sqrt{w(z)}\exp\{-i\Theta(z)\}$ yields

$$ww'' = \tfrac{1}{2}(w')^2 - 2b_1w^2w' - 2(c - \tfrac{1}{4}b_2^2)w^4 + 2(\beta z + \gamma - b_2\delta)w^2 + 2\delta^2, \qquad (2.12)$$

with δ an arbitrary constant. If $c = \frac{1}{4}(b_1^2 + b_2^2)$ then (2.12) is linearizable, otherwise it is not of Painlevé type.

Case V. For $\theta(t) = t^{-1/2}$, $\phi(t) = \lambda t^{1/2}$, $\psi(t) = \frac{1}{2}\lambda^2 t^{1/2} - 4\gamma t^{-1/2}\ln t$ with $\alpha = -\frac{1}{16}$, $\beta = 0$, the similarity reduction is

$$u(x,t) = t^{-1/4}p(z)\exp\left\{i(\tfrac{1}{8}z^2 - \tfrac{1}{2}\lambda z t^{1/2} + \tfrac{1}{4}\lambda^2 t + \mu \ln t)\right\}, \qquad (2.13a)$$

$$v(x,t) = t^{-1/4}q(z)\exp\left\{-i(\tfrac{1}{8}z^2 - \tfrac{1}{2}\lambda z t^{1/2} + \tfrac{1}{4}\lambda^2 t + \mu \ln t)\right\}, \qquad (2.13b)$$

with $z = (x + \lambda t)/t^{1/2}$.

Case VI. For $\theta(t) = 1/(t^2 + 4\alpha)^{1/2}$, $\phi(t) = \lambda t/(t^2 + 4\alpha)^{1/2}$,

$$\psi(t) = \frac{4\alpha\lambda^2 t}{(t^2 + 4\alpha)^{3/2}} - \frac{\gamma}{(t^2 + 4\alpha)^{1/2}}\left\{\frac{2}{\sqrt{\alpha}}\tan^{-1}\left(\frac{t}{2\sqrt{\alpha}}\right) - 1\right\},$$

with $\alpha \neq 0$, $\beta = 0$, $\gamma = 0$, the similarity reduction is

$$u(x,t) = \frac{p(z)\exp\{i\phi(x,t)\}}{(t^2 + 4\alpha)^{1/4}}, \qquad v(x,t) = \frac{q(z)\exp\{-i\phi(x,t)\}}{(t^2 + 4\alpha)^{1/4}}, \qquad (2.14)$$

with

$$\phi(x,t) = \frac{(x^2 t - 8\alpha\lambda x - 4\alpha\lambda^2 t^2)}{4(t^2 + 4\alpha)} - \frac{\gamma}{2\sqrt{\alpha}}\tan^{-1}\left(\frac{t}{2\sqrt{\alpha}}\right), \qquad z = \frac{x + \lambda t}{(t^2 + 4\alpha)^{1/2}}.$$

In Cases V and VI, $p(z)$ and $q(z)$ satisfy the system of equations

$$p'' + (b_1 + ib_2)(pqp' + p^2q') + cp^3q^2 - (\alpha z^2 + \gamma)p = 0,$$
$$q'' + (b_1 - ib_2)(pqq' + q^2p') + cp^2q^3 - (\alpha z^2 + \gamma)q = 0.$$

Setting $p(z) = \sqrt{w(z)}\exp\{i\Theta(z)\}$ and $q(z) = \sqrt{w(z)}\exp\{-i\Theta(z)\}$ yields

$$ww'' = \tfrac{1}{2}(w')^2 - 2b_1w^2w' - 2(c - \tfrac{1}{4}b_2^2)w^4 + 2(\alpha z^2 + \gamma - b_2\delta)w^2 + 2\delta^2, \qquad (2.15)$$

with $\alpha = -\frac{1}{16}$ in Case V and where δ is an arbitrary constant. If $c = \frac{1}{4}(b_1^2 + b_2^2)$ then (2.15) is linearizable, otherwise it is not of Painlevé type.

The similarity reductions in Cases IV–VI are nonclassical. For example, consider the similarity reduction (2.11) with $\beta = 0$, for which an associated one-parameter group of transformations is given by

$$(x,t,u,v) \to \left(e^{2\varepsilon}x, e^{2\varepsilon}t, e^{-\varepsilon}u\exp\{\tfrac{1}{4}i\omega(x,t;\varepsilon)\}, e^{-\varepsilon}v\exp\{-\tfrac{1}{4}i\omega(x,t;\varepsilon)\}\right), \qquad (2.16)$$

with $\omega(x,t;\varepsilon) = (e^{2\varepsilon} - 1)x^2/t$. This maps equation (2.2) to

$$iu_t + u_{xx} + (b_1 + ib_2)u(uv)_x + cu^3v^2 = \tfrac{1}{2}i(1 - e^{-2\varepsilon})\psi_1/t, \qquad (2.17a)$$

$$-iv_t + v_{xx} + (b_1 - ib_2)v(uv)_x + cu^2v^3 = -\tfrac{1}{2}i(1 - e^{-2\varepsilon})\psi_2/t, \qquad (2.17b)$$

where

$$\psi_1 = 2xu_x + 2tu_t + (2t - x^2)u/(2t), \qquad \psi_2 = 2xv_x + 2tv_t + (2t - x^2)v/(2t). \qquad (2.18)$$

$\psi_1 = 0$ and $\psi_2 = 0$ are invariant surface conditions associated with the transformation (2.16). Therefore the group (2.16) may be regarded as a *conditional symmetry* of the GNLS equation (2.2), since it does not map the equation into itself, yet nevertheless provides a reduction to an ODE. Furthermore it is an ordinary (i.e. classical) symmetry of the system of equations (2.2,2.18). (In the terminology of Olver and Rosenau [**7**], equations (2.18) are *side conditions*.)

3. Discussion and Further Examples

In this paper we have obtained new similarity reductions of a generalized nonlinear Schrödinger using the direct method developed by Clarkson and Kruskal [**10**]. This method has been successfully applied to obtain new similarity reductions and exact solutions for several physically significant PDEs (cf. [**14-17**]), which are conditional symmetries of the PDEs. There appears to be a close relationship between the nonclassical method of Bluman and Cole [**5**] and the direct method of Clarkson and Kruskal [**10**]. For both the Boussinesq and Kadomstev-Petviashvili equations, the two different methods yield the same results (cf. [**12,16,29**]), however the precise relationship between the methods has yet to be ascertained. Furthermore, it appears the be the case that for some PDEs one of these methods is simpler to apply and vice-versa for other PDEs.

Consider the PDE $\Delta(x,t,u(x,t))=0$ and suppose that $\mathcal{G}$ is a "symmetry group". For the classical Lie method, $\mathcal{G}$ maps the set of solutions $S := \{u : \Delta(x,t,u)=0\}$ into itself. However for the nonclassical and direct methods, this does not necessarily hold (since for some groups, S is not mapped into itself). It seems that one has to find symmetries of a given PDE *subject to the constraint* (or side condition), $\psi_{\mathcal{G}} = \xi u_x + \tau u_t - \eta = 0$, where ξ, τ and η are the infinitesimals associated with $\mathcal{G}$ (i.e., $\psi_{\mathcal{G}}=0$ is the invariant surface condition). However this supplementary condition does not impose any additional restrictions on the reductions that may be obtained. On the contrary, these conditional symmetries give more reductions than ordinary (or classical symmetries) since $\mathcal{G}$ maps the subset $S_{\mathcal{G}} := \{u : \Delta(x,t,u)=0, \psi_{\mathcal{G}}=0\}$ into S.

To conclude we shall discuss two further examples.

Example 3.1 The Zabolotskaya-Khokhlov (or two-dimensional Burgers) equation

$$(u_t + uu_x + u_{xx})_x + u_{yy} = 0, \tag{3.1}$$

is a model equation in nonlinear acoustics [**30**]; similarity reductions and exact solutions of this equation are discussed in [**31**]. Here, for reasons of simplicity, we discuss reductions of the time-independent Zabolotskaya-Khokhlov equation

$$uu_{xx} + u_x^2 + u_{xxx} + u_{yy} = 0. \tag{3.2}$$

Applying the classical Lie method to equation (3.2) yields three vector fields: $\mathrm{X}_1 = \partial_x$ (x translation); $\mathrm{X}_2 = \partial_y$ (y translation); $\mathrm{X}_3 = 2x\partial_x + 3y\partial_y - 2u\partial_u$ (scaling). There are two associated similarity reductions: (i), a plane wave solution $u(x,y)=w(z)$, with $z = x+\mu y$, and μ an arbitrary constant and where $w(z)$ satisfies

$$w' + \tfrac{1}{2}w^2 + \mu^2 w = Az + B, \tag{3.3}$$

with A and B arbitrary constants, which is linearizable through the transformation $w(z) = 2\phi'(z)/\phi(z)$, where $\phi(z)$ satisfies the Airy equation

$$\phi'' - \tfrac{1}{4}(2Az + 2B + \mu^2)\phi = 0;$$

and (ii), a scaling reduction $u(x,y) = y^{-2/3}w(z)$, with $z = x/y^{2/3}$ and where $w(z)$ satisfies

$$w'' + ww' + \tfrac{4}{9}z^2 w' + \tfrac{10}{9}zw = A, \tag{3.4}$$

with A an arbitrary constant, which is not of Painlevé type.

Using the direct method one obtains two further reductions: (i), an "accelerating solution" $u(x,y) = w(z) - (\kappa y+\mu)^2$, with $z = x+\frac{1}{2}\kappa y^2+\mu y$, κ and μ arbitrary constants and where $w(z)$ satisfies

$$w'' + ww' + \kappa w = 2\kappa^2(z + z_0), \tag{3.5}$$

with z_0 an arbitrary constant, which is not of Painlevé type; and (ii), a second scaling reduction $u(x,y) = y^{-1}w(z)$, with $z = x/y$ and where $w(z)$ satisfies (3.3) with $\mu = 0$.

Example 3.2 The Navier-Stokes equations governing the non-steady flow of an incompressible viscous fluid are

$$\boldsymbol{u}_t + \boldsymbol{u}\bullet\nabla\boldsymbol{u} = -\nabla p + \mu\nabla^2\boldsymbol{u}, \qquad \nabla\bullet\boldsymbol{u} = 0, \tag{3.6}$$

where $\boldsymbol{u} = (u,v,w)$ is the velocity field, p the pressure and μ the viscosity. Applying the classical Lie method yields nine vector fields which generate an infinite dimensional Lie algebra [**32-33**]. For two-dimensional flow, by introducing the stream function ψ such that $\boldsymbol{u} = (-\psi_y, \psi_x)$, equations (3.6) become

$$\nabla^2\psi_t + \psi_x\nabla^2\psi_y - \psi_y\nabla^2\psi_x = \mu\nabla^4\psi, \qquad \nabla^2 \equiv \partial_x^2 + \partial_y^2. \tag{3.7}$$

Similarity reductions and exact solutions of equation (3.7) are discussed in [**34**] using the classical Lie method which gives seven vector fields: $X_1 = \partial_t$; $X_2 = x\partial_x + y\partial_y + 2t\partial_t$; $X_3 = x\partial_x - y\partial_y$; $X_4 = ty\partial_x - tx\partial_x + \frac{1}{2}(x^2+y^2)\partial_\psi$; $X_5 = f(t)\partial_x + f'(t)y\partial_\psi$; $X_6 = g(t)\partial_y - g'(t)x\partial_\psi$; $X_7 = h(t)\partial_\psi$, with $f(t)$, $g(t)$ and $h(t)$ arbitrary functions (these vector fields generate an infinite dimensional Lie algebra). Equation (3.7) also possesses the nonclassical similartity reduction

$$\psi(x,y,t) = \Psi(z) + (x+\kappa y)(\alpha x+\beta y)/t, \qquad z = (x+\kappa y)/t^{1/2}, \tag{3.8}$$

with α, β and κ arbitrary constants and where $\Psi(z)$ satisfies

$$\mu(\kappa^2+1)\Psi'''' - \tfrac{1}{2}z(2\alpha\kappa - 2\beta - 1)\Psi''' + \Psi'' + 2(\alpha+\beta\kappa)/(\kappa^2+1) = 0 \tag{3.9}$$

(a special case of (3.8) is given by Boisvert, Ames and Srivastava [**33**]). Setting

$$\Psi''(z) = \eta(z)\exp\left\{\frac{(2\alpha\kappa - 2\beta - 1)z^2}{8(\kappa^2+1)\mu}\right\} - \frac{2(\alpha+\beta\kappa)}{(\kappa^2+1)},$$

yields

$$\frac{\mathrm{d}^2\eta}{\mathrm{d}z^2} = \left[\frac{(2\alpha\kappa-2\beta-1)^2}{16(\kappa^2+1)^2\mu^2}z^2 - \frac{(2\alpha\kappa-2\beta+3)}{4(\kappa^2+1)\mu}\right]\eta.$$

Therefore $\eta(z) = kD_\nu(\xi)$, with

$$\xi = \gamma z, \qquad \gamma^2 = \pm\frac{2\alpha\kappa-2\beta-1}{2(\kappa^2+1)\mu}, \qquad \nu = \mp\frac{2\alpha\kappa-2\beta+3}{2(2\alpha\kappa-2\beta-1)} - \tfrac{1}{2},$$

and where $D_\nu(\xi)$ is the *Parabolic Cylinder function* satisfying the equation

$$D_\nu''(\xi) = (\tfrac{1}{4}\xi^2 - \nu - \tfrac{1}{2})D_\nu(\xi),$$

together with the boundary condition $D_\nu(\xi) \sim \xi^\nu\exp(-\frac{1}{4}\xi^2)$, as $\xi \to \infty$.

References

[1] G.W. Bluman and J.D. Cole, "*Similarity Methods for Differential Equations*," (Springer-Verlag, Berlin, 1974); P.J. Olver, "*Applications of Lie Groups to Differential Equations*," (Springer-Verlag, Berlin, 1986); G.W. Bluman and S. Kumei, "*Symmetries and Differential Equations*," (Springer-Verlag, Berlin, 1989).

[2] B. Champagne and P. Winternitz, preprint CRM-1278, Montreal (1985); B. Champagne, W. Hereman and P. Winternitz, preprint CRM-1689, Montreal (1990); G.J. Reid, *J. Phys. A: Math. Gen.*, **23** (1990) L853–L859; in "*Lie Theory, Differential Equations and Representation Theory*," ed. V. Hussin (Les Publications de Centre de Recherches Mathématiques, Montreal, 1990) 363–372.

[3] F. Schwarz, *Computing*, **34** (1985) 91–106; *SIAM Rev.*, **30** (1988) 450–481.

[4] A.K. Head, preprint (1990).

[5] G.W. Bluman and J.D. Cole, *J. Math. Mech.*, **18** (1969) 1025–1042.

[6] G.M. Webb, *Physica*, **41D** (1990) 208–218; *J. Phys. A: Math. Gen.*, **23** (1990) 3885–3894.

[7] P.J. Olver and P. Rosenau, *Phys. Lett.*, **114A** (1986) 107–112; *SIAM J. Appl. Math.*, **47** (1987) 263–275.

[8] T. Nishitani and M. Tajiri, *Phys. Lett.*, **89A** (1982) 379–380; P. Rosenau and J.L. Schwarzmeier, *Phys. Lett.*, **115A** (1986) 75–77.

[9] G.R.W. Quispel, F.W. Nijhoff and H.W. Capel, *Phys. Lett.*, **91A** (1982) 143–145; G.R.W. Quispel and H.W. Capel, *Physica*, **117A** (1983) 76–102; A. Oron and P. Rosenau, *Phys. Lett.*, **118A** (1986) 172–176.

[10] P.A. Clarkson and M.D. Kruskal, *J. Math. Phys*, **30** (1989) 2201–2213.

[11] E.L. Ince, "*Ordinary Differential Equations*," (Dover, New York, 1956).

[12] D. Levi and P. Winternitz, *J. Phys. A: Math. Gen.*, **22** (1989) 2915–2924.

[13] H. Airault, *Stud. Appl. Math.*, **61** (1979) 31–53; A.S. Fokas and M.J. Ablowitz, *J. Math. Phys*, **23** (1982) 2033–2042; J.D. Gibbon, A.C. Newell, M. Tabor and Y.B. Zeng, *Nonlinearity*, **1** (1988) 481–490.

[14] P.A. Clarkson, *Euro. J. Appl. Math.*, **1** (1990) 279–300.

[15] P.A. Clarkson, *J. Phys. A: Math. Gen.*, **22** (1989) 2355–2367; *ibid*, **22** (1989) 3821–3848.

[16] P.A. Clarkson and P. Winternitz, *Physica D*, to appear [CRM-1701, Montreal (1990)].

[17] S.-Y. Lou, *J. Phys. A: Math. Gen.*, **23** (1990) L649–L654; *Phys. Lett.*, **151A** (1990) 133–135.

[18] V.E. Zakharov and A.B. Shabat, *Sov. Phys. JETP*, **34** (1972) 62–69.

[19] L. Gagnon and P. Winternitz, *J. Phys. A: Math. Gen.*, **21** (1988) 1493–1511; *ibid*, **22** (1989) 469–497; *Phys. Rev. A*, **39** (1989) 296–306; *Phys. Lett.*, **134A** (1989) 276–281.

[20] P.A. Clarkson and C.M. Cosgrove, *J. Phys. A: Math. Gen.*, **20** (1987) 2003–2024.

[21] D.J. Kaup and A.C. Newell, *J. Math. Phys*, **19** (1978) 798–801.

[22] H.H. Chen, Y.C. Lee and C.S. Liu, *Phys. Scripta*, **20** (1979) 490–492.

[23] V.S. Gerdjikov and M.I. Ivanov, *JINR Preprint E2-82-595* (1982), Dubna, USSR.

[24] M. Wadati, K. Konno and Y.-H. Ichikawa, *J. Phys. Soc. Japan*, **46** (1979) 1965–1966.

[25] F. Calogero and S. De Lillo, *Inverse Problems*, **3** (1987) 633–681; *ibid*, **4** (1988) L33–L37.

[26] J.A. Tuszyński and J.M. Dixon, *J. Phys. A: Math. Gen.*, **22** (1989) 4877–4894; J.M. Dixon and J.A. Tuszyński, *J. Phys. A: Math. Gen.*, **22** (1989) 4895–4920; P.A. Clarkson and J.A. Tuszyński, *J. Phys. A: Math. Gen.*, **23** (1990) 4269–4288.

[27] R.S. Johnson, *Proc. R. Soc. London A*, **357** (1977) 131–141.

[28] M. Florjańczyk and L. Gagnon, *Phys. Rev. A*, **41** (1990) 4478–4485.

[29] P. Winternitz, preprint CRM-1709, Montreal (1990).

[30] E.A. Zabolotskaya and R.V. Khokhlov, *Sov. Phys. Acoustics*, **15** (1969) 35–40; *ibid*, **16** (1970) 39-43; D.G. Crighton, *Ann. Fluid Mech.*, **11** (1979) 11–33; J.K. Hunter, *SIAM J. Appl. Math.*, **48** (1988) 1–37.

[31] P. Barrera and T. Brugarino, *Nuovo Cim.*, **92B** (1986) 142–156; A.T. Cates, *Physica*, **44D** (1990) 303–312; A.T. Cates and D.G. Crighton, *Proc. R. Soc. Lond. A*, **430** (1990) 69–88; G.M. Webb and G.P. Zank, *J. Phys. A: Math. Gen.*, **23** (1990) 5465–5477.

[32] S.P. Lloyd, *Acta Mech.*, **38** (1981) 85–98.

[33] R.E. Boisvert, W.F. Ames and U.N. Srivastava, *J. Eng. Math.*, **17** (1983) 203–221.

[34] B.J. Cantwell, *J. Fluid Mech.*, **85** (1978) 257–271; A. Grauel and W.-H. Steeb, *Int. J. Theo. Phys.*, **24** (1985) 255–265.

Direct Methods in Soliton Theories

R. Hirota

Department of Mathematics, Faculty of Engineering,
Hiroshima University, Higashi-Hiroshima 724, Japan.

§1. Introduction

The basic philosopy underlying the present direct method is not to investigate a given nonlinear patitial differential equation as it stands, but to transform it into a homogeneous form (bilinear, trilinear,etc.) by the dependent variable transformation. We then use a perturbational approach in order to obtain exact solutions to the transformed equation and investigate expressions of the solutions in terms of determinants or/and pfaffians.

Take the Kadomtsev-Petviashvili(KP) equation for example

$$(4u_t - 6uu_x - u_{xxx})_x - 3u_{yy} = 0,$$

which is transformed into the bilinear form

$$(D_x^4 - 4D_xD_t + 3D_y^2)\tau \cdot \tau = 0$$

through the dependent variable transformation

$$u = 2(\log \tau)_{xx}$$

where integration constants are chosen to be zero. In §3 we see that the τ of the KP equation is expressed by a determinant.

The dependent variable transformation is not restricted to the type

$$u = 2(\log \tau)_{xx}$$

but we have several types of transformations. Consider the nonlinear Schrödinger equation, for example,

$$i\psi_t + \psi_{xx} + 2|\psi|^2\psi = 0,$$

which is transformed into the following bilinear form

$$(iD_t + D_x^2)G \cdot F = 0, \quad D_x^2 F \cdot F = 2GG^*, \tag{1.1}$$

through the dependent variable transformation

$$\psi = G/F, \quad F: \quad real.$$

Furthermore we may transform the bilinear equations into other bilinear equations by introducing new dependent variables.

The bilinear equation (1.1) for example are transformed into the following set of bilinear forms

$$D_x(f^* \cdot f + g * \cdot g) = 0, \tag{1.2a}$$

$$(iD_t + D_x^2)f \cdot g^* = 0, \tag{1.2b}$$

$$(iD_t - D_x^2)(f^* \cdot f - g^* \cdot g) = 0, \tag{1.2c}$$

through the dependent variable transformations

$$F^2 = f^* f + g^* g, \quad GF = D_x g \cdot f.$$

On the other hand it is known that the Heisenberg ferromagnet equation

$$\vec{S} = \vec{S} \times \vec{S}_{xx}, \quad \vec{S} = S_1, S_2, S_3,$$

is Gauge equivalent to the nonlinear Schrödinger equation. We note that the Heisenberg ferromagnet equation is transformed into the same bilinear form (1.2a),(1.2b) and (1.2c) as those of the nonlinear Schrödinger equation through the dependent variable transformation

$$S_1 + iS_2 = \frac{2f^* g}{f^* f + g^* g}, \quad S_3 = \frac{f^* f - g^* g}{f^* f + g^* g}.$$

The above relation implies that starting with the given bilinear equation we may construct different types of nonlinear partial differential equations through the different types of dependent variable transformations. In fact, starting with the same set of the bilinear equations (1.2a),(1.2b) and (1.2c) as those of nonlinear Schrödinger equation we have obtained the following nonlinear differential equation[1)]

$$i\phi_t + \phi_{xx} + \frac{2|\phi_x|^2}{1 - |\phi|^2}\phi = 0$$

through the following dependent variable transformations

$$\phi = \frac{g}{F}, \quad F^2 = f^* f + g^* g.$$

§2. The KP and BKP Hierarchies

We have several types of hierarchies of the nonlinear partial differential equations. Among them the KP and BKP hierarchies are of importance. It is known that the KP hierarchy

$$(D_1^4 - 4D_1 D_3 + 3D_2^2)\tau \cdot \tau = 0, \tag{2.1a}$$

$$[(D_1^3 + 2D_3)D_2 - 3D_1 D_4]\tau \cdot \tau = 0, \tag{2.1b}$$

etc., gives the known soliton equations by "reduction". For examples

(i) KdV equation

$$4u_t - 6uu_x - u_{xxx} = 0$$

is obtained by the 2-reduction ($D_2 = 0$) in eq.(2.1a).

(ii) Boussinesq equation

$$u_{tt} - c_0 u_{xx} - 3(u^2)_{xx} - u_{xxxx} = 0$$

is obtained by the 3-pseudo reduction ($D_3 = c_0 D_1$) in eq.(2.1a).

(iii) Coupled KdV equation (Hirota-Satsuma system)

$$u_T - \frac{1}{2}(u_{xxx} + 6uu_x) = 2vv_x,$$

$$v_T + v_{xxx} + 3uv_x = 0,$$

is obtained by the 4-reduction ($D_4 = 0$) in eq.(2.1b).

It is also known that the BKP hierarchy

$$[(D_3 - D_1^3)D_{-1} + 3D_1^2]\tau \cdot \tau = 0, \tag{2.2a}$$

$$[(D_1^6 - 5D_1^3 D_3 - 5D_3^2 + 9D_1 D_5]\tau \cdot \tau = 0, \tag{2.2b}$$

etc., gives the soliton equations by "reduction". For example

(i) Sawada-Kotera equation

$$u_t + 45u^2u_x + 15(u_xu_{xx} + uu_{3x}) + u_{5x} = 0 \tag{2.3a}$$

is the 3-reduction ($D_3 = 0$) in eq.(2.2b).

The Sawada-Kotera equation has the same dispersion relation and nonlinear terms as those of the Lax fifth-order KdV equation

$$u_t + 30u^2u_x + 20u_xu_{xx} + 10uu_{3x} + u_{5x} = 0,$$

except the coefficients.

(ii) A model equation for shallow water waves

$$u_t - u_{xxt} - 3uu_t + 3u_x\int_x^\infty u_t\mathrm{d}x + u_x = 0, \tag{2.3b}$$

is the 3-pseudo reduction ($D_3 = D_1$) in eq.(2.2a), which has the same dispersion relation and nonlinear terms as those of model equation for shallow water waves found and solved by A.K.N.S.(Ablowitz,Kaup,Newell and Segur)

$$u_t - u_{xxt} - 4uu_t + 2u_x\int_x^\infty u_t\mathrm{d}x + u_x = 0.$$

We note that eqs.(2.3a) and (2.3b) have not been successfully solved by the inverse scattering method. We come back to this point after we clarify the structure of the BKP equation.

§3. The Structure of Bilinear Equations

Recent study reveals that the bilinear equations have the extremely simple structure if the solutions are expressed by the determinants or by the pfaffians. We have two types of expressions for the determinants which express the N-soliton solutions for the KP hierarchy. One is a wronskian determinant and another is a gramian type determinant.

$(a.1)$ *Wronski determinants*

Satsuma[2] has shown by using the Laplace expansion theorem that the wronskian determinant gives the solution of the KdV equation. Sato[3] has discovered that the bilinear KP hierarchy reduces to the Plücker relation and that the solutions are expressed by the generalization of the wronskian determinants. Freeman and Nimmo[4,5] have developed the wronskian technique for various soliton equations.The wronskian form of the N-soliton solution to the KP equation is expressed by

$$\tau_{KP} = \det \left| \frac{\partial^{j-1}\phi_i}{\partial x^{j-1}} \right|_{1 \le i,j \le N}$$

where all ϕ_i satisfy the following linear differential equations for $n = 1, 2, \cdots, \infty$

$$\frac{\partial \phi_i}{\partial x_n} = \frac{\partial^n \phi_i}{\partial x^n}.$$

Here we note that the Plücker relation are algebraic identities of determinants among which the simplest one is the following

$$\begin{vmatrix} a_0 & a_1 \\ b_0 & b_1 \end{vmatrix} \cdot \begin{vmatrix} a_2 & a_3 \\ b_2 & b_3 \end{vmatrix} - \begin{vmatrix} a_0 & a_2 \\ b_0 & b_2 \end{vmatrix} \cdot \begin{vmatrix} a_1 & a_3 \\ b_1 & b_3 \end{vmatrix} + \begin{vmatrix} a_0 & a_3 \\ b_0 & b_3 \end{vmatrix} \cdot \begin{vmatrix} a_1 & a_2 \\ b_1 & b_2 \end{vmatrix} = 0,$$

which we write by supressing the rows and using the suffix as follows

$$[0,1][2,3] - [0,2][1,3] + [0,3][1,2] = 0,$$

which is schematically expressed by Sato's Maya diagram[6)]

0 1 2 3   0 1 2 3

[○|○| |] × [| |○|○] − [○| |○|] × [|○| |○]

\+ [○| | |○] × [|○|○|] = 0 .

$(a.2)$ *Gram determinants*

Recently Nakamura[7] has shown that the bilinear KP equation reduces to the Jacobi formula of determinants provided that the solutions are expressed by the gramian type determinant

$$\tau_{KP} = \det \left| M_{ij} \right|_{1 \le i,j \le N},$$

where

$$M_{ij} = c_{ij} + \int^{x} f_i g_j \mathrm{d}x.$$

f_i and g_j satisfy the linear differential equations

$$\frac{\partial f_i}{\partial x_n} = \frac{\partial^n f_i}{\partial x^n}, \quad \frac{\partial g_j}{\partial x_n} = (-1)^{n-1}\frac{\partial^n g_j}{\partial x^n}.$$

Soliton solutions expressed by the gramian type determinants are directly related to the soliton solutions obtained by solving the Gel'fand-Levitan integral equation

$$K(x,z) + F(x,z) + \int_x^{\infty} K(x,y)F(y,z)\mathrm{d}y = 0$$

through the relation

$$K(x,z)|_{z=x} = \frac{\partial}{\partial x}\log\tau_{KP}.$$

(b) Pfaffians

Very recently we have found that soliton solutions to the BKP hierarchy are expressed by pfaffians[8)]

$$\tau_{BKP} = (1,2,3,\cdots,2N),$$

where the (i,j)element is given by

$$(i,j) = c_{ij} + \int_{-\infty}^{x} D_x f_i \cdot f_j \mathrm{d}x,$$

where $c_{ij} = -c_{ji}$ and

$$\frac{\partial f_i}{\partial x_n} = \frac{\partial^n f_i}{\partial x^n},$$

for odd n, and that the BKP equations are reduced to the identity of pfaffians.

We note that the soliton solutions expressed by the pfaffians are related to the solutions of the following integral equation with an antisymmetrix kernel $F(x,z) = -F(z,x)$,[9)]

$$K(x,z) + F(x,z) - \int_x^{\infty} D_y K(x,y) \cdot F(y,z)\mathrm{d}y = 0$$

through the relation

$$\frac{\partial}{\partial x}K(x,z)|_{z=x} = \frac{\partial}{\partial x}\log\tau_{BKP}.$$

This is one of main reasons why the BKP equations such as eqs.(2.3a) and (2.3b) have not been solved yet by the inverse scattering method.

Pfaffians of order $2N$ denoted by $(1,2,\cdots,2N)$ is defined by the expansion rule

$$(1,2,\cdots,2N) = \sum_{j=2}^{2N}(-1)^j(1,j)(2,3,\cdots,\hat{j},\cdots,2N)$$

with antisymmetric elements $(i,j) = -(j,i)$ for all i,j.

We note that a determinant of order N is expressed by the pfaffian of order $2N$

$$\det|a_{ij}|_{1\le i,j\le N} = (1^*,2^*,\cdots,N^*,N,\cdots,2,1)$$

where $a_{ij} = (i^*j), (i,j) = (i^*,j^*) = 0$ .Taking $N=2$, for example, we have

$$\begin{aligned}(1^*,2^*,2,1) &= (1^*,2^*)(2,1) - (1^*,2)(2^*,1) + (1^*,1)(2^*,2)\\ &= \det\begin{vmatrix}(1^*,1) & (1^*,2)\\ (2^*,1) & (2^*,2)\end{vmatrix}.\end{aligned}$$

We have algebraic identities[8] of pfaffians for an arbitray positive integer N:

$$\begin{aligned}&(a_1,a_2,a_3,a_4,1,2,\cdots,2N)(1,2,\cdots,2N)\\ =&(a_1,a_2,1,2,\cdots,2N)(a_3,a_4,1,2,\cdots,2N)\\ -&(a_1,a_3,1,2,\cdots,2N)(a_2,a_4,1,2,\cdots,2N)\\ +&(a_1,a_4,1,2,\cdots,2N)(a_2,a_3,1,2,\cdots,2N),\end{aligned}$$

which we express schematically by Sato's Maya diagram

	0	1	2	3		0	1	2	3
	O	O	O	O	×				
=	O	O			×			O	O
−	O		O		×		O		O
+	O			O	×		O	O	

We rewrite the τ_{KP} of gramian type by using the pfaffian

$$\tau_{KP} = (1^*,2^*,\cdots,N^*,N,\cdots,2,1),$$

where

$$(i^*j) = c_{ij} + \int^x f_i g_j \mathrm{d}x, \quad (i, j) = (i^*, j^*) = 0.$$

Then the Jacobi identiy of determinants is expressed by Sato's Maya diagram

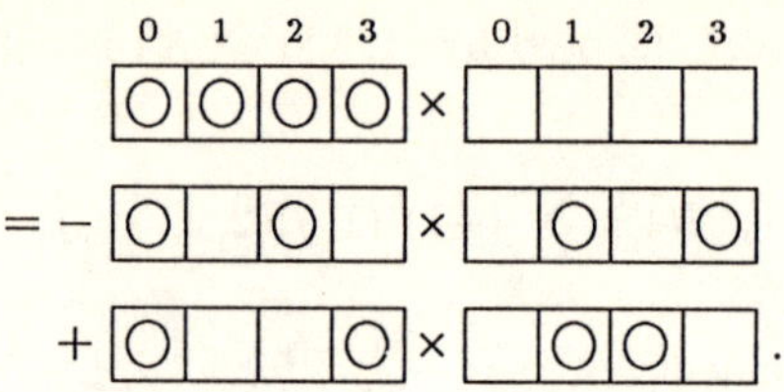

Hence the identity of pfaffians includes the Plücker relation and the Jacobi identity as its special cases.

Now we[10] introduce pfaffians, τ, σ and $\overline{\sigma}$, where τ includes the τ_{KP} and the τ_{BKP} as its special cases,

$$\begin{aligned}
\tau &= (1, 2, \cdots, 2N), \\
\sigma &= (c_0, c_1, 1, 2, \cdots, 2N), \\
\overline{\sigma} &= (d_0, d_1, 1, 2, \cdots, 2N),
\end{aligned}$$

where

$$(i, j) = c_{ij} + \int_{-\infty}^x (\phi_i \overline{\phi_j} - \phi_j \overline{\phi_i}) \mathrm{d}x,$$

$$(c_n, i) = \frac{\partial^n}{\partial x^n} \overline{\phi_i}, \ (d_n, i) = \frac{\partial^n}{\partial x^n} \phi_i,$$

$$(c_m, c_n) = (d_m, d_n) = (c_m, d_n) = 0,$$

for $m, n = 0, 1, 2, \cdots$. ϕ_i and $\overline{\phi_j}$ satisfy the linear differential equations

$$\frac{\partial \phi_i}{\partial x_n} = \frac{\partial^n \phi_i}{\partial x^n}, \quad \frac{\partial \overline{\phi_i}}{\partial x_n} = (-1)^{n-1} \frac{\partial^n \overline{\phi_i}}{\partial x^n}.$$

Substituting τ ,σ and $\overline{\sigma}$ into the KP and second modified KP equations we obtain

$$\begin{aligned}
(D_1^4 - 4D_1 D_3 + 3D_2^2)\tau \cdot \tau &= 24\bar{\sigma}\sigma \ , \\
(D_1^3 + 2D_3 + 3D_1 D_2)\sigma \cdot \tau &= 0 \ , \\
(D_1^3 + 2D_3 - 3D_1 D_2)\bar{\sigma} \cdot \tau &= 0 \ .
\end{aligned}$$

These bilinear equations are transformed into the following coupled nonlinear equations in the ordinary form,

$$(4u_t - 6uu_x - u_{xxx})_x - 3u_{yy} + 24(v\bar{v})_{xx} = 0 \ ,$$
$$2v_t + 3uv_x + v_{xxx} + 3(v_{xy} + v\int^x u_y \mathrm{d}x) = 0 \ ,$$
$$2\bar{v}_t + 3u\bar{v}_x + \bar{v}_{xxx} - 3(\bar{v}_{xy} + \bar{v}\int^x u_y \mathrm{d}x) = 0 \ ,$$

through the dependent variable transformations with $x = x_1$, $y = x_2$, $t = x_3$,

$$u = 2(\log\tau)_{xx}, \quad v = \sigma/\tau, \quad \bar{v} = \bar{\sigma}/\tau \ .$$

We note that Hietarinta[11)] has found coupled bilinear equations which exhibit 4-soliton solutions. One of his equations includes the above as a special case.

Finally we remark again that the present τ is a unified form of the τ_{KP} and τ_{BKP} which have been considered to be in different categories until now.

References

1) R. Hirota:"Bilinear Forms of Soliton Equations",in *Nonlinear Integrable Systems - Classical Theory and Quantum Theory*, Edited by M.Jimbo and T.Miwa, pp.15-37,World Sientific (1983).
2) J.Satsuma:J.Phys.Soc.Jpn.**46**(1979)359.
3) M.Sato:RIMS Kokyuroku **439**(1981)30.
4) N.C.Freeman and J.J.C.Nimmo:Phys.Lett.**95A**(1983)1.
5) J.J.C.Nimmo and N.C.Freeman:Phys.Lett.**95A**(1983)4.
6) R.Hirota:J.Phys.Soc.Jpn.**55**(1986)2137.
7) A.Nakamura:J.Phys.Soc.Jpn.**58**(1989)412.
8) R.Hirota:J.Phys.Soc.Jpn.**58**(1989)2285.
9) R.Hirota:J.Phys.Soc.Jpn.**58**(1989)2705.
10) R.Hirota and Y.Ohta:J.Phys.Soc.Jpn.**60** no.3 (1991).
11) J.Hietarinta:*Partially Integrable Evolution Equations in Physics* ,ed. R.Conte and N.Boccara (Kluwer Academic Publ.Dordrecht,1990) p.459.

Trilinear Form - An Extension of Hirota's Bilinear Form

J. Satsuma, J. Matsukidaira and K. Kajiwara

Department of Applied Physics, Faculty of Engineering,
University of Tokyo, 7-3-1 Hongo, Bunkyo-ku, Tokyo 113, Japan.

Hirota's method has been successfully applied to the soliton equations for obtaining particular solutions, especially those describing interactions of solitons [1]. The key procedure in the method is the bilinearization of the equations through dependent variable transformation. Once an equation is transformed into its bilinear form, we can get explicit solutions by applying a perturbational technique. We now know that there exists a rich algebraic structure behind the trick. It is Sato [2] who first discovered that the hierarchy of bilinear equations which include the Kadomtsev-Petviashvili (KP) equation as the simplest one is nothing but the Plücker relation appeared in the theory of the Grassmann manifold.

Let us briefly describe the above story. We have the KP equation,

$$(4u_{x_3} - 12uu_x - u_{xxx})_x - 3u_{x_2x_2} = 0\ . \tag{1}$$

By introducing the dependent variable transformation,

$$u = (\log \tau)_{xx}\ , \tag{2}$$

eq.(1) is reduced to its bilinear form,

$$(4D_xD_{x_3} - D_x^4 - 3D_{x_2}^2)\,\tau\cdot\tau = 0\ . \tag{3}$$

The solution of eq.(3) is expressed by the Wronskian,

$$\tau = \begin{vmatrix} f^{(1)} & \partial_x f^{(1)} & \dots & \partial_x{}^{m-1} f^{(1)} \\ f^{(2)} & \partial_x f^{(2)} & \dots & \partial_x{}^{m-1} f^{(2)} \\ \vdots & \vdots & \dots & \vdots \\ f^{(m)} & \partial_x f^{(m)} & \dots & \partial_x{}^{m-1} f^{(m)} \end{vmatrix}, \tag{4}$$

where $f^{(j)}$ satisfies

$$\partial f^{(j)}/\partial x_k = \partial^k f^{(j)}/\partial x^k, \tag{5}$$

for $j = 1, 2, \dots, m$ and $k = 1, 2, 3, \dots (x_1 = x)$. For the latter convenience, we have introduced an infinite number of independent variables $x_1, x_2, x_3, \dots$. Let us introduce the microdifferential operator,

$$W = 1 + w_1\partial_x{}^{-1} + w_2\partial_x{}^{-2} + \cdots + w_m\partial_x{}^{-m}\ , \tag{6}$$

where w_j $(j = 1, 2, \ldots)$ are functions of x and $\partial_x{}^{-n}$ is formally defined by

$$\partial_x{}^{-n} = (\mathrm{d}/\mathrm{d}x)^{-n} \tag{7}$$

(although the general theory is developed for $m = \infty$, this simplification keeps its essence). By considering the ordinary differential equation,

$$W\partial_x{}^m f(x) = (\partial_x{}^m + w_1\partial_x{}^{m-1} + \cdots + w_m)f(x) = 0\ , \tag{8}$$

we find that w_j is given by

$$w_j = \frac{\begin{vmatrix} f^{(1)} & \cdots & -\partial_x{}^m f^{(1)} & \cdots & \partial_x{}^{m-1} f^{(1)} \\ \vdots & \cdots & \vdots & \cdots & \vdots \\ f^{(m)} & \cdots & -\partial_x{}^m f^{(m)} & \cdots & \partial_x{}^{m-1} f^{(m)} \end{vmatrix}}{\begin{vmatrix} f^{(1)} & \cdots & \partial_x{}^{m-j} f^{(1)} & \cdots & \partial_x{}^{m-1} f^{(1)} \\ \vdots & \cdots & \vdots & \cdots & \vdots \\ f^{(m)} & \cdots & \partial_x{}^{m-j} f^{(m)} & \cdots & \partial_x{}^{m-1} f^{(m)} \end{vmatrix}}\ , \tag{9}$$

where the denominator is eq.(4) itself and $f^{(1)}(x)$, $f^{(2)}(x)$, ..., $f^{(m)}(x)$ are m linearly independent solutions of eq.(8) which are assumed to be analytic. We now suppose that w_j $(j = 1, 2, \ldots)$ in eq. (8) are also the functions of an infinite number of independent variables $x_1, x_2, x_3, \ldots$. Then imposing the condition (5), we find that w_j is given by

$$w_j = \frac{1}{\tau} p_j(-\tilde{\partial}_x)\,\tau\ , \tag{10}$$

where $p_j, j = 1, 2, \cdots$ are polynomials defined by

$$\exp\left(\sum_{n=1}^{\infty} x_n \lambda^n\right) = \sum_{j=0}^{\infty} p_j(x)\lambda^j, \tag{11}$$

and $\tilde{\partial}_x$ is the differential operator given by

$$\tilde{\partial}_x = (\partial_x, \partial_{x_2}/2, \partial_{x_3}/3, \cdots). \tag{12}$$

Moreover, it is shown that W satisfies

$$\partial W/\partial x_n = B_n W - W\partial_x{}^n\ , \tag{13}$$

$$B_n = (W\partial_x{}^n W^{-1})^+\ , \tag{14}$$

where $(\ \ )^+$ denotes the differential part of the operator. Equations (13) and (14) are called the Sato equation.

If we define an operator L by

$$L = W\partial_x W^{-1} = \partial_x + u_2\partial_x{}^{-1} + u_3\partial_x{}^{-2} + u_4\partial_x{}^{-3} + \cdots , \qquad (15)$$

we see that L and B_n constitute the generalized Lax form. The Lax form generates an infinite series of equations in which the simplest and nontrivial one is the KP eq.(1) with $u = u_2$. The bilinear identities such as eq.(3) are derived by considering the algebraic structure of the τ function (4) with the condition (5) (see [3] for more detail).

The KP equation is an evolution system with 1 time and 2 space variables. It is an interesting question whether the higher-dimensional extension is possible or not in this formalism. Recently we have proposed the soliton equations expressed by trilinear form [4]. The main result is summarized as follows: An infinite number of trilinear equations,

$$\begin{vmatrix} p_i(-\tilde{\partial}_x)p_l(\tilde{\partial}_y)\tau & p_i(-\tilde{\partial}_x)p_m(\tilde{\partial}_y)\tau & p_i(-\tilde{\partial}_x)p_n(\tilde{\partial}_y)\tau \\ p_j(-\tilde{\partial}_x)p_l(\tilde{\partial}_y)\tau & p_j(-\tilde{\partial}_x)p_m(\tilde{\partial}_y)\tau & p_j(-\tilde{\partial}_x)p_n(\tilde{\partial}_y)\tau \\ p_k(-\tilde{\partial}_x)p_l(\tilde{\partial}_y)\tau & p_k(-\tilde{\partial}_x)p_m(\tilde{\partial}_y)\tau & p_k(-\tilde{\partial}_x)p_n(\tilde{\partial}_y)\tau \end{vmatrix} = 0, \qquad (16)$$

for arbitrary nonnegative integers i, j, k, l, m, n, admits solutions written by the two-directional Wronskian,

$$\tau = \begin{vmatrix} f & \partial_x f & \dots & \partial_x{}^{m-1} f \\ \partial_y f & \partial_x\partial_y f & \dots & \partial_x{}^{m-1}\partial_y f \\ \vdots & \vdots & \dots & \vdots \\ \partial_y{}^{m-1} f & \partial_x\partial_y{}^{m-1} f & \dots & \partial_x{}^{m-1}\partial_y{}^{m-1} f \end{vmatrix}, \qquad (17)$$

where $\tilde{\partial}_y$ is defined by

$$\tilde{\partial}_y = (\partial_y, \partial_{y_2}/2, \partial_{y_3}/3, \cdots), \qquad (18)$$

and f is considered to be the function of $x_1, x_2, x_3, \ldots$ and $y_1, y_2, y_3, \ldots$. Moreover, f satisfies

$$\partial f/\partial x_k = \partial^k f/\partial x^k, \quad \partial f/\partial y_k = \partial^k f/\partial y^k. \qquad (19)$$

The simplest case, $(i, j, k) = (l, m, n) = (0, 1, 2)$, of eq.(16) reduces to a coupled system,

$$w_{1x_2} = w_{1xx} - 2w_1 w_{1x} + w_{2x}, \qquad (20a)$$

$$(w_{1y_2} + w_{1yy})w_{2y} - (w_{2y_2} + w_{2yy})w_{1y} = 0, \qquad (20b)$$

by introducing the dependent variables as

$$w_1 = -\partial_x \log \tau, \qquad (21a)$$

$$w_2 = \frac{1}{\tau}\, p_2(-\tilde{\partial}_x)\, \tau\, . \qquad (21b)$$

Since the coupled equation (20) is an evolution system in 2+2 dimension, the trilinear equations may be considered to give a nontrivial extension of the bilinear equations. We here show how this result is explained by the formalism developed by Sato (see [5] for more detail). The function f in the τ function (17) is regarded to be the function which is imposed an additional y-dependence on $f^{(1)}, f^{(2)}, \ldots$ in eq.(4). This means that W composed of w_j given by eq.(10) with the τ function (17) also satisfies the Sato eq.(13) and (14). However, because of the y-dependence, there is another equation which W obeys. Let us derive the equation.

From the dispersion relation (19), we have

$$p_j(\tilde{\partial}_y) f = \partial_y{}^j f \ , j = 1, 2, 3, \ldots . \tag{22}$$

It is shown by using the property of $p_j(\tilde{\partial}_y)$ and W that the ordinary differential equation of $(m-1)$-th degree with respect to x,

$$[p_j(\tilde{\partial}_y) W] \partial^m f = 0, \quad j = 1, 2, \ldots, m-1, \tag{23}$$

has the *same* fundamental solutions, $f, \partial_y f, \ldots, \partial_y{}^{m-2} f$, for all of j. On the other hand, from the assumption,

$$[p_0(\tilde{\partial}_y) W] \partial^m f = W \partial^m f = 0 \tag{24}$$

has the fundamental solutions, $f, \partial_y f, \ldots, \partial_y{}^{m-1} f$. This means that the rank of the matrix,

$$\begin{pmatrix} w_0 & w_1 & \cdots & w_m \\ p_1(\tilde{\partial}_y) w_0 & p_1(\tilde{\partial}_y) w_1 & \cdots & p_1(\tilde{\partial}_y) w_m \\ \vdots & \vdots & \vdots & \vdots \\ p_{m-1}(\tilde{\partial}_y) w_0 & p_{m-1}(\tilde{\partial}_y) w_1 & \cdots & p_{m-1}(\tilde{\partial}_y) w_m \end{pmatrix},$$

whose entries are the coefficients of eqs.(23) and (24), should be 2. Then, since the arbitrary 3×3 minor is zero, we have

$$\begin{vmatrix} p_i(\tilde{\partial}_y) w_l & p_i(\tilde{\partial}_y) w_m & p_i(\tilde{\partial}_y) w_n \\ p_j(\tilde{\partial}_y) w_l & p_j(\tilde{\partial}_y) w_m & p_j(\tilde{\partial}_y) w_n \\ p_k(\tilde{\partial}_y) w_l & p_k(\tilde{\partial}_y) w_m & p_k(\tilde{\partial}_y) w_n \end{vmatrix} = 0, \tag{25}$$

from which we obtain eq.(16) by using eq.(10) and a property of $p_j(\tilde{\partial}_y)$.

We here comment a relationship between the trilinear form and the usual bilinear form. If we introduce the 2-component extension of W in which each coefficient is given by a 2×2 matrix, we obtain a hierarchy in which the Davey-Stewartson equation is the simplest nontrivial system. In this case its solution is given by the double Wronskian [6]. It has been shown that the double Wronskian is also expressed in terms of the two-directional Wronskian [7]. This fact suggests that the trilinear equation can be written in some bilinear form if an auxiliary dependent variable is introduced. However, because of its simplicity the trilinear form may be useful for considering further extension.

In fact we have recently presented semi-discrete [8] and full-discrete [9] trilinear equations which are considered to be four-dimensional coupled systems. For example, we can show that the trilinear equation,

$$\begin{vmatrix} \partial_y \tau_{m,n-1} & \tau_{m,n-1} & \tau_{m+1,n-1} \\ \partial_y \tau_{m,n} & \tau_{m,n} & \tau_{m+1,n} \\ \partial_y \partial_x \tau_{m,n} & \partial_x \tau_{m,n} & \partial_x \tau_{m+1,n} \end{vmatrix} = 0, \tag{26}$$

admits a solution written by the two-directional Casorati determinant,

$$\tau_{m,n} = \det|f(m+i-1, n+j-1)|_{1\leq i,j\leq N} \ , \tag{27}$$

where $f(m,n)$ satisfies

$$\partial_x f(m,n) = f(m,n+1) \ , \quad \partial_y f(m,n) = f(m+1,n) \ . \tag{28}$$

From eq.(26), we obtain an interesting system: Let us introduce the dependent variables by

$$\psi_{m,n} = \log \frac{\tau_{m,n}}{\tau_{m,n-1}}, \quad \phi_{m,n} = \log \frac{\tau_{m,n}}{\tau_{m+1,n}} \ . \tag{29}$$

Then eq.(26) reduces to the coupled system,

$$\partial_x \partial_y \phi_{m,n} = \frac{\partial_x \phi_{m,n} \partial_y \psi_{m,n}}{e^{\phi_{m,n}-\phi_{m,n-1}} - 1} - \frac{\partial \phi_{m+1,n} \partial \psi_{m+1,n}}{e^{\phi_{m+1,n}-\phi_{m+1,n-1}} - 1} \ , \tag{30a}$$

$$\partial_x \partial_y \psi_{m,n} = \frac{\partial_x \phi_{m,n} \partial_y \psi_{m,n}}{e^{\psi_{m,n}-\psi_{m+1,n}} - 1} - \frac{\partial \phi_{m+1,n} \partial \psi_{m+1,n}}{e^{\psi_{m,n-1}-\psi_{m+1,n-1}} - 1} \ , \tag{30b}$$

with a constraint,

$$\psi_{m+1,n} - \psi_{m,n} = \phi_{m,n-1} - \phi_{m,n} \ . \tag{31}$$

If the reduction, $\phi(x,y)_{m,n} = q(x+y)_{m+n}$, $\psi(x,y)_{m,n} = -q(x+y)_{m+n-1}$, is taken, eqs.(30) reduce to

$$\partial_x^2 q_n = -\partial_x q_n \left(\frac{\partial_x q_{n-1}}{e^{q_n - q_{n-1}} - 1} - \frac{\partial_x q_{n+1}}{e^{q_{n+1}-q_n} - 1} \right) \ , \tag{32}$$

which is nothing but the relativistic Toda equation proposed by Ruijsenaars. Hence eqs.(29) is considered to be an 2+2 dimensional extension of the relativistic Toda equation [10].

References

[1] R.Hirota, in *Solitons*, ed. by R.K.Bullough and P.J.Caudrey (Springer, Berlin, 1980).

[2] M.Sato and Y.Sato, in *Nonlinear Partial Differential Equations in Applied Science*, ed. by H.Fujita, P.D.Lax and G.Strang (Kinokuniya / North Holland,Tokyo, 1983) 259.

[3] Y.Ohta, J.Satsuma, D.Takahashi and T.Tokihiro, Prog. Theor. Phys. Suppl. **94** (1988) 210.

[4] J.Matsukidaira, J.Satsuma and W. Strampp, Phys. Lett. A. **147** (1990) 467.

[5] K.Kajiwara, J.Matsukidaira and J.Satsuma, in preparation.

[6] N.C.Freeman, IMA J. Appl. Math. **32** (1984) 125.

[7] R.Hirota, Y.Ohta and J.Satsuma, Prog. Theor. Phys. Suppl. **94** (1988) 59.

[8] J.Matsukidaira and J.Satsuma, J. Phys. Soc. Jpn. **59** (1990) 3413.

[9] J.Matsukidaira and J.Satsuma, to appear in Phys. Lett. A.

[10] J.Hietarinta and J.Satsuma, in preparation.

On the Use of Bilinear Forms for the Search of Families of Integrable Nonlinear Evolution Equations

R. Willox* and F. Lambert

Vrije Universiteit Brussel,
Dienst Theoretische Natuurkunde,
Pleinlaan 2, B-1050 Brussel.
*Research Assistant, National Foundation for Scientific Research, Belgium.

Integrable nonlinear evolution equations with KdV-like two-soliton solutions:

$$2\partial_x^2 \, ln f_2, \qquad f_2 = 1 + \exp\theta_1 + \exp\theta_2 + A_{12} \; \exp(\theta_1 + \theta_2),$$

$$\theta_i = -k_i x + \omega_i t + \tau_i \tag{1}$$

and with corresponding N-soliton solutions, appear in infinite families with a characteristic form of the coupling factor A_{12}.

To find such families one may look for sequences of equations of increasing order admitting two-soliton solutions with the same coupling, and check whether the lower members of a sequence pass the N-soliton test.

Here we are interested in soliton-potential equations with a linear part:

$$V_{x,t} + V_{2x} - V_{(2m+1)x,t} \quad \text{where} \quad V_{rx,st} = \partial_x^r \partial_t^s V \quad \text{and } m = \text{integer}, \tag{2}$$

admitting solutions $V_2 = -2\partial_x ln f_2$ and corresponding potential N-soliton solutions. We discuss the way in which multidimensional bilinear forms [1)] may be used to obtain such equations, and we derive a sequence of candidate soliton equations two of which are found to pass the 3-soliton test.

The procedure relies on an auxiliary family of first-order equations in t, with single power dispersion laws, known to admit KdV-like N-soliton solutions. The simplest such family is the KdV-hierarchy [2)] itself. The KdV-equation can be cast into a two-dimensional bilinear form leading to the "primary equation":

$$f^{-2} D_1(D_3 + D_1^3) f \cdot f \Big|_{f = \exp\frac{q}{2}} \equiv q_{x,t_3} + q_{4x} + 3q_{2x}^2 = 0, \tag{3}$$

(where D_1 stands for D_x and $D_{k>1}$ stands for D_{t_k}) with solutions:

$$q_2 = 2ln f_2\left(\theta_1, \theta_2; A_{12}^{KdV}\right), \qquad \theta_i = -k_i x + k_i^3 t_3 + \tau_i,$$

$$A_{12}^{KdV} = \left(\frac{k_1 - k_2}{k_1 + k_2}\right)^2. \tag{4}$$

Another family is the Sawada-Kotera (SK) hierarchy [3] the lowest member of which corresponds to the sixth-order primary equation:

$$f^{-2}D_1\left(D_5+D_1^5\right)f\cdot f\Big|_{f=\exp\frac{q}{2}} \equiv q_{x,t_5}+q_{6x}+15q_{2x}q_{4x}+15q_{2x}^3=0, \tag{5}$$

$$\text{with solutions}: \quad q_2 = 2ln f_2\left(\theta_1,\theta_2;A_{12}^{SK}\right), \quad \theta_i=-k_ix+k_i^5t_5+\tau_i,$$

$$A_{12}^{SK}=\left(\frac{k_1^2-k_1k_2+k_2^2}{k_1^2+k_1k_2+k_2^2}\right)A_{12}^{KdV}. \tag{6}$$

Higher-order primary equations for both hierarchies can be derived successively from multidimensional bilinear operators, consisting of a leading term $D_1(D_{2m+1}+D_1^{2m+1})$ and of a linear combination of counter-terms of order $2m+2$, of the form:

$$B_{2m+2}\left(D_1,...D_{2p+1},...;\alpha_p\right)=$$

$$D_1\left(D_{2m+1}+D_1^{2m+1}\right)+\sum_p\alpha_p\left(D_{2p+1}+D_1^{2p+1}\right)P_{2m-2p+1}\left(D_{2r+1}\right), \tag{7}$$

with $1\leq p\leq m-1$, and where $P_{2m-2p+1}(D_{2r+1})$ denotes a $(2m-2p+1)$-order product of operators D_{2r+1} with the understanding that D_{2r+1} is of order $2r+1$.
The procedure (which applies also to other families) can be summarized as follows. One introduces a multidimensional expression:

$$\overline{f}_2=f_2\left(\overline{\theta}_1,\overline{\theta}_2;A_{12}\right) \quad \text{with} \quad \overline{\theta}_i=-k_ix+k_i^3t_3+...+k_i^{2p+1}t_{2p+1}+..., \tag{8}$$

such that

$$B_{2m+2}\left(D_1,...D_{2p+1},...;\alpha_p\right)\overline{f}_2\cdot\overline{f}_2$$

$$=2\left[B_{2m+2}\left(k_1,-k_2\right)+A_{12}B_{2m+2}\left(k_1,k_2\right)\right]\exp\left(\overline{\theta}_1+\overline{\theta}_2\right), \tag{9}$$

where $B_{2m+2}\left(k_1,k_2\right)$ denotes the polynomial $B_{2m+2}\left(-k_1-k_2,...k_1^{2p+1}+k_2^{2p+1},...;\alpha_p\right)$ One looks, at successive values of m, for counter-terms which involve only integer values of p and r for which primary equations (of order $2p+2$ and $2r+2$) have previously been obtained, and for appropriate coefficients $\overline{\alpha}_p$, such that:

$$B_{2m+2}\left(D_1,...D_{2p+1},...;\overline{\alpha}_p\right)\overline{f}_2\cdot\overline{f}_2=0. \tag{10}$$

It then follows that the corresponding equation for the field $q(x,...t_{2p+1},...t_{2m+1})$:

$$f^{-2}B_{2m+2}\left(D_1,...D_{2p+1},...;\overline{\alpha}_p\right)f\cdot f\big|_{f=\exp\frac{q}{2}}=0 \tag{11}$$

has solutions $\overline{q}_2=2ln\overline{f}_2$. This equation is a nonlinear partial differential equation with a linear part equal to $B_{2m+2}(\partial_x,...\partial_{t_{2p+1}},...;\overline{\alpha}_p)q$ and with a polynomial nonlinearity, in terms of even-order partial derivatives of q with respect to x and t_{2p+1}, which obeys a simple combinatorial principle [4].

Finally, one considers the equation obtained by differentiating the l.h.side of equ.(11) with respect to x, and one eliminates all partial derivatives of q containing auxiliary variables t_{2p+1} by means of available lower order primary equations in the family (dimensional reduction).

It is worth emphasizing that the resulting primary equations for $q(x, t_{2m+1})$ can be rewritten in terms of the potential field $V = -q_x$ in the form:

$$V_{x,t_{2m+1}} + V_{(2m+2)x} = K_{2m+2}(V) \tag{12}$$

where $K_{2m+2}(V)$ denotes a nonlinearity with a quadratic part:

$$K^{(2)}_{2m+2}(V) = \sum_i \beta_i V_{p_i x} V_{q_i x}, \quad p_i + q_i = 2m+1. \tag{13}$$

Each such evolution equation admits, by construction, potential two-soliton solutions $\overline{V}_2 = -2\partial_x ln f_2(\overline{\theta}_1, \overline{\theta}_2; A_{12})$, with a coupling factor A_{12} which is related to the linear and the quadratic terms through the formula [5]:

$$\begin{aligned} &\frac{(k_1+k_2)^2}{2k_1k_2}(A_{12}-1)\left[(k_1+k_2)^{2m+1} - k_1^{2m+1} - k_2^{2m+1}\right] \\ &\qquad = \sum_i (-)^{p_i+q_i} \beta_i \left(k_1^{p_i} k_2^{q_i} + k_1^{q_i} k_2^{p_i}\right). \end{aligned} \tag{14}$$

This relation can only be satisfied when its l.h.side can be reduced to a polynomial. It may therefore restrict the set of values of m for which a primary equation can be expected.

When $A_{12} = A_{12}^{KdV}$ there is no restriction on m : $A_{12}B_{2m+2}(k_1, k_2)$ can always be reduced to a polynomial. It turns out that the above procedure produces (up to $m = 4$) primary KdV equations of the form:

$$q_{x,t_{2m+1}} + q_{(2m+2)x} + Q_{2m+2}(q_{2x}, \ldots q_{2mx}) = 0, \tag{15}$$

where Q_{2m+2} stands for a polynomial nonlinearity of degree $m+1$ which contains as many different terms as there are different partitions of $2m+2$ into two or more parts larger than one.

When $A_{12} = A_{12}^{SK}$ we see that the condition (14) restricts m to the integers $\{\overline{m}\}$ which are such that $2\overline{m}+1$ is not a multiple of 3. This fact agrees with the lacunary nature of the SK-hierarchy [3]. The first three primary SK equations (m = 2, 3 and 5) take also the form (15).

Let us now try to construct soliton-potential equations with linear parts of type (2). We consider bilinear operators of the form:

$$\widehat{B}_{2m+1,1}(D_1, D_t, \ldots D_{2p+1}, \ldots; \alpha_p) =$$

$$D_1(D_t + D_1 - D_t D_1^{2m}) + \sum_p \alpha_p D_t(D_{2p+1} + D_1^{2p+1}) P_{2m-2k}(D_{2r+1}) \tag{16}$$

and multi-dimensional expressions:

$$\widehat{f}_2 = f_2(\widehat{\theta}_1, \widehat{\theta}_2; A_{12}),$$

$$\widehat{\theta}_i = -k_i x + \omega_i t + \cdots + k_i^{2p+1} t_{2p+1} + \cdots, \quad \omega_i = \frac{k_i}{1 - k_i^{2m}}, \tag{17}$$

such that

$$\widehat{B}_{2m+1,1}(D_1, D_t, ... D_{2p+1}, ...; \alpha_p) \widehat{f}_2 \cdot \widehat{f}_2 =$$
$$2\left[\widehat{B}_{2m+1,1}(k_1, -k_2)) + A_{12} \widehat{B}_{2m+1,1}(k_1, k_2)\right] \exp(\widehat{\theta}_1 + \widehat{\theta}_2) \tag{18}$$

where $\widehat{B}_{2m+1,1}(k_1, k_2)$ stands for the fraction:

$$\widehat{B}_{2m+1,1}\left(-k_1 - k_2, \frac{k_1}{1 - k_1^{2m}} + \frac{k_2}{1 - k_2^{2m}}, ... k_1^{2p+1} + k_2^{2p+1}, ...; \alpha_p\right).$$

Taking first $A_{12} = A_{12}^{KdV}$ we remark that the denominator of $A_{12}^{KdV} \widehat{B}_{2m+1,1}(k_1, k_2)$ remains the same as that of $\widehat{B}_{2m+1,1}(k_1, k_2)$, for any value of m.
Thus, we may look at successive values of m for appropriate counter-terms with coefficients $\widehat{\alpha}_p$, such that:

$$\widehat{B}_{2m+1,1}(D_1, D_t, ... D_{2p+1}, ...; \widehat{\alpha}_p) \widehat{f}_2 \cdot \widehat{f}_2 \Big|_{A_{12} = A_{12}^{KdV}} = 0. \tag{19}$$

Dimensional reduction of the corresponding equations:

$$f^{-2} \widehat{B}_{2m+1,1}(D_1, D_t, ... D_{2p+1}, ...; \widehat{\alpha}_p) f \cdot f \Big|_{f = \exp \frac{q}{2}} = 0 \tag{20}$$

by means of KdV-primary equations (15) produces (at $m = 1$ and $m = 2$) two-dimensional equations for the field $V(x,t) = -q_x(x,t)$, with linear parts of type (2) and with KdV-coupled two-soliton potential solutions. They are the potential versions of the first two members of a sequence of integrable equations which corresponds to a known extension [6)] of the KdV-hierarchy. Taking on the other hand $A_{12} = A_{12}^{SK}$ we remark that the denominator of $A_{12}^{SK} \widehat{B}_{2m+1,1}(k_1, k_2)$ will be the same as that of $\widehat{B}_{2m+1,1}(k_1, k_2)$ if the integers m and p which appear at the r.h.side of equ.(16) are such that the numerator of $\widehat{B}_{2m+1,1}(k_1, k_2)$ contains a factor $k_1^2 + k_1 k_2 + k_2^2$. It is easy to verify that this is only the case for m-values which coincide with multiples of 3 and for p-values taken among the set $\{\overline{m}\}$.
Thus, the lowest possible operator (16) corresponds to m = 3. It is a straightforward matter to check that:

$$\widehat{B}_{7,1}(D_1, D_t, D_5, D_7; \overline{\alpha}_p) \widehat{f}_2 \cdot \widehat{f}_2 \Big|_{A_{12} = A_{12}^{SK}} = 0 \tag{21}$$

with

$$\widehat{B}_{7,1} = D_1(D_t + D_1 - D_t D_1^6) + \frac{3}{5}D_t D_1^2(D_5 + D_1^5) + \frac{1}{7}D_t(D_7 + D_1^7). \qquad (22)$$

Elimination of the auxiliairy variables t_5 and t_7 from the x-derivative of the corresponding equ.(20) by means of the two lower SK-equations produces the potential equation (for $V = -q_x$):

$$\begin{aligned} &V_{x,t} + V_{2x} - V_{7x,t} + 18V_x V_{5x,t} + 15V_{x,t}V_{5x} + 27V_{2x}V_{4x,t} + 30V_{2x,t}V_{4x} \\ &+33V_{3x}V_{3x,t} + 3V_{6x}V_t - 144V_x V_{3x} V_{x,t} - 81V_x^2 V_{3x,t} - 180V_x V_{2x} V_{2x,t} \\ &-45V_{2x}V_{3x}V_t - 45V_x V_{4x} V_t - 63V_{2x}^2 V_{x,t} + 108V_x^3 V_{x,t} + 135V_x^2 V_{2x} V_t \\ &+9V_{2x}\int_x^\infty \partial_t(V_y V_{3y} - V_x^3)dy = 0 \end{aligned} \qquad (23)$$

This equation is also found to admit three-soliton potential solutions with SK coupling. The next candidate soliton equation with a linear part of type (2) and with SK-coupled soliton solutions corresponds to $m = 6$. One verifies that:

$$\widehat{B}_{13,1}(D_1, D_t, D_5, D_7, D_{11}, D_{13})\widehat{f}_2 \cdot \widehat{f}_2\Big|_{A_{12}=A_{12}^{SK}} = 0 \qquad (24)$$

with

$$\begin{aligned} \widehat{B}_{13,1} = &D_1(D_t + D_1 - D_t D_1^{12}) + \frac{1}{13}D_t(D_{13} + D_1^{13} \\ &+ \frac{3}{11}D_t D_1^2(D_{11} + D_1^{11}) + \frac{6}{35}D_t D_1^6(D_7 + D_1^7) \\ &+ \frac{12}{25}D_t D_1^8(D_5 + D_1^5) - \frac{3}{25}D_t D_1^3 D_5(D_5 + D_1^5) \\ &- \frac{9}{35}D_t D_1 D_5(D_7 + D_1^7). \end{aligned}$$

Also in this case it is found that the corresponding equ.(20) passes the 3-soliton test. These results suggest the possible existence of an infinite sequence of integrable equations, connected with the SK-hierarchy, with linear parts of the form (2) subject to the condition m = multiple of 3.

References

1. M. Ito, J. Phys. Soc. Jpn **49** (1980), 771.
2. P. Olver "Application of Lie Groups to Differential Equations", Springer, Heidelberg (1986), 319.
3. J. Weiss, J. Math. Phys. **25** (1984), 13.
4. F. Lambert, R. Willox in "Inverse Problems" Ed. P.C. Sabatier, Springer, Heidelberg (1990), 525.
5. F. Lambert, R. Willox in "Solitons and Applications" Eds. V.G. Makhankov, V.K. Fedyanin and O.K. Pashaev, World Scientific, Singapore (1990), 73.
6. F. Calogero, A. Degasperis "Solitons and the Spectral Transform I", North Holland, Amsterdam (1982), 126.

From Periodic Processes to Solitons and Vice-Versa

J. Zagrodzinski[1], M. Jaworski[1] and K. Wyser[1,2]

[1] Institute of Physics, Polish Academy of Sciences,
02668 Warsaw, Poland.

[2] Department of Physics, ETH Zurich, Switzerland.

The extreme effectiveness of direct methods in solving nonlinear partial differential equations (NLPDE) can be ascribed to an addition property of a function $f(z)$, by which the solution is usually expressed in a form involving $ln\ f(z)$.

We define the function $f : \mathbb{C}^g \to \mathbb{C}$ as having an "addition" property iff

$$f(x+y)\, f(x-y) = \sum_{\varepsilon} X(x,\varepsilon)\, Y(y,\varepsilon), \tag{1}$$

where the sum is over a finite set and where the functions X and Y map $\mathbb{C}_g \otimes Z_g \to \mathbb{C}$. The point is that for functions having the addition property one can calculate immediately the derivatives of $ln\ f(z)$, [1]. On the other hand, the close relation with the Hirota bilinear formalism is clearly seen, [2].

Below we list a few examples of functions having the addition property:

1. Riemann theta functions

$$\Theta(z \mid B) := \sum_{n \in Z^g} \exp i\pi \left[2\langle z,n\rangle + \langle n,Bn\rangle\right], \tag{2}$$

where B represents the Riemann matrix $g \times g$, and the corresponding solution of the NLPDE is quasi-periodic.

2. Exponential functions

$$E(z \mid A) := \sum_{n \in Z_2^g} \exp i\pi \left[2\langle z,n\rangle + \langle n,An\rangle\right], \tag{3}$$

where A is the off-diagonal complex symmetric matrix, and the sum is over 2^g elements. The E functions usually appear in the soliton solutions, as it follows e.g. from direct methods.

3. T-functions, leading to solutions in a form of solitons on the background of multiphase quasi-periodic processes, and representing an intermediate step between (2) and (3)

$$T(z_s, z_p \mid A, B, C) := \sum_{n \in Z_2^g} \exp i\pi \left[2\langle z_s,n\rangle + \langle n,An\rangle\right] \Theta(z_p + Cn \mid B), \tag{4}$$

where $A \in \mathbb{C}^{s\times s}$ (symmetric, off-diagonal), $B \in \mathbb{C}^{p\times p}$ (Riemannian), $C \in \mathbb{C}^{s\times g}$, and T is the order of $g = s + p$, [1].

4. The "smoothed" theta functions $\widetilde{\Theta}$ defined by the integral of the Riemann Θ-function with respect to its characteristics

$$\widetilde{\Theta}(z \mid B) := \int_{-\infty}^{\infty} \exp\left[i\pi\langle\alpha, Q\alpha\rangle\right] \Theta\begin{bmatrix}\alpha\\0\end{bmatrix}(z \mid B)(d\alpha)^g, \tag{5}$$

where Q is an arbitrary $g \times g$ Riemannian matrix. At the moment, we do not know whether those functions lead to essentially new solutions of NLPDEs.

The function$T$ can be obtained by the following trick: we start with the Riemann theta function of order g and decompose its B-matrix into blocks $s \times s$, $p \times p$ and $s \times p$. Denoting the block $s \times s$ by $\widetilde{A}$, we determine the g-dimensional vector d with nonvanishing first s elements $d_s = \text{Diag Im } \widetilde{A}_{ss}$ and the rest being zero. Then the T-function can be defined as the s-fold limit

$$T(z_s, z_p) := \lim_{all d_s \to \infty} \Theta(z - d/2 \mid B). \tag{6}$$

Now the fundamental question arises: when does the solution, expressed by the Riemann theta functions, have the limit expressed by exponential functions according to (6)? In other words, when can the soliton solution be considered as a limit of the quasi-periodic one?

There are two reasons for such a procedure to be carried out carefully. Firstly, the Riemannian B-matrix being a parameter of quasi-periodic solution is not arbitrary in general, but represents the period matrix of a suitably chosen Riemann surface [3], what means that there are constraints imposed on its elements. Thus, increasing only the diagonal elements in the spirit of (6) leads to a matrix B which remains Riemannian, however it can cease to be a period-matrix, giving rise to the expression which may not be a solution.

Secondly, even in the case of cnoidal waves or two-phase solutions, (genus 1 or 2, respectively), where all the elements of B-matrix are arbitrary, there is a solution as a limit (6), canonical contours must be chosen allowing to branch points to coalesce into a single pole.

For example, in Fig.1. two choices of canonical contours are presented for the KdV eqn. The left configuration admits the soliton limit, in contrast to the right configuration, although both represent the same solution.

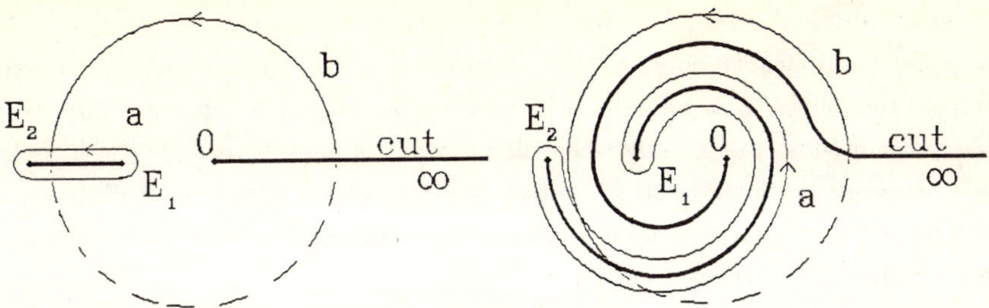

Fig.1. Two examples of the equivalent canonical contours for the KdV eqn

The transformation from one choice of canonical contours to another, for the fixed branch points, is governed by a modular transformation. This transformation is determined by the simplectic matrix $S \in \mathbb{Z}^{2g\times 2g}$, (det$S = 1$), which transforms matrices of Abelian integrals (e.g. for the sG equation), [4,7]

$$H_{ij} = \oint_{a_i^+} dU_J, \quad F_{ij} = \oint_{b_i} dU_j, \quad dU_j = z^{j-1}\left[\prod_{k=0}^{2g}(z - z_k)\right]^{-1/2} \tag{7}$$

into new matrices F', H' of integrals over new contours a_i', b_i', period matrix $B = FH^{-1}$ into new $B' = F'H'^{-1}$ and argument z into z' as follows

$$\begin{bmatrix} F' \\ H' \end{bmatrix} = [S]\begin{bmatrix} F \\ H \end{bmatrix}, \quad S = \begin{bmatrix} a & b \\ c & d \end{bmatrix}, \tag{8a}$$

$$B' = (aB + b)(cB + d)^{-1}, \quad z' = (cB + d)^{-1t}z. \tag{8b}$$

The solution is expressed then by the theta functions of the "new" argument z' with B'-period matrix, which is related to the old one by

$$\Theta\begin{bmatrix} \alpha' \\ \beta' \end{bmatrix}(z' \mid B') = K \ \exp\left[i\pi\langle z, (cB + d)^{-1}cz\rangle\right]\Theta(z \mid B), \tag{9}$$

where K is constant. Since z is linear in x and t, the Gaussian term in (9) gives rise to a constant background (KdV), linear in x background (sG) or only an additional phase shift (NLS). Thus, for an arbitrary choice of canonical contours, sometimes a modular transformation is needed as a starting point, in order to obtain the proper soliton limit of the periodic solution.

In the frame of the traditional finite-gap integration [3]. the estimation how the multisoliton solution approximates (locally) the multi-phase periodic solution is difficult, even for the commonly known equations. On the other hand, if the imaginary parts of diagonal elements of the B-matrix are much greater than the off-diagonal ones, one can expect such an approximation to be good, (see (6)). This is a well-known fact to the people using numerical procedures, particularly when the periodic boundary conditions are imposed in order to evaluate the soliton-type processes. It would be very convenient to have at our disposal a technique for the determination of parameters $(B, \kappa, \omega;\ z = \kappa x + \omega t)$ of a multiphase periodic solution in such a way that the first approximation gives rise to the multisoliton solution. The problem becomes also important from the practical point of view, particularly when the processes of higher order are considered, (genus ≥ 3), since the integration (7) is troublesome and the technique of dispersion equation (effectivization according to the Russian terminology), due to the constraints on the B-matrix is also inconvenient [1].

It turns out that the solution of the above problem exists. Surprisingly, the fundamental results were obtained nearly 100 years ago by Burnside [5] and Baker [6], and only recently rediscovered and developed [7,8].

Let us consider a three-parameter nonabelian group $PSL(2, C)$ generated by a homographic transformation

$$\sigma_n(z) = (\alpha_n z + \beta_n)(\gamma_n z + \delta_n)^{-1}, \quad \alpha_n\beta_n - \gamma_n\delta_n = 1. \tag{10}$$

There are two fixed points A'_n and A''_n of σ_n such that $\sigma_n(A_n) = A_n$. Choosing $A'_n = -A''_n = A_n$, (10) can be represented as

$$\sigma_n = (z\ ch\xi_n + A_n sh\xi_n)/(z\ sh\xi_n/A_n + ch\xi_n), \tag{11a}$$

where

$$th\xi_n = (\mu_n - 1)/(\mu_n + 1), \tag{11b}$$

and μ_n is the second free parameter.

Denoting by G all possible combinations $G = \{\sigma_{n_1}^{j_1}, \ldots, \sigma_{n_k}^{j_k}\}$ with positive and negative powers j, one can give the recipe for calculating all the necessary parameters of a multiphase solution. Each parameter is expressed in the form of a series over that group, and what is the most unexpected fact, the first term of each series corresponds to the soliton solution.

From the mathematical point of view, this result follows from the Schottky uniformization which is determined by the function $H(z,a) := \sum d\{\ln[\sigma_n(z) - a]\}/dz$ with the sum over the discussed group [5,6]. In general, the H function transforms the interior of some circle into the exterior of another one. For suitably chosen $a = \alpha_p/\gamma_p$, (with α_p, γ_p belonging to one of the group generators), this transformation automatically determines the normalized Abelian integrals (if integrated along the circle C_p), leading directly to the period B-matrix of the Riemann surface, (if integrated along some curve connecting the circle and its map), [5,6].

The Riemann surface obtained from the Schottky uniformization is presented in Fig.2 (right). For comparison, an equivalent two-sheet Riemann surface is also shown (Fig. 2 left). It should be noted that the points E and E' are related by the transformations given again in the form of a series over the discussed group, [8].

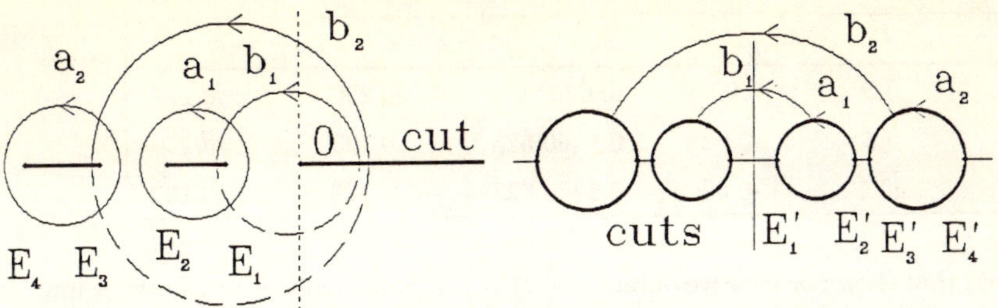

Fig.2. Two equivalent representations of the Riemann surface of genus $g = 2$

Applying the above procedure to the sG equation we obtain

$$B_{mn} = \frac{1}{2\pi i}\left[\delta_{mn}\ln\mu_n + 2\sum_{\sigma\in G_{mn}}\ln\left\{[A_m - \sigma(A_n)][A_m - \sigma(-A_n)]^{-1}\right\}\right] = \qquad (12)$$

$$\sigma \neq I, \text{ if } n = m$$

$$\frac{1}{2\pi i}\left[\delta_{mn}\ln\mu_n + 2(1-\delta_{mn})\ln\left\{[A_m - A_n][A_m + A_n]^{-1}\right\}\right] + o(\mu),$$

and similarly for propagation vectors and angular frequencies

$$\begin{matrix}\kappa_i \\ \omega_i\end{matrix} = \frac{-1}{2\pi i}\sum_{\sigma\in G_n}\left[[\sigma(A_i) - \sigma(-A_i)] \pm \left[\sigma(A_i)^{-1} - \sigma(-A_i)^{-1}\right]/16c\right] \qquad (13)$$

$$= \frac{-1}{\pi i}(A_i \pm 1/16A_i) + o(\mu),$$

$$c = 1 + o(\mu), \qquad (14)$$

where $o(\mu)$ denotes the terms of the order of μ, (being the largest μ_n, $n = 1, 2, \ldots, g$), G_n denotes a subgroup of $G = \{\sigma_{n_1}^{j_1}, \ldots, \sigma_{n_k}^{j_k}\}$ with $n_k \neq n$, while for G_{mn} we have additionally $n_1 \neq m$.

In the soliton limit ($\mu_n \to O$) all the higher terms vanish, while in the leading terms of (12) and (13) one can recognize the multisoliton solution in the Hirota formalism.

In the geometrical interpretation, (see Fig.2) each pair (A_n, μ_n) represents a circle C_n, with the center at $A_n(1+\mu_n)/(1-\mu_n)$, and the radius $A_n \mid \mu_n \mid^{1/2} /(1-\mu_n)$.

Various types of multiperiodic solutions of the sG equation follow from the reality conditions. In particular, for the one-periodic case we have

(i) rotational solution: $A \in \mathbb{R}$, $0 < \mu < 1$, hence B, κ, $\omega \in i\mathbb{R}$,

(ii) oscillatory solution: $A \in \mathbb{R}$, $-1 < \mu < 0$, hence $B = 1/2 + i\beta$; $\beta \in \mathbb{R}$, κ, $\omega \in i\mathbb{R}$,

(iii) oscillatory solution: $A \in i\mathbb{R}$, $-1 < \mu < 0$, hence $B = 1/2 + i\beta$; β, κ, $\omega \in \mathbb{R}$.

Numerical examples corresponding to various types of one-periodic solutions are given below. For clarity, we quote only three decimal digits, however all the quantities have been determined within accuracy of $\pm 10^{-6}$.

type	A	μ	B	κ	ω
(i)	0.5	0.02	i0.623	i0.206	i0.112
(ii)	0.5	-0.02	0.5+i0.623	i0.193	i0.125
(iii)	i0.5	-0.02	0.5+i0.623	-0.125	-0.193

Note that for μ negative we obtain $\text{Re}(B) = 1/2$, while a real (imaginary) A implies imaginary (real) κ and ω, as expected from (12) and (13).

For (i) and (ii) the soliton limit exists and the relevant expressions are given by the leading terms of (12) and (13). On the other hand, for (iii) the limit $\mu \to 0$ corresponds to a small-amplitude quasi-linear (phonon) solution.

Two-periodic solutions can be obtained as various combinations of one-periodic types specified above. However, there exist also solutions which are essentially two-periodic and have no counterparts in the one-periodic case. Indeed, for $A_n \in C$, $-1 < \mu_n < 0$, $(n = 1,2)$, and $A_1 = A_2^*$, $\mu_1 = \mu_2$ we obtain a real solution with $B_{mn} = 1/2 + i\beta_{mn}$, $\beta_m n \in R$, which can be interpreted as a breather wave-train. In the particular case of a stationary breather train we obtain also: κ_n imaginary, ω_n real $(n = 1,2)$, and $\kappa_1 = \kappa_2$, $\omega_1 = -\omega_2$.

Thus, at least for the above mentioned equations, in principle one can construct the multiperiodic solutions, $(\mu_n \neq 0)$, starting from the multisoliton one, $(\mu_n = 0)$, and ascribing a free parameter μ_n to any A_n given by a corresponding soliton. Then, the recipes to calculate B, κ, ω are given by (12-14). The same, in fact, can be done in the frame of traditional finite-gap integration technique by ascribing the finite-length intervals to the soliton (point) spectrum, but the final result is quite illegible in the language of Abelian integrals.

REFERENCES

[1] J. Zagrodzinski, in "Systèmes dynamiques nonlinéaires", ed. P. Winternitz, (Les Presses de l'Univ. de Montréal, 1986); J. Phys. A 17, 3315 (1984).

[2] R. Hirota, in "Solitons", eds. R. Bullough, P. Caudrey (Springer, 1980).

[3] V.E. Zakharov, S.V. Manakov, S.P. Novikov, L.P. Pitaevski, "Theory of solitons" (Plenum, 1984).

[4] J. Zagrodzinski, Phys. Rev. B 29, 1500 (1984).

[5] W. Burnside, Proc. London Math. Soc. 23, 49 (1892).

[6] H.F. Baker, "Abel's theorem and the allied theory including the theory of theta functions" (Cambridge University Press, 1897).

[7] A.I. Bobenko, L.A. Bordag, J. Phys. A *underline*22, 1259 (1989).

[8] A.I. Bobenko, "Uniformization of Riemann surfaces and effectivization of theta-functional formulae", preprint 257 (Tech. Univ. Berlin, 1990).

Part VII

Inverse Methods Related to a Linearization Scheme

The Crum Transformation for a Third Order Scattering Problem

J.J.C. Nimmo

University of Glasgow,
Department of Mathematics,
Glasgow, G12 8QW, UK.

§1 Introduction.

It is well known (see for example Calogero and Degasperis 1982) that the classical Darboux theorem (Darboux 1882) relating solutions of a pair of Schrödinger equations may be exploited in the context of inverse scattering theory. Expressed most simply it is a transformation between a pair of Schrödinger equations where the discrete spectrum of the second contains the discrete eigenvalues of the first together with an extra one.

A generalization of the Darboux theorem due to Crum (1955), obtained by solving the recurrence relation for N applications, allows one to construct solutions of the Schrödinger equation having N more discrete eigenvalues than the original. In soliton theory this result was first used by Wadati, Sanuki and Konno (1975) to obtain a representation of the N-soliton solution of the KdV equation in terms of a Wronskian determinant.

Less well known is the Darboux theorem for a third order scattering problem (Aiyer, Fuchsteiner and Oevel 1986, see also Sabatier 1987, Beals and Coifman 1987 and Sattinger and Zurkowski 1987) which is associated with the Sawada-Kotera equation (Sawada and Kotera 1974). Here we will use the Hirota approach to obtain the analogue of Crum's result for this third order scattering problem. We will show that the nonlinear superposition, given by the Wronskian expression in the case of the Schrödinger equation, is in this case given by a Pfaffian—the square root of the determinant of a skew-symmetric matrix.

Finally, the Darboux theorem will be discussed in terms of factorization of the third-order operator, echoing the work of Adler and Moser (1979) on the Schrödinger operator.

§2 Darboux theorem, Crum transformation and factorization.

The classical Darboux theorem is as follows:- if one has a pair of Schrödinger equations

$$\psi_{xx} + u\psi = k^2\psi, \tag{2.1}$$
$$\phi_{xx} + v\phi = k^2\phi, \tag{2.2}$$

with potentials u and v such that

$$v = u + 2(\ln\theta)_{xx}, \tag{2.3}$$

and θ is a particular solution of (2.1) for any given k then the general solution of (2.2) is

$$\phi(x,k) = \frac{W(\theta(x), \psi(x,k))}{\theta(x)}, \tag{2.4}$$

where $\psi(x,k)$ is the general solution of (2.1) (Darboux 1882).

Crum (1955) showed that if one takes N successive applications of this Darboux theorem from the general solution $\psi_0(x,k)$ of the 'seed' Schrödinger equation

$$\psi_{0_{xx}} + u_0\psi_0 = k^2\psi_0, \tag{2.5}$$

with N particular solutions $\theta_1,\ldots,\theta_N$ then one obtains the general solution

$$\psi_N(x,k) = \frac{W(\theta_1,\ldots,\theta_N,\psi_0(x,k))}{W(\theta_1,\ldots,\theta_N)}, \tag{2.6}$$

of

$$\psi_{N_{xx}} + u_N\psi_N = k^2\psi_N, \tag{2.7}$$

where

$$u_N = u_0 + 2(\ln W(\theta_1,\ldots,\theta_N))_{xx}. \tag{2.8}$$

One way in which the Darboux theorem may be proved is by means of a factorization of the Schrödinger operator (Adler and Moser 1979). One has

$$(\partial^2 + u) = (\partial + v)(\partial - v), \tag{2.9}$$

where $u = -v' - v^2$ is the Miura transformation. This Ricatti equation for v is solved in terms of a solution θ of the zero energy Schrödinger equation to give $v = \theta'/\theta$.

Defining

$$A := \partial + \frac{\theta'}{\theta} \quad \text{and} \quad B := \partial - \frac{\theta'}{\theta}$$

we have the commutator

$$[A,B] = 2(\ln\theta)''. \tag{2.10}$$

In terms of A and B (2.1) is expressed as

$$AB\psi = k^2\psi,$$

while introducing the new wave function $\phi = B\psi$ (c.f. (2.4)) gives

$$BA\phi = k^2\phi,$$

which is (2.2). Equation (2.3) is given by (2.10).

It is also straightforward to prove (2.6,8) by bilinearization of (2.7): we write $\psi_N = G/F$ and $u_N = u_0 + 2(\ln F)_{xx}$ to obtain the Hirota form

$$(\mathrm{D}_x^2 + u_0 - k^2)\, G \cdot F = 0. \tag{2.11}$$

Taking the *ansatz* $F = W(\theta_1,\ldots,\theta_N)$ and $G = W(\theta_1,\ldots,\theta_N,\psi_0)$ one may verify (2.11) by standard techniques (Freeman and Nimmo 1983, Nimmo 1990b).

§3 Third Order Scattering Problem.

The results described in the last section can be reproduced for the third order linear scattering problem

$$\psi_{xxx} + 3u\psi_x = k^3\psi, \tag{3.1}$$

which is appropriate to solving initial value problems for the Sawada-Kotera equation

$$u_t + 15uu_{3x} + 15u_xu_{xx} + 45u^2u_x + u_{5x} = 0 \tag{3.2}$$

(Sawada and Kotera 1974, Dodd and Gibbon 1977, Satsuma and Kaup 1977).

It may be shown (Aiyer et al 1986) that the Darboux theorem relating solutions of (3.1) to those of

$$\phi_{xxx} + 3v\phi_x = k^3\phi, \tag{3.3}$$

where $v = u + 2(\ln\theta)_{xx}$, is given by

$$\phi(x,k) = \frac{\int_{-\infty}^{x}(\theta_y(y)\psi(y,k) - \theta(y)\psi_y(y,k))dy}{\theta(x)}, \tag{3.4}$$

where θ is any solution of (3.1) for a given $k = \zeta$. To prove this result it is necessary that

$$(k^3 - \zeta^3)\psi\theta - 2(\psi_{xx}\theta_x - \psi_x\theta_{xx}), \tag{3.5}$$

vanish at $x = -\infty$ and this is achieved by specifying that θ and its x-derivatives tend to zero in this limit.

The corresponding Crum transformation is as follows: consider

$$\psi_{N_{xxx}} + 3u_N\psi_{N_x} = k^3\psi_N, \tag{3.6}$$

which we write in Hirota form. As for (2.7) we let $\psi_N = G/F$ and $u_N = u_0 + 2(\ln F)_{xx}$ whereupon (3.6) is written as

$$(\mathrm{D}_x^3 + 3u_0\mathrm{D}_x - k^3)G\cdot F = 0. \tag{3.7}$$

The *ansatz* made in this case is

$$F = P(\phi_1,\ldots,\phi_N), \tag{3.8}$$

and

$$G = P(\phi_1,\ldots,\phi_N,\phi_{N+1}), \tag{3.9}$$

where the notation $P(\phi_1,\ldots,\phi_M)$, if M is even, is used to denote the Pfaffian Pf(A) of the $M\times M$ skew-symmetric matrix A defined by

$$A_{ij} = \int_{-\infty}^{x}(\phi_{i_y}\phi_j - \phi_i\phi_{j_y})dy, \tag{3.10}$$

and where

$$\mathrm{Pf}(A) := \sum_{\sigma\in P}\epsilon(\sigma)A_{\sigma(1),\sigma(2)}\cdots A_{\sigma(M-1),\sigma(M)}, \tag{3.11}$$

in which P is the set of permutations of $\{1,\ldots,M\}$ such that the conditions

$$\sigma(1)<\sigma(2),\ \sigma(3)<\sigma(4),\ \ldots,\ \sigma(M-1)<\sigma(M),$$
$$\sigma(1)<\sigma(3)<\ldots<\sigma(M-1),$$

hold and is such that $\mathrm{Pf}(A)^2 = \det(A)$. If M is odd then—since the determinant A would be zero—we define

$$P(\phi_1,\ldots,\phi_M) := P(\phi_1,\ldots,\phi_M,\phi_{M+1}), \tag{3.12}$$

where $\phi_{M+1} = 1$. The functions ϕ_i are chosen to be solutions of the 'seed' equation

$$\phi_{xxx} + 3u_0\phi_x = k_i^3\phi. \tag{3.13}$$

To verify (3.7) requires us to determine the derivatives of the Pfaffians F and G with respect to x. To do this we first consider the skew-symmetric matrix A_λ, where λ is the partition $(\lambda_1, \ldots, \lambda_n)$, defined by

$$A_\lambda = \begin{pmatrix} A & B_\lambda \\ -B_\lambda^T & 0 \end{pmatrix}, \tag{3.14}$$

where A is the $M \times M$ (skew-symmetric) matrix (3.10) and B_λ is the $M \times n$ matrix

$$B_\lambda = \begin{pmatrix} \phi_1^{(\lambda_n)} & \phi_1^{(\lambda_{n-1})} & \ldots & \phi_1^{(\lambda_1)} \\ \vdots & \vdots & \ddots & \vdots \\ \phi_M^{(\lambda_n)} & \phi_M^{(\lambda_{n-1})} & \ldots & \phi_M^{(\lambda_1)} \end{pmatrix}, \tag{3.15}$$

where $\phi_i^{(j)}$ denotes the j-th x-derivative of ϕ_i. Once again one must ensure that the matrix A_λ is of even dimension in order that its Pfaffian be defined. In this case this is done, if necessary, by extending the partition λ by the part $\lambda_{n+1} = 0$. From now on we assume, without loss of generality, that A_λ is of even dimensions and let

$$F_\lambda := \mathrm{Pf}(A_\lambda), \tag{3.16}$$

with $M = N$ and similarly for G_λ with $M = N + 1$.

One may show (Hirota 1989, Nimmo 1990a) that provided the ϕ_i (i=1,...,N) and their x-derivatives tend to zero as $x \to -\infty$

$$F_x = F_{(1)}, \tag{3.17a}$$
$$F_{xx} = F_{(2)}, \tag{3.17b}$$
$$F_{3x} = F_{(3)} + F_{(21)}, \tag{3.17c}$$

with similar expressions for the x-derivatives of G. Also, by virtue of (3.13), we have

$$\left(\sum_{i=1}^{N} k_i^3\right) F = F_{(3)} - 2F_{(21)} + 3u_0 F_{(1)}, \tag{3.18}$$

We may now verify that F and G given by (3.8,9) satisfy (3.7) for $k = k_{N+1}$. Using (3.17,18) we have

$$3\left[G_{(21)}F - G_{(2)}F_{(1)} + G_{(1)}F_{(2)} - GF_{(21)}\right],$$

which may be shown to vanish using a Pfaffian identity (see Nimmo 1990b for details).

§4 Factorization.

We now consider the factorization of the operator

$$(\partial^3 + 3u\partial), \tag{4.1}$$

which defines the scattering problem (3.1). The factorization

$$(\partial^3 + 3u\partial) = (\partial + v)(\partial - v)\partial, \tag{4.2}$$

discussed by Fordy and Gibbons (1981), which is intimately related to the factorization (2.9), gives a Miura-like transformation between (3.2) and a modified Sawada-Kotera equation in v. This, however, does not share the property of (2.9) which corresponds to the Darboux theorem for the Schrödinger equation. Instead we consider a more complicated factorization (Athorne and Nimmo 1991)

$$(\partial^3 + 3u\partial) = (\partial + w)(\partial - w + v)(\partial - v), \tag{4.3}$$

where we may parametrize v and w as $v = \theta'/\theta$ and $w = \theta''/\theta'$, where θ is any solution of $\theta''' + 3u\theta' = 0$. Thus we define

$$A_\theta := \partial + \frac{\theta''}{\theta'}, \quad B_\theta := \partial - \frac{\theta''}{\theta'} + \frac{\theta'}{\theta} \quad \text{and} \quad C_\theta := \partial - \frac{\theta'}{\theta}.$$

The scattering problem (3.1) is now

$$A_\theta B_\theta C_\theta \psi = k^3 \psi. \tag{4.4}$$

If we apply C_θ to both sides of (4.4) to obtain

$$C_\theta A_\theta B_\theta C_\theta \psi = k^3 C_\theta \psi. \tag{4.5}$$

and introduce ϕ such that

$$C_{\theta^{-1}} \phi = C_\theta \psi, \tag{4.6}$$

so that (4.5) gives

$$C_\theta A_\theta B_\theta C_{\theta^{-1}} \phi = k^3 C_{\theta^{-1}} \phi. \tag{4.7}$$

By observing that $C_\theta A_\theta = C_{\theta^{-1}} A_{\theta^{-1}}$, $B_\theta = B_{\theta^{-1}}$ and imposing suitable boundary condition (4.7) gives

$$A_{\theta^{-1}} B_{\theta^{-1}} C_{\theta^{-1}} \phi = k^3 \phi. \tag{4.8}$$

The product of operators on the left hand side of (4.8) is $\partial^3 + 3v\partial$ so that (4.8) corresponds to (3.3). Thus the relation (4.6) defines a Darboux theorem for (3.1).

Finally, since (4.6) may be written as

$$(\phi\theta)' = (\psi/\theta)',$$

from which we get (3.4), this factorization gives the Darboux theorem of (Aiyer et al 1986).

References

Athorne, C and Nimmo, J J C (1991) Factorization and Darboux theorems for second and third order operators (in preparation).

Aiyer, R N, Fuchsteiner, B and Oevel, W (1986) J Phys A **19** 3755.

Beals, R and Coifman, R R (1997) Inv Prob **3** 577.

Calogero, F and Degasperis, A (1982) *Solitons and the Spectral Transform I*, North-Holland, Amsterdam.

Crum, M M (1955) Q J Math **6** 121.

Darboux, G (1882) C R Acad Sci Paris **94** 1456.

Dodd, R K and Gibbon, J D (1977) Proc R Soc Lond **A358** 287.

Fordy, A P and Gibbons J (1981) J Math Phys **21** 2508.

Freeman, N C and Nimmo, J J C (1983) Proc R Soc Lond **A389** 319.
Hirota, R (1989) J Phys Soc Japan **58** 2285.
Nimmo, J J C (1990a) J Phys A **23** 751.
Nimmo, J J C (1990b) Proc R Soc Lond **A431** 361.
Sabatier, P C (1987) in *Inverse Problems* (J R Cannon and U Horning eds), Birkhauser, Basel.
Satsuma, J and Kaup, D J (1977) J Phys Soc Japan **43** 692.
Sattinger, D H and Zurkowski, V D Physica D **26** 225.
Sawada, K and Kotera, T (1974) Prog Theor Phys **51** 1355.
Wadati, M, Sanuki, H and Konno, K (1975) Prog Theor Phys **53** 419.

Darboux Theorems Connected to Dym Type Equations

W. Oevel

Department of Mathematical Sciences,

Loughborough University,

LE11 3TU, UK.

We consider integrable evolution equations defined by Lax equations of the form

$$\frac{d}{dt}L = [P_{\geq 2}(L^q), L] \ , \tag{1}$$

where L is a differential operator parametrized by a number of potentials, L^q is some (usually fractional) power of this operator and $P_{\geq 2}$: $\sum_{i>-\infty}^{N} a_i(x)\partial^i \rightarrow \sum_{i\geq 2}^{N} a_i(x)\partial^i$ is the projection of a pseudo-differential operator to its terms of differential order larger than 1. Equations of this type have been considered in [1]-[4]. It can be shown easily ([2]) that (1) defines an integrable hierarchy of coupled equations for the fields $w_N, w_{N-1},...$ for the following choices of the differential operator L:

$$\begin{aligned} L &= w_N\partial^N + ... + w_3\partial^3 + w_2\partial^2 + w_1\partial + w_0 \ , \\ L &= w_N\partial^N + ... + w_3\partial^3 + w_2\partial^2 + w_1\partial \ , \\ L &= w_N\partial^N + ... + w_3\partial^3 + w_2\partial^2 \ , \end{aligned} \tag{2}$$

for arbitrarily chosen order N. According to [1] even deeper reductions to operators L satisfying the constraint $L^* = \pm\partial^2 L \partial^{-2}$ are admissible, where $(w_0 + w_1\partial + w_2\partial^2 + ...)^* = w_0 - \partial w_1 + \partial^2 w_2 - ...$ is the formal adjoint of the operator.
As will be demonstrated lateron, the choice $L = w^2\partial^2$ leads to the Harry Dym equation for w:

$$L_t = [P_{\geq 2}(L^{3/2}), L] \quad \leftrightarrow \quad w_t = \frac{1}{4}w^3 w_{xxx} \ , \tag{3}$$

other choices for L lead to coupled systems of Dym like equations ([2]).
There are general statements about the Lax hierarchy (1), such as the commutativity of these equations for different choices of the power q or their Hamiltonian structure ([1],[2]).
Further, it is known ([5]) that reciprocal transformations involving a change of coordinates describe certain invariances of the Dym equation (3) and relations to the Korteweg-de Vries hierarchy. As a compact definition of the Dym hierarchy is given by (1), the natural question arises whether general statements about reciprocal transformations can be obtained on the level of Lax representations (1). Indeed, the following theorem yields a very

general statement about Bäcklund transformations involving a change of independent variables. Thus, the known results for the Dym equation can be generalized to the systems of iso-spectral Dym like equations associated to scattering problems of the form (2). The crucial step is the transformation of the Lax operators under a change of independent variables:

Definition: Given a function $\Phi(x,t)$ we introduce new independent variables $\hat{x} := \Phi(x,t)$ and $\hat{t} := t$. To any function $a(x,t)$ we associate the new function $\hat{a}(\hat{x},\hat{t}) := a(x,t)$, to any pseudo-differential operator $A = \sum_{i>-\infty}^{N} a_i(x,t)\partial^i$ we associate the new operator

$$\hat{A} := A = \sum_{i>-\infty}^{N} \hat{a}_i(\hat{x},\hat{t}) \, (\hat{\Phi_x}(\hat{x},\hat{t}) \, \hat{\partial} \,)^i \, . \tag{4}$$

Here $\hat{\partial}$ is the differential symbol linked to the new variable $\hat{x}$ by $\partial = \frac{\partial \hat{x}}{\partial x}\hat{\partial}$. By $\hat{P}_{\geq 2} : \sum_{i>-\infty}^{N} \hat{b}(\hat{x})\hat{\partial}^i \to \sum_{i\geq 2}^{N} \hat{b}(\hat{x})\hat{\partial}^i$ we denote the projection related to the new symbols.

Theorem ([2],[3]) : Let the operator $L(x,t)$ and the functions $\Phi(x,t)$, $\Psi(x,t)$ satisfy the equations

$$L_t = [P_{\geq 2}(L^q), L] \quad , \quad \Phi_t = P_{\geq 2}(L^q)\Phi \quad , \quad \Psi_t = P_{\geq 2}(L^q)\Psi \quad . \tag{5}$$

i) For arbitrary constants $a, b, c, d, \tilde{a}, \tilde{b}$ we introduce the new variables $\hat{x} := \frac{a\Phi+b}{cx+d}$, $\hat{t} := t$ and define

$$\hat{L}(\hat{x},\hat{t}) := \frac{1}{cx+d} L(x,t) \, (cx+d) \quad , \quad \hat{\Psi}(\hat{x},\hat{t}) := \frac{\tilde{a}\Psi(x,t) + \tilde{b}}{cx+d} \, . \tag{6}$$

ii) Alternatively, we introduce $\hat{x} := \Psi_x$, $\hat{t} := t$ and define

$$\hat{L}(\hat{x},\hat{t}) := \partial L(x,t)\partial^{-1} \quad , \quad \hat{\Psi}(\hat{x},\hat{t}) := \Psi_x(x,t) \quad . \tag{7}$$

Then, in both cases, $\hat{L}$ and $\hat{\Psi}$ satisfy the evolution equations

$$\hat{L}_{\hat{t}} = [\hat{P}_{\geq 2}(\hat{L}^q), \hat{L}] \quad , \quad \hat{\Psi}_{\hat{t}} = \hat{P}_{\geq 2}(\hat{L}^q)\hat{\Psi} \, . \tag{8}$$

This theorem characterizes Bäcklund transformations of the iso-spectral equations for the scattering operator L to iso-spectral equations of the new operator $\hat{L}$. These transformations are characterized by "eigenfunctions" Φ satisfying the linear evolution equations $\Phi_t = P_{\geq 2}(L^q)\Phi$. Thus, as "spatial" part of the transformation the eigenvalue problem $L\Phi = \lambda\Phi$ may also be imposed for Φ. In the spirit of the classical Darboux theorem we also have the transformation of further eigenfunctions Ψ to new eigenfunctions $\hat{\Psi}$ of

the transformed operator $\hat{L}$.
To illustrate the procedure we consider the spectral problem $L = w^2\partial^2$, choose $q = 3/2$, calculate $P_{\geq 2}(L^{3/2}) = w^3\partial^3 + \frac{3}{2}w^2 w_x \partial^2$, and find the Harry Dym equation for the field w:

$$L_t = 2w\, w_t\, \partial^2 = [P_{\geq 2}(L^{3/2}), L] = \frac{w}{2}\, w^3 w_{xxx}\partial^2 \ . \tag{9}$$

Note, that the time evolution for the eigenfunctions is trivially solved for constants (and x), i.e. instead of considering the eigenvalue equation $L\Phi = \lambda\Phi$ we may also consider $L(\Phi + \alpha) = \lambda(\Phi + \alpha)$ for arbitrary constant $\alpha = \lambda_0/\lambda$. Thus, introducing the spectral problem

$$w^2\Phi_{xx} = \lambda_0 + \lambda\Phi \ \ , \ \ \Phi_t = w^3\Phi_{xxx} + \frac{3}{2}w^2 w_x \Phi_{xx} \tag{10}$$

we introduce $\hat{x} := \Phi_x, \hat{t} := t$, and calculate

$$\hat{L} := \partial L \partial^{-1} = (w\Phi_{xx})^2\hat{\partial}^2 + 2\lambda\hat{x}\hat{\partial} \ . \tag{11}$$

Putting $\hat{w} = w\Phi_{xx}$ we obtain the time evolution for $\hat{w}$ from the evolution $\hat{L}_{\hat{t}} = [\hat{P}_{\geq 2}(\hat{L}^{3/2}), \hat{L}]$ of the new Lax operator. Calculating $\hat{P}_{\geq 2}(\hat{L}^{3/2}) = \hat{w}^3\hat{\partial}^3 + \frac{3}{2}(\hat{w}^2\hat{w}_{\hat{x}} + \lambda\hat{x}\hat{w})\,\hat{\partial}^2$ one finds the new integrable equation

$$\hat{w}_{\hat{t}} = \frac{1}{4}\ (\ \hat{w}^3\hat{w}_{\hat{x}\hat{x}\hat{x}} + 3\lambda^2(\hat{x} - \frac{\hat{x}^2\hat{w}_{\hat{x}}}{\hat{w}})\ )\ . \tag{12}$$

For $\lambda = 0$ this reduces to the Dym equation (3) and the transformation reads $\hat{w} = w\Psi_{xx} = \lambda_0/w$, i.e. we rediscover a well known invariance for the Dym equation. The above theorem clearly generalizes the results of [5] and gives direct access to similar statements for coupled systems (1) related to higher order spectral problems (2).

References

[1] B.A. Kupershmidt, Commun. Math. Phys. **99**, 51-73 (1985)

[2] B.G. Konopelchenko and W. Oevel, *A r-Matrix Approach to Nonstandard Classes of Integrable Equations*, preprint (1990)

[3] W. Oevel and C. Rogers, *Gauge Transformations and Reciprocal Links in (2+1) Dimensions*, preprint (1990)

[4] K. Kiso, Progr. Theoret. Phys. **83**, 1108-1114 (1990)

[5] C. Rogers and M.C. Nucci, Physica Scripta **33**, 289-292 (1986)

Forced Initial Boundary Value Problems for Burgers Equation

M.J. Ablowitz[1] and S. De Lillo[2]

[1]Program in Applied Mathematics,
University of Colorado,
Boulder, CO 80309, USA..
[2]Dipartimento di Fisica,
Università di Perugia,
and I.N.F.N. Sezione di Perugia,
06100 Perugia, Italy.

In recent years there has been considerable research devoted towards the understanding of nonlinear evolution equations. There are a number of such equations which are exactly solvable and for which significant informations can be obtained. These studies are primarily directed towards autonomous equations with either rapidly decaying or periodic boundary values. On the other hand, forced nonlinear equations are certainly of physical significance. As such, the "forced" Burgers equation may be viewed as a prototypical model. In fact it describes the time evolution of onedimensional nonlinear diffusive systems under the influence of an external driver (see, for example, Ref. [1]).

In Ref. [2] by making use of semiline results of Burgers equation [3], we found the solution of the initial value problem on the whole line for the forced Burgers equation

$$u_t = (u_x+u^2)_x + \delta(x)\, F(t), \tag{1}$$

where $u = u(t,x)$ and $\delta(x)$ is the usual Dirac delta function. $F(t)$ is a given function of time which is assumed to be continuous and bounded.

The main result of Ref [2] was the reduction of the initial value problem for Eq(1) to a linear integral equation of Volterra type in t.

For semplicity we restricted our considerations to a trivial initial datum $u(0,x)=u_0(x) = 0$.

In this contribution we give the solution of the initial value problem for Eq(1) in the general case of a non-trivial initial datum $u_0(x) = 0$.

We will show that also in this case the problem reduces to a linear integral equation of Volterra type in t.

We start by observing that the presence of $\delta(x)$ in Eq(1) implies a connection (through the forcing term) between two initial/boundary value problems on the semiline, i.e. for $x>0$ and $x<0$ respectively, for the Burgers equation

$$u_t = (u_x+u^2)_x \tag{2}$$

We find the connection to the semiline problem to be useful. An alternative approach problem using the standard linearization on the whole line can be employed as well (cf. [2]).

We introduce the generalised Hopf-Cole transformation [3]

$$v(x,t) = C(t)u(x,t)\exp[\int_0^x dx'u(x',t)] \tag{3.a}$$

$$u(x,t) = v(x,t)/[C(t)+\int_0^x dx'v(x',t)] \tag{3.b}$$

$$\text{with } C(0) = 1 \tag{3.c}$$

This transformation implies the relations

$$v(x,0) = u(x,0)\exp[\int_0^x dx'u(x',0)] \tag{4.a}$$

$$v(t,0) = C(t)u(t,0) \tag{4.b}$$

$$v_x(t,0) = C(t)[u_x(t,0)+u^2(t,0)] \tag{4.c}$$

and transforms Eq(2) into the heat equation

$$v_t = v_{xx} \tag{5.a}$$

with

$$\dot{C}(t) = v_x(0,t) \tag{5.b}$$

We then consider the initial/boundary value problem for Burgers equation on the semiline $x\in[0+\infty)$ characterized by the following set of initial and boundary data

$$u(0,x) = u_0(x) \tag{6.a}$$

$$u(t,0) = G(t) \tag{6.b}$$

with

$$G(0) = u_0(0) \tag{6.c}$$

Consequently, once C(t) is determined, then the initial boundary value problem for u(t,x), characterized by (6.a), (6.b) and (6.c), is solved through the following algorithm:

(i) Compute the initial and boundary data for v(x,t) on the semiline

$$v(0,x) = v_0(x) = u_0(x)\exp[\int_0^x dx' u_0(x')] \tag{7.a}$$

$$v(t,0) = C(t)G(t) \tag{7.b}$$

(ii) Evaluate v(t,x) from (5.a), (7.a) and (7.b).
(iii) Recover u(t,x) from v(t,x) via (3.b).

Finally, the last task is to compute C(t).
In order to evaluate v(t,x) it is useful to introduce the Laplace transform

$$\mathcal{L}\,(v(t,x)) \equiv \hat{v}(s,x) = \int_0^\infty dt e^{-st} v(t,x) \tag{8}$$

obtaining from (5.a) and (7.a) the equation

$$\hat{v}_{xx} - s\hat{v} = -v_0(x) \tag{9}$$

Solving this equation with the boundary condition (7.b) for v(t,x) yields

$$\hat{v}(s,x) = \frac{1}{2}\int_0^\infty dy v_0(y)\frac{1}{\sqrt{s}}[e^{-(x+y)\sqrt{s}} + e^{-|x-y|\sqrt{s}}] + \frac{\hat{v}_x(s,0)}{\sqrt{s}}e^{-\sqrt{s}x} \tag{10}$$

where the relation

$$\mathcal{L}\,(\dot{C}(t)) = \hat{v}_x(s,0) = \sqrt{s}\hat{v}(s,0) \tag{11}$$

has also been used.

From (10) and (11) we get the inverse Laplace transform

$$v(t,x) = w(t,x) - \frac{1}{\sqrt{\pi}}\int_0^t dt' \frac{\dot{C}(t')e^{-x^2/4(t-t')}}{(t-t')^{1/2}} \tag{12.a}$$

$$w(t,x)=(\pi t)^{-1/2}\int_0^\infty dy v_0(y)\cosh(xy/2t)\exp[-(x^2+y^2)/4t] \qquad (12.b)$$

Let us now turn our attention to the solution valid on the semiline $x\in(-\infty,0]$. Through the same approach is straightforward to derive for Eq(9) the solution

$$\hat{v}(s,x)=\frac{1}{2}\int_{-\infty}^{0} dy v_0(y)\frac{1}{\sqrt{s}}[e^{-|x-y|\sqrt{s}}+e^{(x+y)\sqrt{s}}]+\frac{\hat{v}_x(s,0)}{\sqrt{s}}e^{-\sqrt{sx}} \qquad (13)$$

which replaces (10). The inverse Laplace transform will now be given by

$$v(t,x)=w(t,x)-\frac{1}{\sqrt{\pi}}\int_0^t dt'\frac{\dot{C}(t')e^{-x^2/4(t-t')}}{(t-t')^{1/2}} \qquad (14.a)$$

$$w(t,x)=(\pi t)^{-1/2}\int_{-\infty}^{0} dy v_0(y)\cosh(xy/2t)\exp[-(x^2+y^2)/4t] \qquad (14.b)$$

which replace (12.a) and (12.b) respectively.

We are now ready to solve the forced problem described by Eq(1). The two solutions of Eq(2) in the quarter planes $\{x>0,\ t\geq0\}$ and $\{x<0,\ t\geq0\}$ will be denoted by $u_R(t,x)$ and $u_L(t,x)$ respectively. They are taken to satisfy the continuity condition

$$u_L(t,0)=u_R(t,0)=G(t) \qquad (15)$$

as $x\to0_-$ and $x\to0_+$, respectively.

Integrating Eq(1) across the discontinuity on the x axis yields, via (15),

$$[u_{xR}(t,0)+u_R^2(t,0)]-[u_{xL}(t,0)+u_L^2(t,0)]=-F(t) \qquad (16)$$

Relation (16) induces a discontinuity at x=0 both in the solution v(t,x) of the linear problem and in the x-derivative $v_x(t,x)$.

In fact, via (4.b), (4.c) and (5.b) there obtains

$$C_L(t)=v_{xL}(t,0)=C_L(t)[u_{xL}(t,0)+u_L^2(t,0)] \qquad (17.a)$$

$$C_R(t)=v_{xR}(t,0)=C_R(t)[u_{xR}(t,0)+u_R^2(t,0)] \qquad (17.b)$$

and

$$v_L(t,0) = C_L(t)G(t) \tag{18.a}$$
$$v_R(t,0) = C_R(t)G(t) \tag{18.b}$$

We can now derive the connection between $u_L(t,x)$ and $u_R(t,x)$. By using (17.a), (17.b) and (16) we get

$$\frac{\dot{C}_L(t)}{C_L(t)} - \frac{\dot{C}_R(t)}{C_R(t)} = F(t) \tag{19}$$

which then integrated with the initial condition (3.c) ($C_R(0)=C_L(0)=1$), yields

$$C_L(t)=C_R(t)\exp[\int_0^t dt'F(t')] \tag{20}$$

The above relation establishes the connection between $u_L(t,x)$ and $u_R(t,x)$ as a factorization in terms of the functions $C_L(t)$, $C_R(t)$ and of the forcing term.

The last task is to obtain the evolution equation for $C_R(t)$ (or equivalent for $C_L(t)$).

We observe that (18.a) and (18.b) imply

$$\dot{C}_L(t)v_R(t,0) = \dot{C}_R(t)v_L(t,0) \tag{21}$$

which, when (12.a) and (16.a) are used, yields

$$C_R(t)\Big[\int_{-\infty}^{0} \frac{dy v_0(y)e^{-y^2/4t}}{\sqrt{t}} + \int_0^t dt' \frac{\dot{C}_L(t')}{(t-t')^{1/2}}\Big] =$$

$$= C_L(t)\Big[\int_0^{\infty} \frac{dy v_0(y)e^{-y^2/4t}}{\sqrt{t}} + \int_0^t dt' \frac{\dot{C}_R(t')}{(t-t')^{1/2}}\Big] \tag{22}$$

substituting (20) in (22) and introducing the function C(t) as

$$C_R(t) = 1 + \hat{C}(t) \tag{23}$$

We finally obtain the following integral equation

$$\hat{C}(t) = h(t) + \int_0^t dt'\hat{C}(t)K(t,t') \tag{24.a}$$

where

$$h(t) = \int_0^{\infty} \frac{dy v_0(y) e^{-y^2/4t}}{\sqrt{t}} - \exp\left[-\int_0^t dt' F(t')\right] \int_{-\infty}^{0} \frac{dy v_0(y) e^{-y^2/4t}}{\sqrt{t}} -$$

$$- \frac{1}{2\pi} \int_0^t dt_1 \int_0^{t_1} dt_2 (t-t_1)^{-1/2} (t_1-t_2)^{-1/2} \frac{d}{dt_2} \left[\exp \int_{t_1}^{t_2} F(y) dy - 1 \right] \qquad (24.b)$$

$$K(t,t') = \frac{1}{4\pi} \int_{t'}^{t} dt'' (t-t'')^{-1/2} (t''-t')^{-1/2} \left[\frac{\exp \int_{t''}^{t'} F(y) dy - 1}{(t''-t')} \right] \qquad (24.c)$$

Eq(24.a) is a linear integral equation of Volterra type in t; it admits a unique continuous solution under the assumption that F(t) is a continuous, bounded function of its argument.

Aknowledgments

M.J.A. was partially supported by the Air Force Office of Scientific Research under Grant No. AFOSR-90-0039, the NSF under Grant No. DMS-8916182, and the Office of Naval Research under Grant No. N00014-90-J-1218.

S.D.L. was supported by a special Grant for Scambi Internazionali from the Istituto Nazionale di Fisica Nucleare.

References

[1] M.Kercher, G. Parisi and Y.C. Zheng, *Physical Review Letters* 56, 889 (1986).

[2] M.J. Ablowitz and S. De Lillo, "Forced and Semiline Solutions of Burgers Equation", Preprint Program in Applied Mathematics, University of Colorado, Boulder, PAM #69, Jan 1991.

[3] F. Calogero and S. De Lillo, *Nonlinearity* 2, 37 (1989).

Creation and Annihilation of Solitons in Nonlinear Integrable Systems

V.K. Mel'nikov

Laboratory of Theoretical Physics,

Joint Institute for Nuclear Research,

141980 Dubna Moscow Region, USSR.

Creation and annihilation of solitons in nonlinear integrable systems is probably the most unexpected fact in the theory of these systems which itself is rich in various surprises. Though the creation and annihilation of solitons is one of the most widespread processes in nature, it was thought until very recently that this phenomenon could hardly be observed in nonlinear integrable systems, i.e. the latter cannot describe processes of that kind. Now, it became clear that this is not so and nonlinear integrable systems are most appropriate for this purpose.

In the present report, we exemplify this phenomenon by two nonlinear integrable systems. The first is the modified Korteweg-de Vries equation with a self-consistent source, i.e. the following system of equations:

$$\frac{\partial u}{\partial t}+6u^2\,\frac{\partial u}{\partial x}+\frac{\partial^3 u}{\partial x^3}=2\sum_{n=1}^{N}\left(\varphi_n p_n+\overline{\varphi}_n\overline{p}_n-\psi_n q_n-\overline{\psi}_n\overline{q}_n\right)\ , \tag{1}$$

$$\frac{\partial \varphi_n}{\partial x}+u\psi_n-i\zeta_n\varphi_n=\frac{\partial \psi_n}{\partial x}-u\varphi_n+i\zeta_n\psi_n=0\ , \tag{2}$$

$$\frac{\partial p_n}{\partial x}+uq_n-i\zeta_n p_n=\frac{\partial q_n}{\partial x}-up_n+i\zeta_n q_n=0\ . \tag{3}$$

Here and hereafter the bar means complex conjugation. We shall be interested in the case when the solution $u=u(x,t)$ of this system is a real, rapidly decreasing function of x, i.e. the condition

$$\sum_{r=0}^{3}\int_{-\infty}^{\infty}\left|\frac{\partial^r u(x,t)}{\partial x^r}\right|dx<\infty \tag{4}$$

is valid at any $t\geq 0$. In accordance with this condition we require that solutions $\varphi_n=\varphi_n(x,t)$, $\psi_n=\psi_n(x,t)$ and $p_n=p_n(x,t)$, $q_n=q_n(x,t)$

of the corresponding linear Dirac system at any $t \geq 0$ should satisfy the conditions

$$|\varphi_n(x,t)| + |\psi_n(x,t)| \to \infty, \quad \text{if} \quad x \to \pm\infty, \tag{5}$$

$$|p_n(x,t)| + |q_n(x,t)| \to 0, \quad \text{if} \quad x \to \pm\infty. \tag{6}$$

Hence, it follows that points $\zeta = \zeta_n$, $n = 1, \ldots, N$, are the points of the discrete spectrum of the Dirac operator

$$L = \Lambda\partial + U, \quad \partial = \frac{\partial}{\partial x}, \tag{7}$$

where

$$\Lambda = diag(1, -1), \quad U = \begin{vmatrix} 0 & u \\ u & 0 \end{vmatrix}. \tag{8}$$

It turns out that this nonlinear evolution system may be integrated by the inverse scattering method for the Dirac operator L. There are the evolution equations for the S-matrix elements

$$\frac{\partial S_{11}(\zeta)}{\partial t} + 2i\zeta \sum_{n=1}^{N} \left(\frac{W_n}{\zeta^2 - \zeta_n^2} + \frac{\overline{W}_n}{\zeta^2 - \overline{\zeta}_n^2} \right) S_{11}(\zeta) = 0, \tag{9}$$

$$\frac{\partial S_{12}(\zeta)}{\partial t} - 8i\zeta^3 S_{12}(\zeta) = \frac{\partial S_{21}(\zeta)}{\partial t} + 8i\zeta^3 S_{21}(\zeta) = 0, \tag{10}$$

$$\frac{\partial S_{22}(\zeta)}{\partial t} - 2i\zeta \sum_{n=1}^{N} \left(\frac{W_n}{\zeta^2 - \zeta_n^2} + \frac{\overline{W}_n}{\zeta^2 - \overline{\zeta}_n^2} \right) S_{22}(\zeta) = 0, \tag{11}$$

where

$$W_n = \varphi_n(x,t)\, q_n(x,t) - \psi_n(x,t)\, p_n(x,t). \tag{12}$$

From these equations it follows that the functions $S_{11}(\zeta)$ and $S_{22}(\zeta)$ admit the representation

$$S_{11}(\zeta) = S_1(\zeta) \prod_{n=1}^{N} \left[\frac{(\zeta - \zeta_n)(\zeta + \overline{\zeta}_n)}{(\zeta + \zeta_n)(\zeta - \overline{\zeta}_n)} \right]^{r_n}, \tag{13}$$

$$S_{22}(\zeta) = S_2(\zeta) \prod_{n=1}^{N} \left[\frac{(\zeta + \zeta_n)(\zeta - \overline{\zeta}_n)}{(\zeta - \zeta_n)(\zeta + \overline{\zeta}_n)} \right]^{r_n}, \tag{14}$$

where

$$\frac{\partial S_1(\zeta)}{\partial t} = \frac{\partial S_2(\zeta)}{\partial t} \equiv 0, \tag{15}$$

$$\frac{d\zeta_n}{dt} = \frac{i}{(1+\gamma_n)r_n}\left(W_n + \gamma_N \overline{W}_n\right), \tag{16}$$

$$\gamma_n = \begin{cases} 0, & \text{if} \quad \zeta_n + \overline{\zeta}_n \neq 0, \\ 0, & \text{if} \quad \zeta_n + \overline{\zeta}_n \equiv 0. \end{cases} \tag{17}$$

Note that according to these equations in the process of evolution the imaginary part of any of the eigenvalues, say ζ_{n_0}, can vanish, i.e. ζ_{n_0} itself can fall on the real axis in the process of evolution. This results in that a soliton corresponding to that eigenvalue disappears. Moreover, if then the quantity ζ_{n_0} will leave the real axis, the soliton that disappeared for a while will appear again, i.e. as though regenerated. It is obvious that the appearance of the new eigenvalue and the creation of the corresponding soliton may proceed without their preliminary disappearance.

Evolution equations for the normalization constants B_m and $\hat{B}_m$ in this case have the following form:

$$\frac{\partial B_m}{\partial t} + \left[8i\zeta_m^3 - i(W_m + \gamma_m \overline{W}_m)b_m\right] B_m = 0, \tag{18}$$

$$\frac{\partial \hat{B}_m}{\partial t} - \left[8i\overline{\zeta}_m^3 - i(\overline{W}_m + \gamma_m W_m)\overline{b}_m\right] \hat{B}_m = 0, \tag{19}$$

where the quantities b_m depend on the choice of the solution φ_m, ψ_m of the corresponding linear Dirac system, $m = 1, \ldots, N$.

As a second example we shall consider the Sine-Gordon equation with a self-consistent source, i.e. consider the system of equations of the form

$$\frac{\partial^2 \theta}{\partial t \partial x} + sin\theta = 4\sum_{n=1}^{N}(\varphi_n p_n + \overline{\varphi}_n \overline{p}_n - \psi_n q_n - \overline{\psi}_n \overline{q}_n), \tag{20}$$

$$\frac{\partial \varphi_n}{\partial x} + \frac{1}{2}\frac{\partial \theta}{\partial x}\psi_n - i\zeta_n \varphi_n = \frac{\partial \psi_n}{\partial x} - \frac{1}{2}\frac{\partial \theta}{\partial x}\varphi_n + i\zeta_n \psi_n = 0, \quad n = 1, \ldots, N, \tag{21}$$

$$\frac{\partial p_n}{\partial x} + \frac{1}{2}\frac{\partial \theta}{\partial x}q_n - i\zeta_n p_n = \frac{\partial q_n}{\partial x} - \frac{1}{2}\frac{\partial \theta}{\partial x}p_n + i\zeta_n q_n = 0, \quad n = 1, \ldots, N. \tag{22}$$

We are interested in the case when the solution $\theta = \theta(x,t)$ as $x \to \pm\infty$ tends to some quantities multiple of 2π so quickly that at any $t \geq 0$ the inequality holds

$$\int_{-\infty}^{\infty}\left\{|sin\theta(x,t)| + \left|\frac{\partial \theta(x,t)}{\partial x}\right|\right\} dx < \infty. \tag{23}$$

Moreover, we assume that the solutions $\varphi_n = \varphi_n(x,t)$, $\psi_n = \psi_n(x,t)$ and $p_n = p_n(x,t)$, $q_n = q_n(x,t)$ of the corresponding linear Dirac system in this case also satisfy the conditions (1) at any $t \geq 0$. Then, the evolution equations for the S-matrix elements have the form

$$\frac{\partial S_{11}(\zeta)}{\partial t} + 2i\zeta \sum_{n=1}^{N} \left(\frac{W_n}{\zeta^2 - \zeta_n^2} + \frac{\overline{W}_n}{\zeta^2 - \overline{\zeta}_n^2} \right) S_{11}(\zeta) = 0, \tag{24}$$

$$\frac{\partial S_{12}(\zeta)}{\partial t} - \frac{i}{2\zeta} S_{12}(\zeta) = \frac{\partial S_{21}(\zeta)}{\partial t} + \frac{i}{2\zeta} S_{21}(\zeta) = 0, \tag{25}$$

$$\frac{\partial S_{22}(\zeta)}{\partial t} - 2i\zeta \sum_{n=1}^{N} \left(\frac{W_n}{\zeta^2 - \zeta_n^2} + \frac{\overline{W}_n}{\zeta^2 - \overline{\zeta}_n^2} \right) S_{22}(\zeta) = 0, \tag{26}$$

where the quantities W_n were determined earlier by (2) and $\zeta = \zeta_n$ are the points of the discrete spectrum of the corresponding Dirac operator, $n = 1, \ldots, N$. The evolution equations for the normalisation constants B_m and $\hat{B}_m$ in this case have the following form:

$$\frac{\partial B_m}{\partial t} + \left[\frac{i}{2\zeta_m} - i(W_m + \gamma_m \overline{W}_m) b_m \right] B_m = 0, \tag{27}$$

$$\frac{\partial \hat{B}_m}{\partial t} - \left[\frac{i}{2\zeta_m} - i(\overline{W}_m + \gamma_m W_m) \overline{b}_m \right] \hat{B}_m = 0. \tag{28}$$

In conclusion I should like to remark that the method I am using here is very general. This method permits one to get the solution of the Cauchy problem in the case of the source including the Fourier integral over eigenfunctions of the continuous spectrum of the Dirac operator in a very general form.

Part VIII

Nonlinear Excitations in more than one Space Dimension

Multidimensional Nonlinear Schrödinger Equations Showing Localized Solutions

P.C. Sabatier

Département de Physique Mathématique,
Université de Montpellier II,
34095 Montpellier Cedex 5, France.

Summary. There exists non linear generalizations of the Schrödinger equation, for an arbitrary number of space dimensions, showing exponentially localized solutions that propagate at constant velocity. The equations also show a number of properties that propagate at constant velocity. The equations also show a number of properties similar to those of the linear Schrödinger equation, e.g. scale invariance and center-of mass separation. Other properties are not so good but one may have obtained here a good direction for generalizing the Schrödinger equation.

Main Equations and their localized solutions

For an arbitrary number N of space dimensions, the equation

$$i \frac{\partial q}{\partial t} + \Delta q - sq \, \Delta \log |q| = 0 \tag{1}$$

with $s > 1$, and a class of related equations, show solutions which are exponentially confined solitary waves and propagate as particles (Sabatier, 1990 a,b). They are related to one-dimensional solitons, or to the two-dimensional solitons (or dromions ?) discovered by Boiti et al. (1988), but here the properties are intrinsic, and thus one can impose to the solutions any desired symmetry in the moving frame, and the confinement needs not any driving force at ∞. We may call the solutions which are spherically symmetric in the moving frame "ballons" or "bullons", according to our personal memories of balls or bull's heads.

Proving the claimed solutions is trivial. By setting $q = F \exp[-i\alpha]$, we reduce (1) to

$$\frac{\partial F}{\partial t} - 2 \,\mathrm{grad}\, \alpha \,.\, \mathrm{grad}\, F - F\Delta\alpha = 0 \tag{2}$$

$$F^{-1} \Delta F + \left(\frac{\partial \alpha}{\partial t} - \mathrm{grad}\, \alpha \,.\, \mathrm{grad}\, \alpha - s \,\Delta \log F \right) = 0 \quad . \tag{3}$$

where grad without special index denotes the gradient in the x-space. Assume a linear phase α, say, $\beta, \gamma \in \mathbb{R}$, $\eta \in \mathbb{R}^n$:

$$\alpha(x,t) = \beta + \gamma t + \eta . x \quad . \tag{4}$$

The solution of (2) is then a function F depending only on the moving coordinates :

$$X = x - x^{(0)} + 2\eta t \tag{5}$$

where $x^{(0)}$ is a reference point. Assume $s \neq 1$ and set $F = f^{1-s}$, the equation (3) then reduces to the Helmholtz equation

$$\begin{cases} \Delta_x f - \kappa^2 f = 0 \\ \kappa^2 = (1-s)^{-2}\, \eta.\eta - \gamma \end{cases} \tag{6}$$

The conditions ($s > 1$, $\kappa^2 > 0$) guarantee the existence of exponentially confined solutions (e.g. $f(X) = \prod_{i=1}^{n} \cosh \kappa_i x_i$, with $\sum_i \kappa_i^2 = \kappa^2$) without any driving force, and with symmetries one may require, e.g., for $N = 3$, $s = 2$, the solution which is spherically symmetric in the moving frame (bullon, or ballon ?) :

$$F(x,t) = \frac{\|x-x^{(0)} + 2\eta t\|}{\sinh\left[\kappa \|x-x^{(0)} + 2\eta t\|\right]} \quad . \tag{7}$$

A similar study holds for the equation

$$i\frac{\partial q}{\partial t} + \Delta q + (\mu - V)q - sq\,\Delta\,(\log |q|) = 0 \tag{8}$$

where $\mu \in \mathbb{R}$, V is a real function of x and t depending on X only (as given by (5)) if a solution with phase α (as given by (4)) should exist. Then the Helmholtz equation (6) is replaced by

$$(1-s)^2\, \Delta_x f + (\gamma - \eta.\eta + \mu - V)f = 0 \tag{9}$$

and convenient conditions on s,γ,η,μ,V, can guarantee the confinement. Also if a few simple conditions are satisfied by the parameters, the results can be extended to equations of the form

$$i\,\theta \frac{\partial q}{\partial t} + (\Delta - V)q - sqD\left[\log |q| + \rho \log(q/q^*)\right] \tag{10}$$

where D is the operator defined by

$$Df = \Delta f + b.\mathrm{grad}\, f + c\, \mathrm{grad}\, f.\mathrm{grad}\, f + d\frac{\partial f}{\partial t} \tag{11}$$

θ,s,ρ,c,d are real numbers, $b \in \mathbb{R}^N$. The parameter ρ is zero in the cases corresponding to a "real" interaction, which we call the "modular" cases, and is

1 in the cases where (10) involves q without ever "separating" modulus and phase, which we call "logarithmic cases". We also define the "basic" cases as those when D reduces to Δ.

Other Properties

It is remarkable that the equation (1) and many equations of the form (8) and (10) share with the linear Schrödinger equation a number of properties. Locality and scale invariance are the most obvious ones - in contrast with the Bialynicki-Birula & Mycielski logarithmic extension of the Schrödinger Equation(1976).The equations are homogeneous and time reversal invariant. They can also exhibit Euclidean and Galilean invariance (when V(x) does). Let us now write down the multiparticle extension of the equation (1) (the other cases are easy to study in the same way), with the usual physical quantities, say :

$$i\hbar \frac{\partial\Psi}{\partial t} + \frac{h^2}{2}\sum_{i=1}^{n} \frac{\Delta_i}{m_i}\Psi - V\Psi - s\frac{\hbar^2}{2}\Psi\sum_{i=1}^{n}\frac{\Delta_i}{m_i}\log|\Psi| = 0 \qquad (12)$$

where the "wave function" Ψ depends on the n particle position vectors $x_1, x_2, \ldots, x_n$, $\left(x_i \in \mathbb{R}^N\right)$, and $\Delta_i \equiv \Delta_{x_i}$. If we introduce the Jacobi coordinates

$$\xi_j = \frac{\sum_{i=1}^{j} m_i x_i}{\sum_{i=1}^{j} m_i} - x_{j+1} \qquad \xi_n = \frac{\sum_{i=1}^{n} m_i x_i}{\sum_{i=1}^{n} m_i} = X$$

$$\mu_j^{-1} = \left(\sum_{i=1}^{j} m_i\right)^{-1} + (m_{j+1})^{-1} \qquad M = \sum_{i=1}^{n} m_i = \mu_n \qquad (15)$$

there is a solution separating the center-of mass motion and the relative motion, as in the linear case :

$$\begin{cases} \Psi = \psi(\xi_1,\xi_2,\ldots,\xi_{n-1})\,\varphi(X) \\ i\,\hbar\,\psi^{-1}\dfrac{\partial\psi}{\partial t} - V + \displaystyle\sum_{i=1}^{n-1}\frac{\hbar^2}{2\mu_i}\left(\psi^{-1}\Delta_{\xi_i}\psi - s\,\Delta_{\xi_i}\log|\psi|\right) = 0 \\ i\,\hbar\,\varphi^{-1}\dfrac{\partial\varphi}{\partial t} + \dfrac{\hbar^2}{2M}\left(\varphi^{-1}\Delta_X\varphi - s\,\Delta_X\log|\varphi|\right) = 0 \end{cases} \qquad (14)$$

In the two particle case, once the center-of mass motion has been separated, the relative motion is that of a bullon if V confines a bound state which is then completely identified to a breather ($\eta = 0$) in the moving frame.

The existence of such interesting properties led us to seek conservation laws, and here we must admit that usual generalizations of the Schrödinger equation do better. The usual current conservation law holds for all modular cases - say - for equation (11) and $\rho = 0$:

$$i\,\theta \frac{\partial}{\partial t}(qq^*) + \operatorname{div}[q^* \operatorname{grad} q - q \operatorname{grad} q^*] = 0 \qquad (15)$$

but we were able to derive a lagrangian density for one modular case only (Sabatier 1990b)

$$i \frac{\partial q}{\partial t} + \Delta q - Vq - sq \frac{\Delta|q|}{|q|} = 0 \qquad (16)$$

which does not show bullons ! This is a pity, because a nice energy functional is conserved :

$$E(t) = \int_{\mathbb{R}^N} dx\, [\operatorname{grad} q . \operatorname{grad} q^* + V\,|q|^2 + s|q|\,\Delta|q|] \ . \qquad (17)$$

In the logarithmic case, there may exist more complicated conservation laws. In particular, the basic logarithmic case on one hand is reduced to the linear case by setting $q = p^{1-s}$, so that it shows all the corresponding conservation laws, and on the other hand shows the linear phase solutions of the basic modular case. Let us write down its lagrangian density and its energy :

$$L = (qq^*)^{\frac{1}{1-s}} \left[\frac{1}{2}\, i \left(\frac{\dot{q}}{q} - \frac{\dot{q}^*}{q^*}\right) - \frac{\operatorname{grad} q}{q} . \frac{\operatorname{grad} q^*}{q^*} - V\right]$$

$$E = \int_{\mathbb{R}^N} dx\, [\operatorname{grad} q . \operatorname{grad} q^* + V\,qq^*]\,(qq^*)^{\frac{s}{1-s}}\,dx \qquad (18)$$

where the upper dots denote time derivatives. Unfortunately, the physical interest of these results is unclear.Furthermore, the stability of a localized solution Φ of (1) against a small perturbation $\epsilon\,\Phi\left(e^{-i\omega t}\,\varphi_1 + e^{i\omega t}\,\varphi_2\right)$ has been studied by G. Auberson in the one-dimensional case who obtained a negative result : perturbed normalizable solutions are obtained with $\operatorname{Im}\omega \neq 0$.

Further generalizations

Since improving the results is necessary, one may hope to do it either by understanding the physical meaning of completed cases or by further generalizations. We tried two directions. The first one (Degasperis & Sabatier 1990) is devoted to what we call "real" non linear Schrödinger equations, i.e.

$$\left(i\frac{\partial q}{\partial t} + \Delta + v\right)q = 0 \tag{19}$$

where v is a function of q and x with real values only. Setting $q = r\exp[i\varphi]$ yields a solution of (19) iff

$$\frac{\partial}{\partial t}\log r + 2\,\mathrm{grad}\,\varphi.\mathrm{grad}\,\log r + \Delta\varphi = 0 \tag{20}$$

$$r^{-1}\,\Delta r + \left(v + \frac{\partial \varphi}{\partial t} - \mathrm{grad}\,\varphi.\mathrm{grad}\,\varphi\right) = 0 \quad . \tag{21}$$

For a fixed function $\varphi(x,t)$, any other solution $R(x,t)$ of (20) is of the form $R = r/F(W)$ where F is a differentiable function of W and

$$\frac{\partial W}{\partial t} + 2\,\mathrm{grad}\,\varphi.\mathrm{grad}\,W = 0 \quad . \tag{22}$$

Hence, all the equations which share with (19) a solution whose phase is $\varphi(x,y,t)$ are obtained from (19) by substituting V to v, with

$$V = v + \Delta\log F(W) + \mathrm{grad}\,\log F(W).\mathrm{grad}\,\log[r^2/F(W)] \tag{23}$$

and the modulus of this solution is R. Hence all the "real" generalizations could be solved if (19) and (22) are solved for all φ. Now (22) is solved if the characteristic equations

$$2\,dt = \left(\frac{\partial\varphi}{\partial x_1}\right)^{-1} dx_1 = \ldots \left(\frac{\partial\varphi}{\partial x_n}\right)^{-1} dx_n \tag{24}$$

are solved. In the one-dimensional case, the input solution r of (20) readily yields the integrating factor $(-r^2)$ of the (unique) characteristic equation, since (20) also reads :

$$\frac{\partial}{\partial t}r^2 = -2\frac{\partial}{\partial x}r^2\frac{\partial\varphi}{\partial x} \quad . \tag{25}$$

Thus $W(x,t)$ is given by $\mathrm{grad}_{x,t}W = \begin{pmatrix} r^2 \\ -2r^2\dfrac{\partial\varphi}{\partial x}\end{pmatrix}$, or

$$W = \int_a^x r^2(\xi,t)d\xi - 2\int_{t_0}^t r^2(a,\tau)\left[\frac{\partial\varphi}{\partial\xi}(\xi,\tau)\right]_{\xi=a} d\tau + b \tag{26}$$

where r, φ refer to the input solution of (19), a, t_0, b, are real numbers. If (19) is integrable, the ansatz yields equations at least partially integrable. In the multidimensional case, only separable solutions of (19) yield solutions of the characteristic equations and a class of equations that are partially integrable.

The same ansatz applies to the Klein Gordon equation - the only tractable case being the 1+1 equation

$$\frac{\partial^2 q}{\partial x^2} - \frac{\partial^2 q}{\partial t^2} + vq = 0 \tag{27}$$

whose solution is $q = r \exp[i\varphi]$ iff

$$2\, r^{-1} \left(\frac{\partial r}{\partial x}\frac{\partial \varphi}{\partial x} - \frac{\partial r}{\partial t}\frac{\partial \varphi}{\partial t} \right) + \frac{\partial^2 \varphi}{\partial x^2} - \frac{\partial^2 \varphi}{\partial t^2} = 0 \tag{28}$$

$$r^{-1} \left(\frac{\partial^2 r}{\partial x^2} - \frac{\partial^2 r}{\partial t^2} \right) + v - \left(\frac{\partial \varphi}{\partial x} \right)^2 + \left(\frac{\partial \varphi}{\partial t} \right)^2 = 0 \quad . \tag{29}$$

Again, the equation (28) does not depend on v, and, for fixed φ, we can construct from the input solution $r(x,t)$ all the others as $R = r/F(w)$ provided

$$\frac{\partial w}{\partial t}\frac{\partial \varphi}{\partial t} - \frac{\partial w}{\partial x}\frac{\partial \varphi}{\partial x} = 0 \tag{30}$$

Again, in this 1+1 case, the characteristic equation

$$\left(\frac{\partial \varphi}{\partial t} \right)^{-1} dt = - \left(\frac{\partial \varphi}{\partial x} \right)^{-1} dx \tag{31}$$

has the integrating factor r^2 since (28) also reads

$$\frac{\partial}{\partial t} \left(r^2 \frac{\partial \varphi}{\partial t} \right) = \frac{\partial}{\partial x} \left(r^2 \frac{\partial \varphi}{\partial x} \right) \tag{32}$$

and w is given by $\mathrm{grad}_{x,t}\, w = r^2\, \mathrm{grad}\, \varphi$.

The second direction for generalizing solvable non linear Schrödinger equations is by using Darboux transformations of the equation and a coupling condition. Of course, the form of (1) strongly recalls a Darboux transformation, but those of the linear Schrödinger equation in the multidimensional case have never been published as yet, up to my knowledge. Yet, assume f and u are arbitrary solutions of the two following equations

$$i \frac{\partial f}{\partial t} + (\Delta - V) f = 0 \tag{33}$$

$$i \frac{\partial u}{\partial t} + (\Delta - V) u = \gamma^2 u \tag{34}$$

where V is an arbitrary function of x and t, γ is a number, and consider the vector

$$\underset{\sim}{q} = \text{grad } f - u^{-1} f \text{ grad } u \tag{35}$$

It is not difficult to show that $\underset{\sim}{q}$ obeys the vectorial Schrödinger equation

$$\Delta \underset{\sim}{q} + i \frac{\partial q}{\partial t} - V \underset{\sim}{q} + 2(\underset{\sim}{q} \; . \; \text{grad}) \frac{\text{grad } u}{u} = 0 \tag{36}$$

and if V is allowed to depend on u, this result can be used together with a limiting process in order to obtain solvable non linear equations.

Preliminary results are complicated.

References

I. Bialynicki-Birula & J. Mycielski (1976). "Nonlinear wave mechanics". Ann. Phys. 100, 62-93.

M. Boiti, J. J-P. Léon, L. Martina and F. Pempinelli (1989). "Scattering of localized solitons in the plane". Phys. Lett. A 632, 432-438.

A. Degasperis & P.C. Sabatier (1990). "Localized solutions of (N+1) dimensional evolution equations". Phys. Lett. A.

P.C. Sabatier (1990a). "Quest of multidimensional non linear equations with exponentially confined solutions" Inv. Prob. 6 L29-L32.

P.C. Sabatier (1990b). "Multidimensional non linear Schrödinger equations with exponentially confined solutions" Inv. Prob. 6 L47-L53.

New Soliton Solutions for the Davey-Stewartson Equation*

F. Pempinelli, M. Boiti, L. Martina, O.K. Pashaev† and D. Perrone

Dipartimento di Fisica dell'Università and Sezione INFN,
73100 Lecce, Italy.

1.- The N-soliton solution.

Let us consider the hyperbolic Davey-Stewartson equation (DSI) written in characteristic coordinates u=x+y and v=x-y

$$iQ_t + \sigma_3(Q_{uu}+ Q_{vv}) + [A,Q] = 0 \tag{1}$$

where Q is a 2x2 off diagonal matrix field

$$Q = \begin{pmatrix} 0 & q(u,v,t) \\ r(u,v,t) & \end{pmatrix} \tag{2}$$

and A is a 2x2 diagonal matrix field A, so called auxiliary field, which is chosen to have arbitrary boundary values a_1 and a_2 according to the formula

$$A = \frac{1}{2}\begin{pmatrix} -\int_{-\infty}^{u} du'\,(Q^2)_v + a_1(v,t) & 0 \\ 0 & \int_{-\infty}^{v} dv'\,(Q^2)_u + a_2(u,t) \end{pmatrix}. \tag{3}$$

The DSI admits a Lax representation $[T_1,T_2]=0$ where

$$T_1 = \partial_x+\sigma_3\partial_y+Q, \quad T_2 = i\partial_t+\sigma_3\partial_y^2+Q\,\partial_y- \frac{1}{2}\,\sigma_3 Q_x+ \frac{1}{2}\,Q_y+A\ . \tag{4}$$

Therefore it belongs to the hierarchy of equations related to the spectral problem of Zakharov-Shabat (ZS) in the plane $T_1\psi = 0$. Of interest is the reduced case $r = \varepsilon\bar{q}$ (the overbar means complex conjugation and $\varepsilon^2= 1$) and specifically the so called focusing case $\varepsilon=-1$, the one we consider in the following.

Recently BLMP (Boiti, Leon, Martina and Pempinelli) [1],

---

* Work supported in part by M.U.R.S.T.
† Permanent address: JINR, Dubna (Moscow), USSR

by using the Bäcklund transformations, found that bidimensional localized (exponentially decaying) soliton solutions do exist for the DSI equation and for all the equations of the ZS hierarchy. This result has been reobtained by defining a new Spectral Transform (ST) [2,3], which satisfies the following requirements:

i) for vanishing boundaries values a_1 and a_2 it reduces to the ST introduced by Fokas and Ablowitz [4];

ii) its time evolution can be explicitly integrated;

iii) the discrete part of the spectrum corresponds to solitons and the continuous part to the radiation.

For an alternative Spectral Transform which does not satisfies the last two requirements see ref. [5].

In order to linearize the initial value problem for the DSI equation one needs to give the ST of Q considered as potential in the Zakharov-Shabat spectral problem

$$T_1\psi \equiv (\partial_x + \sigma_3\partial_y + Q)\psi = 0 \tag{5}$$

and of the boundaries a_1 and a_2 considered as potentials in two different time-dependent Schrödinger equations

$$(i\partial_t + \partial_v^2 + a_1)\phi_1 = 0 \tag{6}$$

$$(i\partial_t - \partial_u^2 + a_2)\phi_2 = 0 \tag{7}$$

The complex spectral parameter k is introduced by requiring that the eigenfunctions ψ and ϕ_i satisfy, as $k\to\infty$, the asymptotic properties

$$\psi \exp[-ik(\sigma_3 x - y)] = \mathbf{1} + O\left(\frac{1}{k}\right) \tag{8}$$

$$\phi_1 \exp[-ikv + ik^2 t] = 1 + O\left(\frac{1}{k}\right) \tag{9}$$

$$\phi_2 \exp[iku - ik^2 t] = 1 + O\left(\frac{1}{k}\right). \tag{10}$$

Then the ST is defined as the measure of the departure from the analyticity of the eigenfunctions ψ and ϕ_i

$$\frac{\partial\psi}{\partial\bar{k}} = \iint d\ell\wedge d\bar{\ell}\ \psi(\ell)R(k,\ell) \tag{11}$$

$$\frac{\partial\phi_i}{\partial\bar{k}} = \iint d\ell\wedge d\bar{\ell}\ \phi_i(\ell)R_i(k,\ell) \qquad i = 1,\ 2. \tag{12}$$

Integral equations for ψ and ϕ_i can be written [2,3,6]. The integral equation for ψ is of Volterra type and therefore ψ has simple poles only if the homogeneous solution ψ_o has simple poles. Because ψ_o is related to the eigenfunctions ϕ_i of eqs. (6), (7) by

$$\psi_o = \mathrm{diag}\,(\phi_1 \exp[ik^2t],\ \phi_2 \exp[-ik^2t]). \qquad (13)$$

it has poles for potentials a_i which are wave solitons of the Kadomtsev-Petviashvili (KP) equation. Therefore the discrete spectrum of ψ is due to a_1 and a_2. If we use the usual N-soliton solution for the KP [7] we get an N-soliton solution for the DSI that describes the interaction of N solitons which exponentially decay in all directions in the plane and which, after the interaction, do not change shape and velocity. The only effect of the interaction is a shift in the position and in the phase. The solution depends on 8N real parameters, the discrete complex eigenvalues λ_n, μ_n, the complex normalization coefficients ρ_n, η_n (related by the reduction condition $\rho_n(\bar{\mu}_n-\mu_n)=\varepsilon\bar{\eta}_n(\bar{\lambda}_n-\lambda_n)$) and the initial positions of the solitons u_{on}, v_{on} $(n=1,2,\ldots,N)$ [3,8]. One can define the mass (energy, charge or number of particles according to the physical context) of the solution q as $M = \iint |q|^2 du\, dv$ and the mass of the nth soliton at $t=\pm\infty$ as $M_n^{(\pm)} = \iint |q_n^{(\pm)}|^2 du\, dv$, where $q_n^{(\pm)}$ is the asymptotic behaviour of q at $t=\pm\infty$ computed in the rest reference frame of the soliton. Every soliton preserves its mass and M equals the total mass of solitons.

2.- Bifurcation and related solutions.

One can show [8] that the N-soliton solution is structurally unstable at $t=-\infty$ and at $t=+\infty$ when any couple of discrete eigenvalues have the same real part or are equal. The explicit expression of q and A is given in [8].

In the first case ($\lambda_{i\mathcal{R}}=\lambda_{j\mathcal{R}}$ or $\mu_{i\mathcal{R}}=\mu_{j\mathcal{R}}$) the solitons after the interaction:

i) have a bidimensional shift,

ii) change their form but do not change their mass.

Fig. 1 shows the evolution of $|q|$ and $\mathrm{Tr}\,\sigma_3 A$ in the two soliton case. Note the form of these new localized solitons which have two bumps.

In the second case ($\lambda_i=\lambda_j$ or $\mu_i=\mu_j$) the solitons after the interaction:

i) have a bidimensional shift,

ii) change their form,

iii) change their mass but the total mass is conserved.

Moreover, the form and the mass of solitons depend on the initial position u_{on}, v_{on} of every soliton [9]. Fig. 2 shows the evolution of $|q|$ and $Tr\sigma_3 A$ in the two soliton case.

3.- More general N-soliton solution.

By choosing more general boundaries than those used before an N-soliton solution can be derived which depends on 4N(N+1) real parameters, in contrast with the previous one depending only on 8N real parameters (see ref.[10] for details). Of these parameters 2N(N+2) are determined by the choice of the boundaries and fix the velocity and the possible location of the solitons in the plane, while the remaining $2N^2$ govern the dynamics of the solitons during the interaction. In general, neither the mass of the single soliton nor the total mass of the solitons is preserved by the interaction (only M is conserved) and solitons can be created and annihilated (some special effects previously obtained by using direct methods [11,12] are also recovered).

More precisely, this solution is obtained by choosing for $r_i(k,\ell)$ the most general discrete ST compatible with the requirement that the a_i are real, i.e.

$$R_1(k,\ell) = \sum_{n,m} r^{(1)}_{nm} \exp[-i(k^2-\ell^2)t]\, \delta(\ell-\bar{\lambda}_m)\, \delta(k-\lambda_n) \tag{14}$$

$$R_2(k,\ell) = \sum_{n,m} r^{(2)}_{nm} \exp[i(k^2-\ell^2)t]\, \delta(\ell-\bar{\mu}_m)\, \delta(k-\mu_n). \tag{15}$$

where the constant matrices $r^{(i)}$ are hermitian.

Of special interest are the cases in which one or more masses are zero. The two soliton solution which displays all the richness of the general case is considered in details in ref. [10]. The solution in this case can simulate relativistic quantum effects as inelastic scattering ($M \neq \sum_n M^{(-)}_n \neq \sum_n M^{(+)}_n$), fission ($M^{(-)}_2 = 0$), fusion ($M^{(+)}_2 = 0$), pair creation ($M^{(-)}_1 = M^{(-)}_2 = 0$), pair annihilation ($M^{(+)}_1 = M^{(+)}_2 = 0$), self interaction ($M^{(-)}_1 = M^{(+)}_2 = 0$), creation from the vacuum of one soliton ($M^{(-)}_1 = M^{(\pm)}_2 = 0$), annihilation of one soliton ($M^{(+)}_1 = M^{(\pm)}_2 = 0$), vacuum fluctuation ($M^{(\pm)}_1 = M^{(\pm)}_2 = 0$).

REFERENCES

[1] M Boiti, J Léon, L Martina and F Pempinelli, Phys. Lett. A 132, 432 (1988)

[2] M Boiti, J Léon and F Pempinelli, J. Math. Phys. 31, 2612 (1990)

[3] M Boiti, J Léon and F Pempinelli, Phys. Lett. A 141, 96 and 101 (1989)

[4] A S Fokas, Phys. Rev. Lett. 51, 3 (1983), A S Fokas and M J Ablowitz, J. Math. Phys. 25, 2494 (1984)

[5] A S Fokas and P M Santini, Phys. Rev. Lett. 63, 1329 (1989); A S Fokas and P M Santini, Physica D 44, 99 (1990)

[6] M Boiti, J Léon, F Pempinelli and A K Pogrebkov, "Solitons and Spectral Transform for DSI and KPI equations" in "Solitons and Applications", Eds V G Makhanhov, V K Fedyanin and O K Pashaev (World Scientific, Singapore, 1990)

[7] S V Manakov, V E Zakharov, L A Bordag, A R Its and V B. Matveev, Phys. Lett. A 63, 205 (1977)

[8] M Boiti, J Léon and F Pempinelli, Inverse Problems 6, 715 (1990)

[9] M Boiti, J Léon, L Martina, F Pempinelli and D Perrone, "Asymptotic bifurcations of multidimensional solitons" in Proceedings of NEEDS '90 Workshop, Dubna USSR (Springer Verlag)

[10] M Boiti, L Martina, O K Pashaev, F Pempinelli, "Dynamics of multidimensional solitons", Preprint (Lecce, December 1990)

[11] J Hietarinta and R Hirota, Phys. Lett. A 145, 237 (1990)

[12] R Hernandez Heredero, L Martinez Alonso and E Medina Reus, "Fusion and fission of dromions in the Davey-Stewartson equation", Preprint, April 1990 (Madrid)

FIGURE CAPTION

Figures describe time sequence of the level contours of $|q|$, on the left, and of $\mathrm{Tr}\sigma_3 A$, on the right, for the two soliton solution ,in the frame of reference of the soliton 1. In Fig. 1 is considered the case $\mu_{1\Re}=\mu_{2\Re}$ (λ_1=-0.1-i, μ_1=i , ρ_1=2.004, λ_2=-1.01i, μ_2=1.3i, ρ_2=2.108) and in Fig. 2 the case $\mu_1=\mu_2$ (λ_1=-0.1-i, μ_1=i , ρ_1=2.004, λ_2=-1.01i, μ_2=i, ρ_2=2.108).

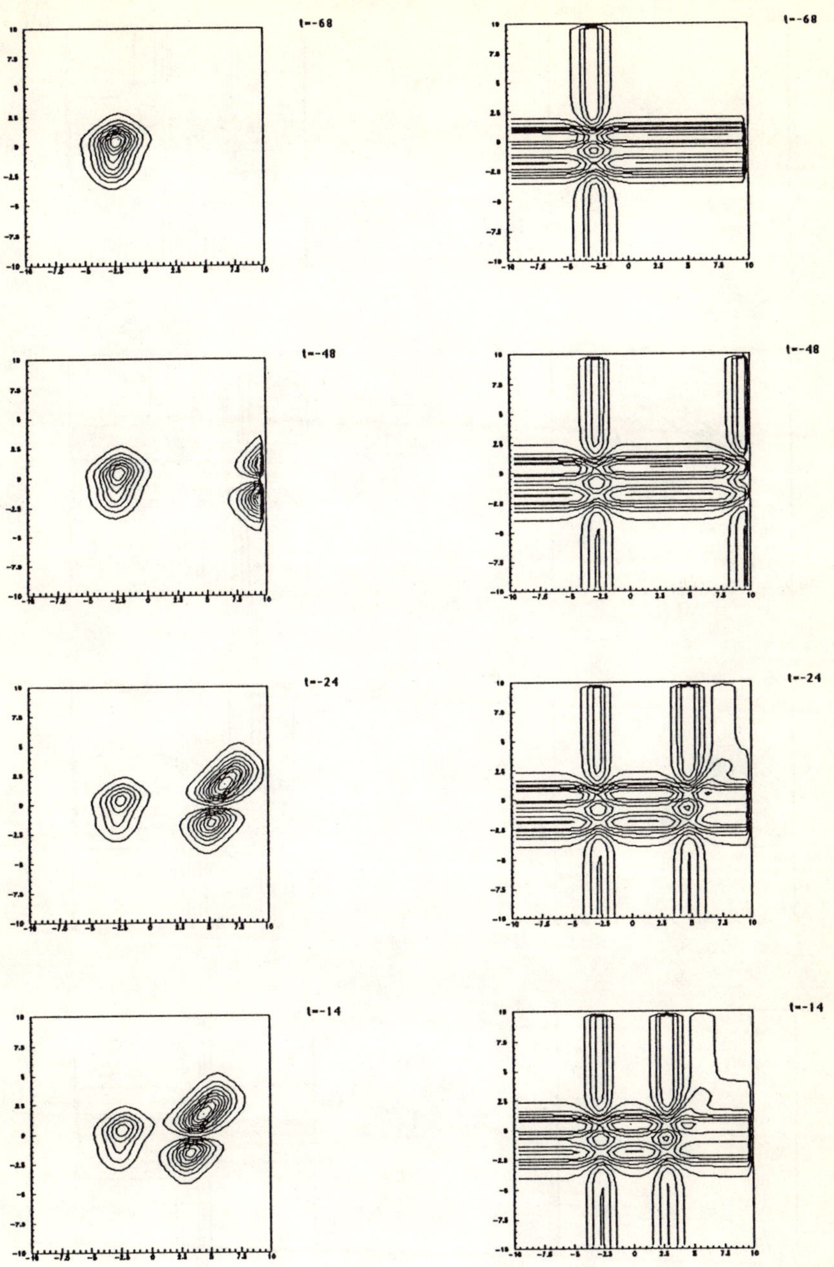

Fig. 1a

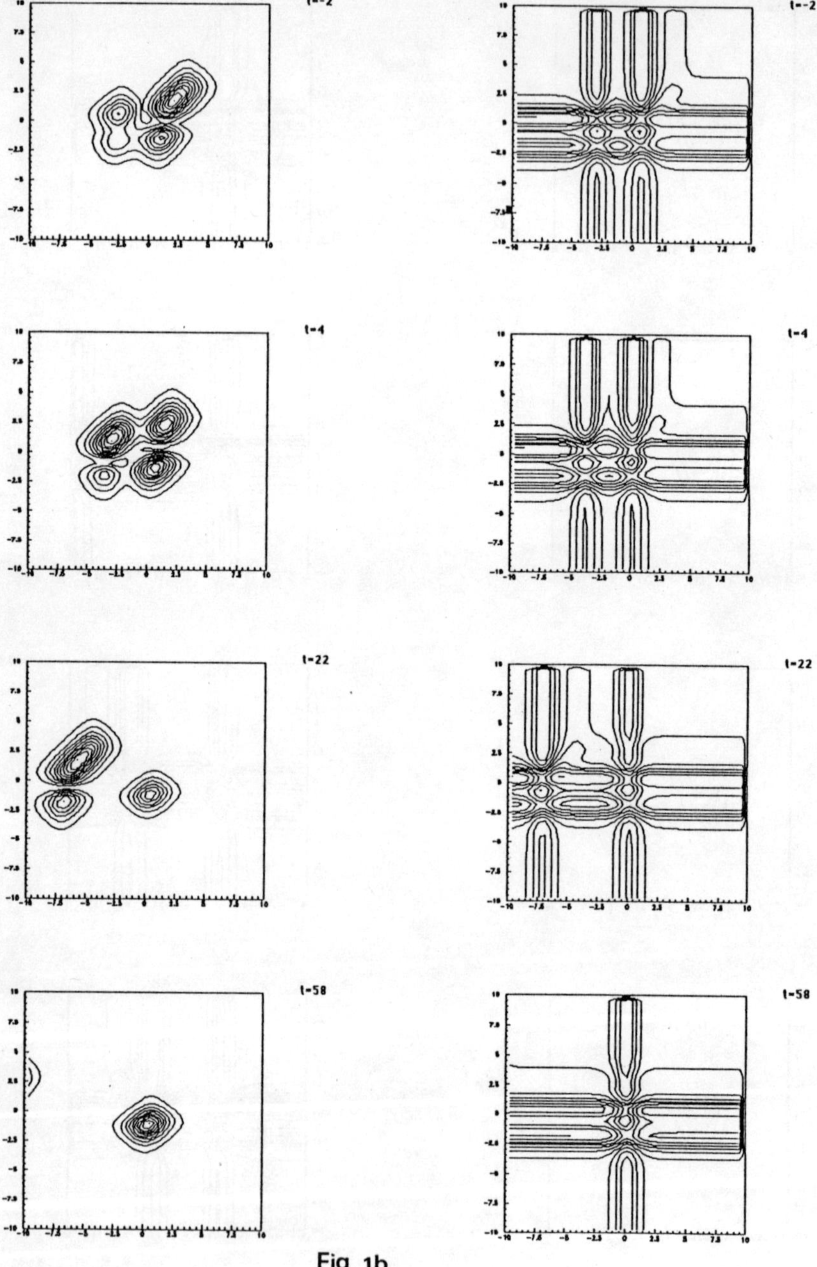

Fig. 1b

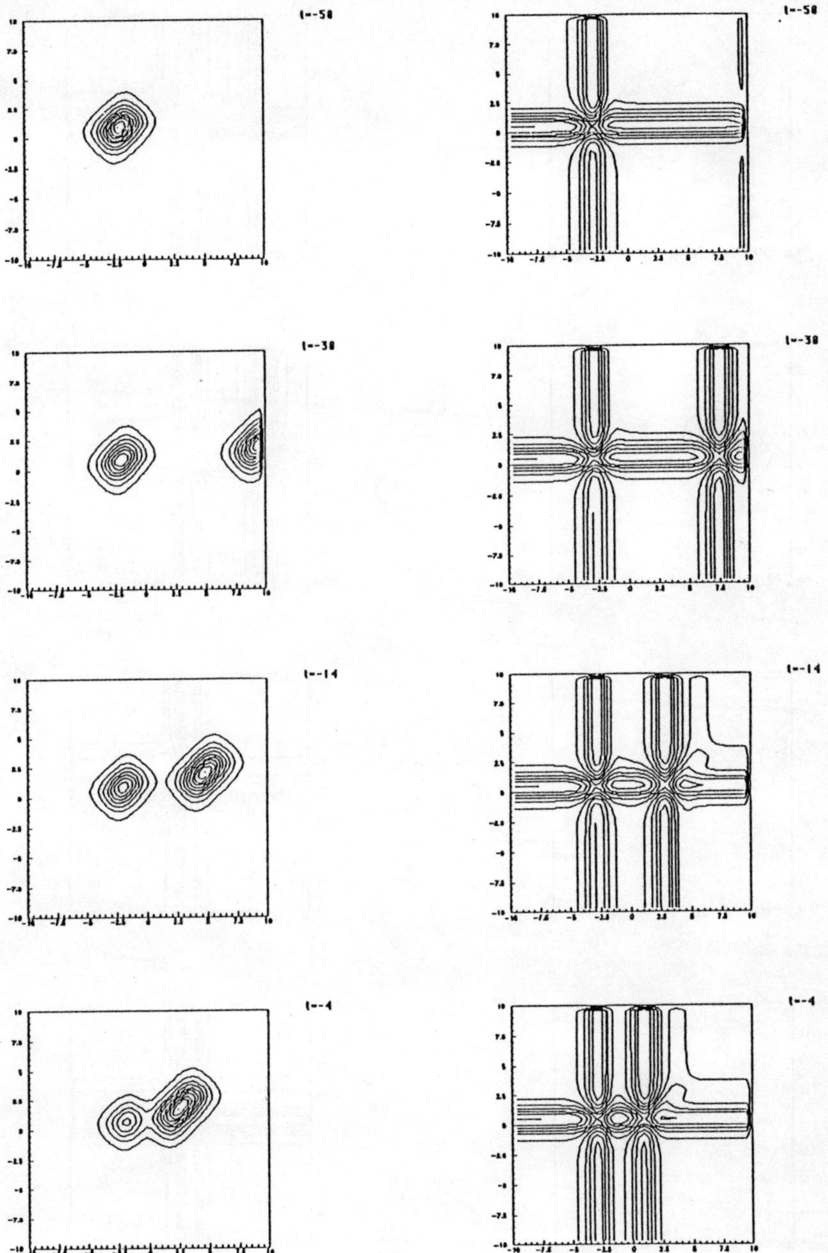

Fig.2a

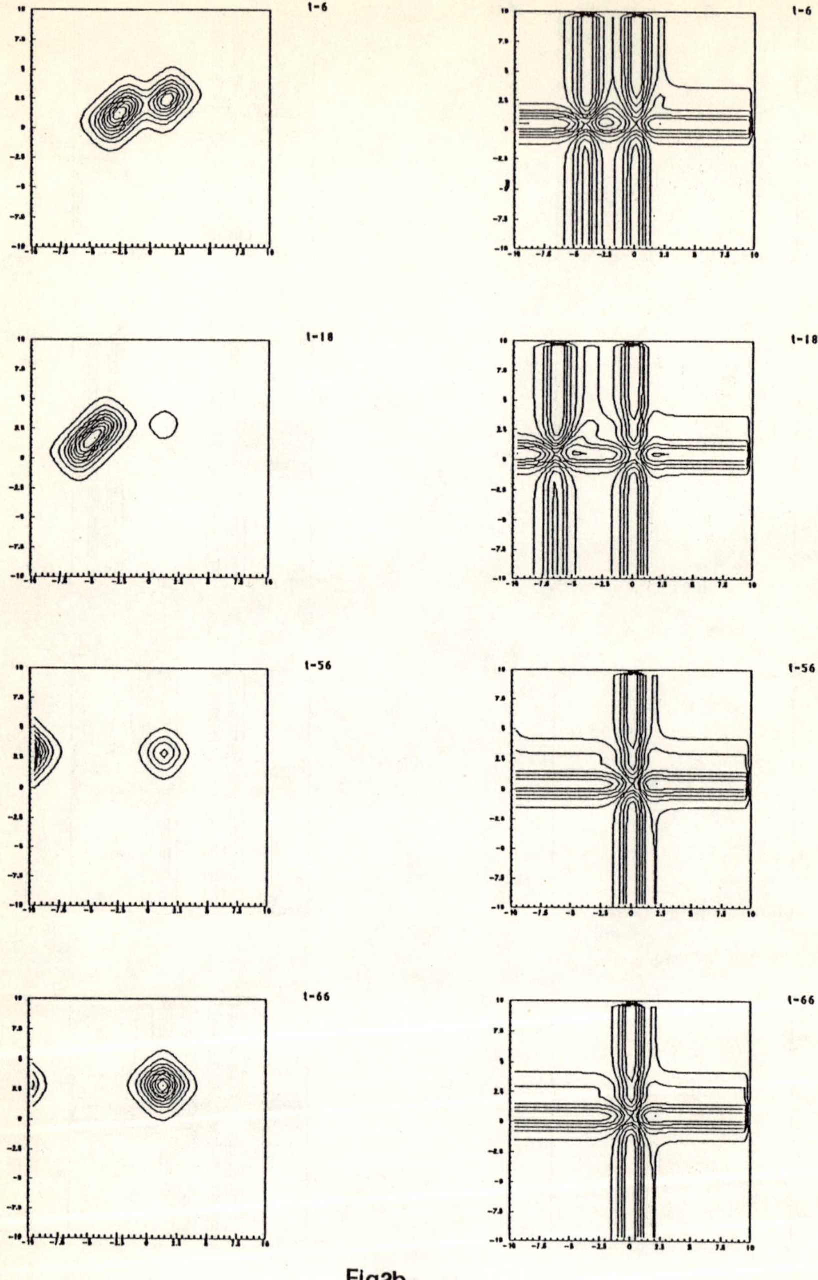

Fig.2b

$2+1$ Dimensional Dromions and Hirota's Bilinear Method

J. Hietarinta

Department of Physics, University of Turku
20500 Turku, Finland.

1. Introduction. Hirota's bilinear formalism is perhaps the best method for constructing solutions of integrable nonlinear evolution equations [1,2]. In this lecture we show how using this method one can easily construct one-dromion solutions for generic equations of nonlinear Schrödinger (nlS) and Korteweg–de Vries (KdV) type [3,4].

In order to construct dromion solutions it is important to make a clear distinction between the physical function and all the other functions that appear in the equation. This is because it is only the physical function that displays localized solutions. The basic concept is the following:

1) The building blocks are plane wave solitons, however, these plane waves must be chosen so that the corresponding *physical function vanishes.* The individual plane-wave-solitons are therefore kind of 'ghost' solitons.
2) A one-dromion solution is made out of *two intersecting ghost solitons.* Outside the interaction region the physical function vanishes.

In the following we show how these principles work for the nlS- and KdV-type equations.

2. Nonlinear Schrödinger-type equations. Let us first look at the Davey-Stewartson-equation (DS) for which the first dromion solutions were constructed. Its bilinear form is [3]

$$(iD_t + D_x^2 + D_y^2)G\cdot F = 0, \tag{1a}$$

$$D_x D_y F\cdot F = 2|G|^2, \tag{1b}$$

where the D-operators are defined by [1]

$$P(D_{\vec{x}})G\cdot F = P(\partial_{\vec{y}})G(\vec{x}+\vec{y})F(\vec{x}-\vec{y})\big|_{\vec{y}=0}, \tag{2}$$

In this case the physical function is $u = G/F$. To construct a ghost soliton we take therefore

$$G = 0, \quad F = 1 + c\,e^{\eta+\eta^*}, \quad \eta = \vec{p}\cdot\vec{x} + \text{const.}\ , \tag{3}$$

which clearly yields $u = 0$ and solves (1a). It solves also (1b), if the parameter vector $\vec{p}$ is given by

$$\vec{p}_1 = (p, 0, \Omega), \ \text{ or } \ \vec{p}_2 = (0, q, \omega) \tag{4}$$

To construct a dromion solution we take a combination of two ghost plane waves as an ansatz

$$F = \delta + \alpha\, e^{\eta_1+\eta_1^*} + \beta\, e^{\eta_2+\eta_2^*} + \gamma\, e^{\eta_1+\eta_2+\eta_1^*+\eta_2^*}. \tag{5}$$

When this is substituted into (1b) we obtain

$$G = \rho e^{\eta_1+\eta_2}, \quad |\rho|^2 = (p+p^*)(q+q^*)(\gamma\delta - \alpha\beta) \tag{6}$$

Next we must look at the conditions following from (1a), but it is as easy to do the computations for the generic equation,

$$B(D_{\vec{x}})G\cdot F = 0, \tag{7}$$

where B is a polynomial with the property that $B(i\vec{p})$ is real. Substituting (5,6) into (7) yields in the following conditions

$$B(\vec{p}_1 + \vec{p}_2) = 0,\; B(-\vec{p}_1{}^* + \vec{p}_2) = 0. \tag{8}$$

These are the dispersion relations from which Ω and ω in (4) are determined.

The above shows that a one-dromion solution can be constructed for any pair of bilinear equations (1b,7). In fact we can generalize also the second equation, i.e. the above process works for the generic nlS-type equation

$$B(D_{\vec{x}})G\cdot F = 0, \tag{9a}$$

$$A(D_{\vec{x}})F\cdot F = 2|G|^2. \tag{9b}$$

where A is an even polynomial with $A(0) = 0$. The previous ansatz (5,6) implies that the $\vec{p}_i$ must now be chosen so that

$$A(\vec{p}_i + \vec{p}_i{}^*) = 0,\; i = 1, 2. \tag{10}$$

This equation must have two *nonparallel* solutions $\vec{p}_i$, which means that A cannot be too simple, e.g. just a power of a linear polynomial. In addition (9b) implies

$$\gamma\delta A(\vec{p}_1 + \vec{p}_1{}^* + \vec{p}_2 + \vec{p}_2{}^*) + \alpha\beta A(\vec{p}_1 + \vec{p}_1{}^* - \vec{p}_2 - \vec{p}_2{}^*) = |\rho|^2. \tag{11}$$

from which ρ can be solved; from (9a) we get the dispersion relations (8) as before.

3. KdV-type equations. The dromion solutions have been previously discussed only in the context of the DS-equation. However, when one uses the bilinear formalism the concept given before can be applied to other equations as well, e.g. for generic KdV-type bilinear equations [4]

$$P(D_{\vec{x}})F\cdot F = 0, \tag{12}$$

where P is an even function of its variables and $P(0) = 0$.

The first problem is to choose the proper physical function, let us try

$$u = (\alpha\partial_x + \beta\partial_y)(\gamma\partial_x + \delta\partial_y) \log F, \tag{13}$$

where $\Delta := \alpha\delta - \gamma\beta \neq 0$. Ghost solutions are then obtained with

$$F = 1 + e^{\eta},\ \eta = \vec{p}\cdot\vec{x} + \eta_0, \quad \vec{p} = (\beta p, -\alpha p, \Omega),\ \text{or}\ \vec{p} = (\delta q, -\gamma q, \omega), \tag{14}$$

where Ω and ω are fixed by the dispersion relation $P(\vec{p}) = 0$.

Equation (12) is of the generic type for which one can always construct two-soliton solutions:

$$F = 1 + e^{\eta_1} + e^{\eta_2} + K\, e^{\eta_1+\eta_2},\ K = -\frac{P(\vec{p}_1 - \vec{p}_2)}{P(\vec{p}_1 + \vec{p}_2)}. \tag{15}$$

(To avoid singularities we must restrict the parameters $\vec{p}_i$ so that $K > 0$.) Let us take one p-type soliton and one q-type soliton as given in (14), then from (13) we get the physical function u as:

$$u = \Delta^2 pq(1-K)\frac{e^{\eta_1+\eta_2}}{(1 + e^{\eta_1} + e^{\eta_2} + K\, e^{\eta_1+\eta_2})^2}, \tag{16}$$

which has the standard dromion form. Note that in contrast to the nlS-case there are no free parameters other than p and q, in particular the amplitude (=maximum of u) is fixed to $\frac{1}{4}\Delta^2 pq(1-\sqrt{K})/(1+\sqrt{K})$.

If P is one of the integrable polynomials we can construct multidromion solutions in a straightforward manner, we just take any multisoliton solution composed of solitons of type (14).

As an example let us take a closer look at the Kadomtsev–Petviashvili-equation (KP). Let us start with its bilinear form

$$(D_x^4 + 3D_y^2 - 4D_xD_t)F\cdot F = 0. \tag{17}$$

A possible choice for the physical function is $u = \partial_y(\partial_y - \partial_x)\log F$, and then we can use $\eta = px + \frac{1}{4}p^3 t$ and $\eta = qx + qy + \frac{1}{4}(q^3 + 3q)t$ as the two non-parallel ghost solitons. Any multisoliton solution constructed with these two kinds of plane waves (with different p's and q's) will yield a multidromion solution in terms of u [4,5].

Note, however, that in the standard treatment of the KP-equation one takes $w = \partial_x^2 \log F$ as the physical function. The underlying bilinear equation is the same (equation (17)) for both u and w, but obviously the corresponding nonlinear equations have to be different.

To get from (17) to a nonlinear form we use first the canonical substitution $F = e^v$, which yields

$$v_{xxxx} + 3(v_{xx})^2 + 3v_{yy} - 4v_{xt} = 0. \tag{18}$$

The equation for w is obtained from this by taking two x-derivatives of the equation and using the new variable $w = v_{xx}$:

$$w_{xxxx} + 3(w^2)_{xx} + 3w_{yy} - 4w_{xt} = 0. \tag{19}$$

In order to get the nonlinear equation for u we first operate on (18) by $\partial_y(\partial_y - \partial_x)$, and introduce an auxiliary function $\phi = v_{xx}$, then (18) transforms to the pair

$$\begin{gathered} u_{xxxx} + 6\phi_y(\phi_y - \phi_x) + 6\phi u_{xx} + 3u_{yy} - 4u_{xt} = 0, \\ u_{xx} = \phi_{yy} - \phi_{xy}. \end{gathered} \tag{20}$$

This is therefore the nonlinear equation that has multidromion solutions. Since it is based on the same KP-bilinear form it may be considered as variant of the KP-equation.

4. Conclusions. We have shown here how the dromion concept can be generalized to all bilinear equations of type (9) or (12). The basic idea behind our approach is to *build dromions from intersecting plane waves.* The component plane waves must be such that in isolation they produce only a vanishing physical function. This means that a physical contribution is obtained only at the points where the plane waves interact. The method that has been presented here can be further generalized, e.g. for multicomponent nonlinear Schrödinger-type systems.

One question that has not yet been studied sufficiently is the stability of dromions. The following heuristic argument suggests that the dromions are in fact stable: The construction discussed here shows that dromions exist only at the intersection of plane waves. At this intersection there must also be a phase shift, that is, a bend in the plane wave (this is clearly shown in the graphical illustrations [5]). Thus to eliminate a dromion we must eliminate this phase shift in both of the intersecting plane waves. However, this is not a local effect, as it means that we must make a simultaneous parallel transport on two semi-infinite plane waves. It seems to be difficult to induce such a long range effect by a small perturbation.

References.

[1] R. Hirota, in "Solitons", R.K. Bullough and P.J. Caudrey (eds.), p. 157, (Springer, 1980).

[2] J. Hietarinta, in "Partially integrable equations in Physics", R. Conte and N. Boccara (eds.), p. 459, (Kluwer, 1990).

[3] J. Hietarinta and R. Hirota, Phys. Lett. A **145**, 237 (1990).

[4] J. Hietarinta, Phys. Lett A **149**, 113 (1990).

[5] J. Hietarinta and J. Ruokolainen: "Dromions - The Movie", 15min video (1990)

Skyrmions Scattering in $(2+1)$ Dimensions

B. Piette and W.J. Zakrzewski

Department of Mathematical Sciences,
Univerity of Durham, Durham DH1 3LE, England.
20500 Turku, Finland.

Abstract. We consider instanton solutions of the $O(3)\sigma$-model in two Euclidian dimensions modified by the addition of appropriate potential and skyrme-like terms as static solitons - skyrmions of the same model in $(2+1)$ dimensions. We find that the addition of the potential and skyrme terms stabilises the skyrmions and that the force between them is repulsive. In the scattering process initiated at low relative velocities the skyrmions bounce back while at larger velocities they scatter at right angles. The scattering is quasi-elastic and the skyrmions preserve their shape after the collision. We study the time evolution of a family of two soliton configurations corresponding to different relative orientations of the skyrmions in the $O(3)$ space.

1. Introduction

Over the last few years sigma models in low dimensions have become an increasingly important area of research. Although the σ-models are integrable in two dimensions ,[1-3] it appears that only very special models are integrable[4] in (2+1) dimensions. In particle physics we are interested primarily in Lorentz invariant models, but as all such σ-models in (2+1) dimensions appear to be nonintegrable, it is natural to consider numerical evolutions in these cases.

The simplest Lorentz invariant (2+1) dimensional σ model is the $O(3)$ model, which involves one real vector field of 3 components, $\vec{\phi} \equiv (\phi^1, \phi^2, \phi^3)$. In (2+1) dimensions $\vec{\phi}$ is a function of the space-time coordinates (t, x, y) which we also write as (x^0, x^1, x^2). The model is defined by the Lagrangian density

$$\mathcal{L} = \tfrac{1}{4}(\partial^\mu \vec{\phi}) \cdot (\partial_\mu \vec{\phi}) \,, \tag{1.1}$$

together with the constraint $\vec{\phi} \cdot \vec{\phi} = 1$, *i.e.* $\vec{\phi}$ lies on a unit sphere S^2_{ϕ}. In (1.1) the Greek indices take values $0, 1, 2$ and label space-time coordinates, and ∂_μ denotes partial differentiation with respect to x^μ. Note that we have set the velocity of light, c, equal to unity, so that in all our calculations we can use dimensionless quantities. The Euler-Lagrange equations derived from (1.1) are

$$\partial^\mu \partial_\mu \vec{\phi} + (\partial^\mu \vec{\phi} \cdot \partial_\mu \vec{\phi}) \vec{\phi} = \vec{0} \,. \tag{1.2}$$

For boundary conditions we take

$$\vec{\phi}(r, \theta, t) \rightarrow \vec{\phi}_0(t) \quad \text{as} \quad r \rightarrow \infty \,, \tag{1.3}$$

where (r, θ) are polar coordinates and where $\vec{\phi}_0$ is independent of the polar angle θ. In two Euclidean dimensions (*i.e.* taking $\vec{\phi}$ to be independent of time) this condition ensures finiteness of the action, which is precisely the requirement for quantisation in terms of path integrals. In (2+1) dimensions it leads to a finite potential energy.

It is convenient to express the $\vec{\phi}$ fields in terms of their stereographic projection onto the complex plane W

$$\phi^1 = \frac{W + W^\star}{1 + |W|^2}, \quad \phi^2 = i\frac{W - W^\star}{1 + |W|^2}, \quad \phi^3 = \frac{1 - |W|^2}{1 + |W|^2}. \tag{1.4}$$

The W formulation is very useful, because it is in this formulation that the static solutions take their simplest form; namely, as originally shown by Belavin and Polyakov [5] and Woo,[6] they are given by W being any rational function of either $x + iy$ or of $x - iy$.

The instanton solutions can be considered as soliton configurations for the (2+1) dimensional model only if they are stable and do not desintegrate when we consider their time evolution. We have analysed this problem in some detail and have found [7] [8] [9] that the solitons of the $O(3)$ σ model are unstable. This is due to the fact that the model has no intrinsic scale and so admits the existence of solitons of arbitrary size. Hence under small perturbations the solitons can either expand indefinitely or shrink to become infinitely tall spikes of zero width. Our simulations have shown that this is exactly what happens in this model. In fact, as soon as the solitons are purturbed, *e.g.* start moving, they start shrinking. This is true not only in the full simulation of the model but also [10] in the approximation to the full simulation provided by the so-called "collective coordinate" approach in which the evolution is approximated by the geodesic motion on the manifold of static solutions.

A few words about our numerical procedures. Most of our simulations were performed using a 4th order Runge-Kutta method of simulating time evolution. They were performed on fixed lattices which varied from 201×201 to 512×512, with lattice spacing $\delta x = \delta y = 0.02$. The time step was 0.01.

So far as the boundary conditions are concerned most of our simulations were performed with fixed boundary conditions as all the effects associated with the variation of the fields at the boundaries are very small. However, even though small, they are nonzero and so we tested their effects by introducing some absorption or by extrapolating the fields at the boundaries. We have found that the waves coming from the boundaries or the waves reflected from the boundaries can effect our results quite significantly. In particular, some preliminary results obtained on smaller size lattices were not confirmed in our bigger lattice simulations. Having tested our results by changing the lattice size and varying the boundary conditions we are reasonably confident of our results; although we believe some more work is required to be absolutely certain.

2. Skyrme Model

To stabilise the $O(3)$ model we introduced a scale into the model that would prevent the instantons from both shrinking and expanding. Guided by the ideas of Skyrme[11][12] we chose to add to our Lagrangian density the following extra terms

$$\mathcal{L}_e = -\tfrac{1}{4}\Big(\theta_1\big((\partial^\mu\vec{\phi}.\partial_\mu\vec{\phi})^2 - (\partial^\mu\vec{\phi}.\partial^\nu\vec{\phi})(\partial_\mu\vec{\phi}.\partial_\nu\vec{\phi})\big) + \theta_2(1+\phi^3)^4\Big), \tag{2.1}$$

where θ_1 and θ_2 are two new (real) parameters of the model. It is clear that the model based on the Lagrangian with these terms is still Lorentz invariant and for positive values of θ's its Hamiltonian is positive definite. Moreover, despite the appearance to the contrary, the Lagrangian does not contain time derivatives higher than two and so its equation of motion takes the conventional form.

$$\begin{aligned}
&\partial_\mu\partial^\mu\phi^i - (\vec{\phi}.\partial_\mu\partial^\mu\vec{\phi})\phi^i + 2\theta_2(1+\phi^3)^3\,(\delta_{i3}-\phi^i\phi^3)\\
&-2\theta_1\Big[\partial_\mu\partial^\mu\phi^i(\partial_\nu\vec{\phi}.\partial^\nu\vec{\phi}) + \partial_\nu\phi^i(\partial_\mu\partial^\nu\vec{\phi}.\partial^\mu\vec{\phi}) - \partial_\nu\partial_\mu\phi^i(\partial^\nu\vec{\phi}.\partial^\mu\vec{\phi})\\
&\quad - \partial_\mu\phi^i(\partial^\nu\partial_\nu\vec{\phi}.\partial^\mu\vec{\phi}) + (\partial_\mu\vec{\phi}.\partial^\mu\vec{\phi})(\partial_\nu\vec{\phi}.\partial^\nu\vec{\phi})\phi^i - (\partial_\nu\vec{\phi}.\partial_\mu\vec{\phi})(\partial^\nu\vec{\phi}.\partial^\mu\vec{\phi})\phi^i\Big] = 0.
\end{aligned} \tag{2.2}$$

The equation (2.2) is rather difficult to solve, but if we restrict ourselves to looking for static solutions and then consider $\vec{\phi}$ which corresponds to W being analytical (*i.e.* $W = W(x+iy)$) then it is easy to check that

$$W = \lambda(x+iy) \tag{2.3}$$

is a static solution if

$$\lambda = \sqrt[4]{\frac{\theta_2}{2\theta_1}}. \tag{2.4}$$

This is a particular case of the one instanton solution of the $O(3)$ σ model, but with the fixed "size" (determined by λ).

It is easy to show that this solution is stable with respect to any perturbations. In fact, if we evolve it with (2.4) different from λ we find that the system has an excess of energy which it uses to alter its size up or down to the correct value and at the same time it sends out a wave of radiation.

So what are the scattering properties of our skyrmions? First we considered the field configuration described by

$$W(x,y,t) = \frac{(x+iy-a+vt)(x+iy+a-vt)}{2\mu(a-vt)} \tag{2.5}$$

and calculating from it $W(x,y,0)$ and $\partial_t W(x,y,0)$. This configuration describes[9] two skyrmions (located at $\pm a$ at $t=0$); their widths are the same and given by $\lambda = \frac{1}{\mu}$. We chose $a = 1.0$ and considered the dependence of the evolution on the values of v.

When $v = 0$, the two skyrmions (initially at rest) start repelling and move apart from each other. If v is small the skyrmions move towards each other, slow down and then start moving backwards. If v is taken large enough, the skyrmions collide with each other, form a ring and eventually scatter at 90°, again moving away from each other. There is a critical velocity in-between these two cases and as we found that its value increases with θ.

We observed also that as the skyrmions move towards each other with their velocities approaching the critical value, the time during which the skyrmions stay close together increases, implying the trapped nature of the quasi bound-state formed by the skyrmions.

3. Modified Initial Condition

We also looked at the evolution of the modified initial condition for W:

$$W(x,y,t) = \frac{\sqrt{1+R^2}}{2\mu} \frac{(x+iy-a+vt)(x+iy+a-vt)}{(a-vt+iR(x+iy))} \tag{3.1}$$

where R is a parameter and μ is related to the θ's as above.

This configuration is very much like the previous one: it corresponds to two skyrmions located initially at $x = \pm a$, moving towards each other. However, the two skyrmions point in different directions in the $O(3)$ space and their relative direction depends on R. When $R = 0$ (3.1) is actualy equal to (2.5). Notice also that for $R \neq 0$, (3.1) is not symmetric with respect to the reflection along the x axis. Moreover, the shape of the energy density of the system also depends on R; as R increases, the energy of the solitons increases slightly, but at the same time the energy is more spread out. For example, when $\theta_2 = 1.$, $\mu = 0.25$ and $a = 1.0$ the energy contained in the domain, $x, y \in [-4,4][-4,4]$ when $R = 10$ is slightly smaller than twice the energy contained in the same domain for one soliton. Nevertheless, the total energy of the system is always larger than the energy of two separated skyrmions. It is easy to see that as $r = \sqrt{x^2+y^2}$ goes to infinity, the energy density goes to zero as r^{-6} when $R = 0$ but only like r^{-4} for other values of R.

As the local excess of energy decreases when R increases, the critical velocity increases and the scattering is much slower. For example, for the above values of θ's and a, $v_{cr} = 0.37$ for $R = 0$ and $v_{cr} = 0.54$ for $R = 1$. For larger values of v, as they collide, the skyrmions merge to form a ring from which they emerge as two skyrmions of different sizes. As they separate the skyrmions oscillate in size until they both reach the correct size for a single skyrmion.

4. Conclusions

We saw that the skyrmions behave very much like real solitons. In the scattering they preserve their shape and although some radiation effects are present these effects are always very small.

In conclusion we see that the modified $O(3)$ model, although non-integrable, is almost integrable in that it has many features in common with many integrable models. Most differences or deviations are rather small. As most physically relevant models are not integrable our results suggest that the results found in some integrable models should not be dissmissed as not relevant; it is quite likely that some of these results may also hold in models which, strictly speaking, are not integrable but whose deviations from integrability are rather small.

Finally, our results suggest that the modified $O(3)$ σ model is a good candidate for being a toy model of solitons in (2+1) dimensions.

REFERENCES

1. Y.Y. Goldschmidt and E. Witten, *Phys. Lett*, **91 B**, 392 (1980)
2. V.E. Zakharov and A.V. Mikhailov, *Sov. Phys. JETP* **47**, 1017 (1979)
3. J. Harnad, Y. Saint-Aubin and S. Shnider, *Comm. Math. Phys.*, **92**, 329 (1984)
4. R.S. Ward, *Nonlinearity*, **1**, 671 (1988)
5. A.A. Belavin and A.M. Polyakov, *JETP Lett*, **22**, 245 (1975)
6. G. Woo*J. Math. Phys.*, **18**, 1264 (1977)
7. R.A. Leese, M. Peyrard and W.J. Zakrzewski - *Nonlinearity*, **3**, 387 (1990)
8. W.J. Zakrzewski - Soliton-like Scattering in the $O(3)$ σ model in (2+1) Dimensions - *Nonlinearity* to appear (1990)
9. M. Peyrard, B. Piette and W.J. Zakrzewski - Soliton Scattering in the Skyrme Model in (2+1) Dimensions 1. and 2., Durham University preprints DTP-90/37 and 39 (1990)
10. R.A. Leese, Low-energy Scattering of Solitons in the CP^1 model, *Nucl. Phys.* **B**, to appear (1990)
11. T.H.R. Skyrme - *Proc. Roy. Soc.***A 260**, 127 (1961)
12. E. Witten - *Nucl. Phys.* **B 223**, 433 (1983), G. Adkins, C. Nappi and E. Witten - *Nucl. Phys.* **B228**, 552 (1983)

[illegible]

[illegible]

[illegible]

REFERENCES

1. [illegible]

2. [illegible]

3. [illegible]

4. [illegible]

5. [illegible]

6. [illegible]

7. [illegible]

8. [illegible]

9. [illegible]

10. [illegible]

11. [illegible]

12. [illegible]

Subject Index

Reference is made to the *first* page of relevant articles

Index of Contributors

Printing and binding: Druckerei Triltsch, Würzburg